RENEWALS 45

EXS 98

Neurotransmitter Interactions and Cognitive Function

Edited by Edward D. Levin

Birkhäuser Verlag
Basel • Boston • Berlin

Edward D. Levin
Duke University Medical Center
Department of Psychiatry and Behavioral Sciences
Neurobehavioural Research Laboratory
Box 3412
Durham NC 27710
USA

Library of Congress Cataloging-in-Publication Data
Neurotransmitter interactions and cognitive function / edited by Edward D. Levin.
p. cm. – (EXS 98)
Includes bibliographical references and index.
ISBN-13: (invalid) 978-3-7643-7171-7 (alk. paper)
ISBN-10: 3-7643-7771-2 (alk. paper)
1. Neurotransmitters. 2. Cognition. 3. Cognitive neuroscience.

QP364.7.N4685 2006
612.8'2–dc22 2006047644

Bibliographic information published by Die Deutsche Bibliothek
Die Deutsche Bibliothek lists this publication in the Deutsche Nationalbibliografie; detailed bibliographic data is available in the Internet at <http://dnb.ddb.de>.

ISBN 10: 3-7643-7771-2 Birkhäuser Verlag, Basel – Boston – Berlin
ISBN 13: 978-3-7643-7771-7
The publisher and editor can give no guarantee for the information on drug dosage and administration contained in this publication. The respective user must check its accuracy by consulting other sources of reference in each individual case.
The use of registered names, trademarks etc. in this publication, even if not identified as such, does not imply that they are exempt from the relevant protective laws and regulations or free for general use.
This work is subject to copyright. All rights are reserved, whether the whole or part of the material is concerned, specifically the rights of translation, reprinting, re-use of illustrations, recitation, broadcasting, reproduction on microfilms or in other ways, and storage in data banks. For any kind of use, permission of the copyright owner must be obtained.

© 2006 Birkhäuser Verlag, P.O. Box 133, CH-4010 Basel, Switzerland
Part of Springer Science+Business Media
Printed on acid-free paper produced from chlorine-free pulp. TCF ∞
Cover illustration: See page 114. Reprinted from Buxbaum-Conradi H, Ewert J-P (1999) Responses of single neurons in the toad's caudal ventral striatum to moving visual stimuli and test of their efferent projection by extra-cellular antidromic stimulation/recording techniques. *Brain Behav Evol* 54: 338–354. With kind permission of Karger.
Typesetting: PTP-Berlin Protago-TeX-Production GmbH, Germany

Printed in Germany
ISBN 10: 3-7643-7771-2 e-ISBN 10: 3-7643-7772-0
ISBN 13: 978-3-7643-7771-7 e-ISBN 13: 978-3-7643-7772-4

9 8 7 6 5 4 3 2 1 www.birkhauser.ch

Library
University of Texas
at San Antonio

Contents

List of contributors

Patrizio Blandina, Dipartimento di Farmacologia Preclinica e Clinica, Università di Firenze, 50139 Firenze, Italy; e-mail: patrizio.blandina@unifi.it

John P. Bruno, Departments of Psychology and Neuroscience, Ohio State University, Columbus, OH 43210, USA

Jamie G. Bunce, Department of Psychology, University of Connecticut, Storrs, CT 06269, USA

James J. Chrobak, Department of Psychology, University of Connecticut, Storrs, CT 06269, USA; e-mail: james.chrobak@uconn.edu

Hans C. Dringenberg, Department of Psychology and Centre for Neuroscience Studies, Queen's University, Kingston, Ontario, K7L 3N6, Canada; e-mail: dringenb@post.queensu.ca

Jörg-Peter Ewert, Department of Neurobiology, Faculty of Natural Sciences, University of Kassel, 34132 Kassel, Germany; e-mail: ewertjp@gmx.de

Adi Guterman, Department of Psychology, Faculty of Social Sciences and the Brain and Behavior Research Center, University of Haifa, Israel

Rouba Kozak, Department of Psychology, University of Michigan, Ann Harbor, MI 48109, USA

Min-Ching Kuo, Department of Psychology and Centre for Neuroscience Studies, Queen's University, Kingston, Ontario, K7L 3N6, Canada

José Larrauri, Duke University, Department of Psychology and Neuroscience, Durham, NC 27708, USA

Edward D. Levin, Department of Psychiatry and Behavioral Sciences, Box 3412, Duke University Medical Center, Durham, NC 27710, USA; e-mail: edlevin@duke.edu

Vicente Martinez, Department of Psychology, University of Michigan, Ann Harbor, MI 48109, USA

Robert D. Oades, Biopsychology Research Group, University Clinic for Child and Adolescent Psychiatry, Virchowstr. 174, 45147 Essen, Germany; e-mail: oades@uni-essen.de

Vinay Parikh, Department of Psychology, University of Michigan, Ann Harbor, MI 48109, USA

Maria Beatrice Passani, Dipartimento di Farmacologia Preclinica e Clinica, Università di Firenze, Italy

Jos Prickaerts, Department of Psychiatry, RED Europe, Johnson & Johnson Pharmaceutical Research and Development, Turnhoutseweg 30, 2340 Beerse, Belgium

Amir H. Rezvani, Department of Psychiatry and Behavioral Sciences, Box 3412, Duke Unveristy Medical Center, Durham, NC 27710, USA

Jerry B. Richards, Research Institute on Addictions, Buffalo University, NY 14203, USA

Gal Richter-Levin, Department of Psychology, Faculty of Social Sciences and the Brain and Behavior Research Center, University of Haifa, 31905 Haifa, Israel; e-mail: gal.r-l@psy.haifa.ac.il

Helen R. Sabolek, Department of Psychology, University of Connecticut, Storrs, CT 06269, USA

Martin Sarter, Department of Psychology, University of Michigan, Ann Harbor, MI 48109, USA; e-mail: msarter@umich.edu

Nestor Schmajuk, Duke University, Department of Psychology and Neuroscience, Durham, NC 27708, USA; e-mail: nestor@duke.edu

Wolfgang W. Schwippert, Department of Neurobiology, Faculty of Natural Sciences, University of Kassel, Germany; e-mail: wschwip@uni-kassel.de

Thomas Steckler, Department of Psychiatry, RED Europe, Johnson & Johnson Pharmaceutical Research and Development, Turnhoutseweg 30, 2340 Beerse, Belgium

Geoff Warnock, Department of Psychiatry, RED Europe, Johnson & Johnson Pharmaceutical Research and Development, Turnhoutseweg 30, 2340 Beerse, Belgium; e-mail: gwarnock@prdbe.jnj.com

Mohammad R. Zarrindast, Department of Pharmacology, School of Medicine, Tehran University of Medical Sciences, Tehran, Iran; e-mail: zarinmr@ams.ac.ir

Neurotransmitter Interactions and Cognitive Function
Edited by Edward D. Levin
© 2006 Birkhäuser Verlag/Switzerland

The rationale for studying transmitter interactions to understand the neural bases of cognitive function

Edward D. Levin

Department of Psychiatry and Behavioral Sciences, Box 3412, Duke University Medical Center, Durham, NC 27710, USA

The brain is an organ of communication. Neurons within the brain connect in networks that communicate with each other to provide behavioral function. This network organization is particularly evident with regard to cognitive function, from simple sensorimotor plasticity to attentional, learning and memory processes. Cognitive function involves the participation of diverse brain areas including parts of the limbic system, such as the hippocampus and the amygdala, as well as the frontal cortex, portions of the thalamus and the basal forebrain cholinergic and midbrain monoaminergic nuclei, which project to the more dorsal and rostral brain regions. The interactions of these systems can be characterized by the different neurotransmitters used to communicate between the different brain regions and neuronal types. A variety of drugs are available which more or less selectively stimulate or block receptors for these transmitters, which can be used to manipulate the activity of these systems to discover their functional interactions. Studying the interactions among these drugs can provide a window through which neural communication underlying cognitive function can be studied and new therapeutic treatments for cognitive dysfunction can be developed.

The chapters of this book were written by international experts in the field of behavioral neuroscience and provide a background review of the literature in the area of neurotransmitter interactions and cognitive function. They also provide insight into the experimental methods used for discovering new information concerning these neural interactions. This is a vigorous and wide-ranging field of research. A portion of that research is covered in this particular book. The current book is a follow-up to the previous volume with the same name published in 1992 [1]. In the past decade and a half substantial progress has been made in the area with considerable improvements in our understanding of how neural systems interact in the basis of cognitive function. The orientation underlying this field of endeavor was insightfully discussed in the earlier book by the late Dr. Roger Russell who emphasized the vital consideration of the integrated organism in the full understanding of neurobehavioral function [2]. It is not the understanding of the critical parts of the system that underlie function that are particularly important, but how these parts interact and work in concert to synthesize behavior that provides a more complete and accurate understanding of behavioral function. This is the philosophy that underlies the current volume as well.

A variety of international experts using a variety of experimental approaches have contributed to this volume. Neural systems using acetylcholine have been very well characterized in the basis of cognitive function. The chapter by Dr. Zarrindast provides an overview of the neural systems involved in cognitive function, particularly cholinergic systems. Specific interactions of cholinergic systems are examined in detail in the following chapters. Dr. Warnock and colleagues describe the interactions between acetylcholine and corticotropin releasing factor (CRF) in the modulation of cognitive behavior. Dr. Sarter and colleagues have characterized forebrain dopaminergic-cholinergic interactions in the basis of attention, shedding light on both psychostimulant addiction and schizophrenia. Chrobak et al. have characterized another principal branch of the forebrain cholinergic innervation, the septohippocampal cholinergic system. Intraseptal cholinergic infusions alter memory in the rat: method and mechanism.

Certainly the cholinergic system is not alone in the neural basis of cognitive function. Blandina and Passani demonstrated that central histaminergic system interactions are important for cognitive function. Dringenberg and Kuo investigated cholinergic, histaminergic, and noradrenergic regulation of LTP stability and induction threshold: cognitive implications. Our group has investigated the interactions of nicotinic acetylcholinergic systems with antipsychotic drug effects on dopaminergic and serotonergic systems and cognitive function in intact systems as well as dysfunction produced by NMDA glutamate blockade. The relative importance of limbic system interactions has been investigated by Guterman and Richter-Levin who evaluated the effects of neuromodulators of long-term potentiation in the amygdala and hippocampus in response to stress.

While the majority of research on neurotransmitter interactions and cognitive function has investigated these functional interactions in rodents, it is important to explore other experimental models to determine the generality and specificity of the functional interactions. Drs. Ewert and Schwippert showed that modulation of visual perception and action by forebrain structures and their interactions in amphibians. It is also vital to conduct human studies to determine how the animal model studies match clinical reality. Dr. Oades has investigated function and dysfunction of monoamine interactions in children and adolescents with attention deficit hyperactivity disorder (ADHD).

An important developing technology is the emergence of computer models of neurocognitive function. Because the neurotransmitter interactions are quite complex, it is important to develop computational tools to organize in a systematic fashion the discoveries concerning the parts of the systems investigated into an understanding of the function whole. Larrauri and Schmajuk have developed computer models of an elementary form of neurobehavioral plasticity, prepulse inhibition. Computer models will be of increasing use for the understanding of complex interacting neural systems forming the basis of cognitive function.

Neurotransmitter interactions in the basis of cognitive function is critical for the basic understanding of how neural systems produce functions that comprise cognition. As different parts of the neural networks work together the respective neurotransmitters and their receptors play important links in the system. These transmitter interactions are also critical to the understanding of cognitive dysfunction. Cognitive disorders such as ADHD and Alzheimer's disease as well as the cognitive impairments of schizophrenia are characterized by disrupted interactions of a variety of neural systems using a variety of neurotransmitters and receptor systems. Finally, understanding neural interactions underlying cognitive function is essential to the rational development of therapeutic treatments for cognitive dysfunction. Chronic dysfunction of one neural system would inevitably cause adaptive changes in other related neural systems. Drug therapy for cognitive dysfunction is most often directed at modulating neurotransmitter systems by direct receptor agonist or antagonist actions or indirect actions on the synthesis, degradation or sequestration of neurotransmitters. Like chronic dysfunction chronic drug therapy will inevitably cause adaptive changes in the target system as well as interacting neural systems. Understanding the neurotransmitter interactions underlying cognitive function is essential for both the appreciation of the complexities of chronic neural dysfunction as well their therapeutic treatment.

References

1. Levin ED, Decker MW, Butcher LL (eds) (1992) *Neurotransmitter interactions and cognitive function*. Birkhäuser, Basel
2. Russell RW (1992) Interactions among neurotransmitters: Their importance to the "Integrated Organism". In: Levin ED, Decker MW, Butcher LL (eds): *Neurotransmitter interactions and cognitive function.* Birkhäuser, Basel

Neurotransmitter Interactions and Cognitive Function
Edited by Edward D. Levin
© 2006 Birkhäuser Verlag/Switzerland

Neurotransmitters and cognition

Mohammad R. Zarrindast

Department of Pharmacology, School of Medicine, Tehran University of Medical Sciences, Tehran, Iran

Introduction

Cognition deficits have received much attention over the last two decades. The severe impairment of cholinergic function in dementias, particularly in age-related cognitive decline and Alzheimer's disease has been indicated. However, loss of cholinergic activity may play a key role in the cognitive symptoms but it cannot clearly demonstrate the entire mechanism involved. Increase in acetylcholine or administration of direct cholinergic agonists are not able to combat such cognitive impairments. Moreover, stimulation of monoaminergic activity, in conjunction with cholinergic therapies, may induce effective treatment of Alzheimer's disease [1]. Furthermore, effects of other neurotransmitters, such as cathecholamines, serotonin, GABA, histamine, adenosine, nitric oxide and cholecystokinin and their role in learning and memory in animals have been tested. Decrease or increase in levels of the neurotransmitters or activation or blockade of different receptors related to the neurotransmitters indicate that other mechanisms may alter learning and memory. The role of a network consisting of different neurotransmitter systems may be important for learning and memory processing. Therefore, the aim of this section is to simply show the involvement of different neurotransmitter systems in cognitive behavior.

Cholinergic system and cognition

Both human and animals studies have shown that the cholinergic system, particularly muscarininc acetylcholine receptors, may have a role in memory [2–5]. Moreover, functional imaging studies revealed that cholinergics increase and anticholinergics depress activity measures in subcortical regions such as thalamus that are responsible for maintaining arousal and attention [6].

A reliable relationship between the status of basal forebrain cholinergic neurons and severity of age-related impairment has been indicated [7, 8], and an extensive literature has also demonstrated an age-dependent decline in various aspects of learning and memory [9].

Cholinergic function could contribute to both the cognitive deficits and dementia. Although acetylcholine could be considered to be a neurotransmitter that is highly involved in learning and memory processes, the validity of experimental data from pharmacological and lesion studies, which were interpreted in terms of cholinergic mechanisms, has been seriously questioned by some authors [10, 11].

The latest picture of muscarinic receptors shows five subtypes, named M1-M5, which are typical members of the superfamily of G-protein coupled receptors. Memory impairment caused by scopolamine may be due to blockade of M1 receptors [2].

Brain cholinergic innervation comes from five major nuclei: (a) the basal forebrain, which innervates the cortex and hippocampus; (b) the diencephalus, which gives rise to local circuits and innervates the cortex; (c) the striatum, which also gives rise to local circuits; (d) the brain stem, which innervates the thalamus, the basal forebrain, the hindbrain, and the cerebellar cortex; and (e) the spinal cord, which innervates the cranial and somatic muscles and secretory glands. The system is extensively interconnected, leading to the coordinated firing of neurons and different cholinergic subsystems [12–14].

Damage to the basal forebrain (CBF) region can result in global cognitive impairments; for instance, aneurysms of the anterior communicating artery that injure the basal forebrain are associated with amnesia and impairments in executive function [15–17].

Damage to components of the cholinergic basal forebrain (CBF) with electrolesion, excitotoxins, or the cholinergic neurotoxin AF64A, produces deficits in a variety of cognitive tasks. These deficits have typically been ascribed to impairments in working/episodic memory or attention [18–21]. In addition, the pathological hallmarks of Alzheimer's disease include the extensive degeneration of cholinergic neurons in the CBF as well as neurofibrillary targets and amyloid plaques in the target of the CBF such as the cerebral cortex and hippocampus [22, 23].

The cholinergic nuclei of the forebrain have a diffuse distribution, thus traditional approaches to removing these structures (i.e. excitotoxic, electrolytic, radiofrequency lesions) in experimental animals invariably include damage to non-cholinergic neurons and only incomplete destruction of cholinergic cells. The development of an animal model to examine the cholinergic hypothesis therefore requires more selective neuropathology than is possible with traditional lesion strategies [10, 24].

Nicotine acetylcholine receptor mechanism and cognition

Nicotine is the only chemical available in a biologically significant quality in tobacco that has been shown to meet criteria for an abusable drug.

Nicotine receptors exist as a variety of subtypes, a hetrogenicity that is due to the diversity of the genes encoding acetylcholine nicotininc receptor subunit. Sixteen acetylcholine nicotinic receptor subunit genes have been cloned from vertebrates ($\alpha 1$ to $\alpha 8$, $\beta 1$ to $\beta 4$, γ, ε and δ). These receptors are cationic channels that belong to ligand-gated ion channels, which are key molecules in the cholinergic nicotinic transmission in a number of areas in brain peripheral nervous system and can be opened by nicotine and acetylcholine [25].

Nicotinic receptors are mainly located in various cortical areas, the periacqueductal grey matter, the basal ganglia, the thalamus, the hippocampus, the cerebellum, the retina, and in chickens the optic lobes [14].

Most of nicotine effects have shown to be mediated by changes in the release of a number of neurotransmitters [26]. The agent produces several behavioral changes through different neurotransmitter systems.

Nicotine induces purposeless chewing through dopaminergic or nicotinic mechanisms in rats [27], grooming in rats by activation of cholinergic and dopaminergic mechanisms [28], hypothermia through indirect dopaminergic mechanism [29], and anxiogenesis in mice through adrenergic and cholinergic systems [30]. The drug increases apomorphine-induced licking [31] and sniffing [32] behaviors in rats, and attenuates naloxone-induced jumping behavior in morphine-dependent mice by central nicotinic receptors [33]. It potentiates morphine analgesic effect [34], induces antinociception by cholinergic and opioid mechanisms [35], and potentiates sulpiride-induced catalepsy through cholinergic and nicotininc mechanism [36]. There is cross-tolerance between morphine- and nicotine-induced hypothermia in mice [37]. Therefore, one may expect that nicotine affects learning and memory through different neurotransmitters.

Activation of neuronal nicotinic acetylcholine receptors (nAChRs) has been shown to maintain cognitive function aging or the development of dementia. Nicotinic receptor agonists may improve cognitive function in aged or impaired subjects. Epidemiological and also both *in vitro* and *in vivo* animal studies have shown that smoking may be protective against the development of neurodegenerative diseases. However, nonsmokers may have twice the risk for Alzheimer's disease (AD) or Parkinson's disease, smoking may have more association with Parkinson's disease (PD) than Alzheimer's disease. Negative association between cigarette smoking and AD or PD shown by several epidemiological studies have also been suggested. The epidemiological data suggest that smoking protects against development of some forms of PD, however, there is a study that suggests that smoking increases the likelihood of PD along with other factors, including old age and family history of PD. Controversy also exists about the benefit of smoking in AD, and epidemiological data are less consistent for a protective effect of nicotine in AD; one study indicates that nicotine intake may protect against neurochemical markers of neurodegeneration related to AD (for review see [38]).

Nicotine has been shown to have both facilitating and impairing effects on learning and memory in animals. The dose of nicotine used may play an important role in the drug effect. The drug has been shown to improve recall in humans [39] and produce a retrieval deficit in mice [40]. The interaction of nicotine with postsynaptic nicotine receptors should play an important part in the production of its effects. However, many central effects of the drug have been shown to be attributed to changes in the release of a number of neurotransmitters [26] including acetylcholine [41]; the precise mechanisms involved in its responses are not clear.

Gilliam and Schlessinger [40] indicated retrieval deficits for nicotine in mice, however, in a passive avoidance learning task, it increased step-down latencies in mice, indicating increase in memory [42]. Locomotor activity is a major problem in testing the effect of different agents on learning and memory and it has been suggested that this task is more reliable than other methods of memory and learning assessment in this respect [43]. Nicotine in the dose used (0.5 mg/kg), which was effective, did not alter locomotion and thus the drug may improve memory retrieval [42]. Although cholinergic mechanism has been shown to be involved in memory [44], antimuscarinic, atropine failed to alter the nicotine-induced improvement of retrieval. Thus the involvement of muscarinic mechanism in the response of nicotine seems unlikely. It has been suggested that both peripheral and central mechanisms are involved in learning and memory processes [45, 46]. However, the improvement of retrieval induced by nicotine was decreased by the nicotine receptor antagonist mecamylamine [47] but not peripheral nicotinin receptor antagonist hexamethonium. These results may indicate that central nicotinic receptor sites are involved. There is evidence that dopaminergic neurons possess nicotinic receptors and nicotine enhances dopamine release by increasing neuronal firing and *via* direct presynaptic action on terminals [48]. Dopaminergic mechanisms have been shown to affect learning [49, 50]. The increase in nicotine enhancing response in memory retrieval [42] by a D1 dopamine receptor antagonist, SCH 23390 [51], but not D2 receptor antagonist sulpiride [52], may indicate that D1 dopamine receptor mechanism exerts a negative influence on the improvement of retrieval by nicotine. However, the antagonist alone did not elicit any response. SCH 23390 may also bind with high affinity to 5-HT2 receptors in the brain [53] and antagonize 5-HT2 receptor activation both centrally and peripherally [54, 55]. Although nicotine has been proposed to release catecholamines [26] and adrenergic mechanisms have been shown to affect learning and memory processes [45], the α-adrenoceptor antagonist phenoxybenzamine did not change the step-down latencies, thus α-adrenoceptor mechanisms appear not to be involved in the nicotine-induced improvement of retrieval. The β-adrenoceptor antagonist propranolol increased the nicotine response. The antagonist alone also increased the retrieval of the learned task, which is not in agreement with the report [56] showing that β-adrenoceptor activation enhances memory. It seems unlikely that propranolol actually potentiated the effect of nicotine, and thus further studies are needed to elucidate the precise mechanism involved in the interaction between β-adrenoceptor and nicotine mechanism in the process of learning and memory (see [42]).

Interaction of different neurotransmitter systems with cholinergic system in cognition

However, many clinical [57–59] and experimental [60] studies have also shown that the cholinergic system may be, in part, important in cognition, and there is evidence which shows interactions between cholinergic and other neurotransmitter systems such as adrenergic, dopaminergic, serotonergic, GABA, adenosine CCK, nitric oxide, opioids and histamine systems, which should be considered in memory processes.

Dopamine receptor mechanism and cognition

Dopamine has been suggested to be a potential substrate for synaptic plasticity and memory mechanisms [61]. Direct pharmacological manipulation of dopamine activity by administration of dopamine agonists provides evidence of a role for dopamine in learning and memory [43, 45, 62–64]. It has been suggested that dopamine uptake inhibition improves learning of inhibitory avoidance and increases hippocampal acetylcholine release [65]. However, it is not clear whether stimulation of dopamine receptor sites facilitates or impairs learning and memory. The discrepancy between the results may be due to activation of different receptor subtypes and different experimental design. To date, five dopamine receptor subtypes have been cloned and are differentiated as belonging either to the D1 (D1 and D5 subtypes) or D2 (D2, D3 and D4 subtypes) receptor families [66–68]. D1 dopamine receptor stimulation leads to an increase in the formation of cAMP, while the activation of the D2 dopamine receptor either does not increase cAMP or actually decreases it [69]. The D3 dopamine receptor is not linked to adenylate cyclase and is not influenced by GTP, which regulates binding associated with D1 and D2 receptors [70, 71].

Both D1 and D2 receptors have been implicated in various learning and memory processes [64, 72–74]. Dopamine D1 receptor agonists were found to enhance passive [72] and improve cognitive performance in rats [75], or have no effect on learning [64, 74]. In active avoidance test, a single administration of D1 receptor agonist SKF 38393 in rats [76] improved retrieval, which was antagonized by the D1 antagonist [42]. It has been proposed that dopamine D1 receptors are involved in at least one form of the cognitive processes [77].

Low and high doses of apomorphine, which is a mixed D1/D2 dopamine receptor agonist [78], in one-way active avoidance procedure in mice improved or impaired retrieval, respectively [42]. The opposite effects induced by low and high doses of apomorphine have also been shown for single-trial passive avoidance learning in mice [43], suggesting that the central dopamine systems may play an important role in modulating memory processes. Since D1 antagonist SCH 23390 [51] and a high dose of the D2 antagonist sulpiride [52] reversed the impairment induced by higher doses of apomorphine, it has been concluded that both D1 and D2 receptors are involved in retrieval deficits. However, there is a report indicating that apomorphine attenuates forgetting [45], others [43] suggested that low dose of apomorphine by acting at pre-synaptic dopamine receptors improves memory retrieval, while higher doses of the drug stimulate postsynaptic D2 dopamine receptor and impair memory retrieval. Administration of low and high doses of the D2 receptor agonist bromocriptine in mice [69, 79] also improved or impaired retrieval, respectively. While a low dose of sulpiride (20 mg/kg) antagonized the improvement induced by low dose of bromocriptine, a high dose of the antagonist reversed the impairment induced by higher doses of bromocriptine treatment. When D2 agonist quinpirole [80] was employed, a similar result was obtained. However, the effect was slight and not statistically significant. Low and high doses of the D2 agonists and bromocriptine may act on pre- or post-synaptic dopamine receptors, repectively, and it may be suggested that activation of pre-synaptic D2 receptors will improve, while stimulation of postsynaptic D2 receptors will impair retrieval in trained mice. It has been suggested that D2 receptors in the ventral hippocampus are involved in memory performance, possibly through the regulation of acetylcholine release [81].

Other experiments also suggested that D3 receptors are involved in the modulation of stimulus-reward learning by the mesoamygdaloid dopamine receptors [82].

Noradrenergic system and cognition

Noradrenergic pathways have been suggested to play an important role in the modulation of learning and memory [83–85]. The role of brain norepinephrine in memory processes has been originally assessed through post-training intracerebral administration of the neurotransmitter, as well as of reserpine [86, 87]. Based on these studies, locus coeruleus [88, 89], amygdala [90–92] and hippocampus [93–95] have been implied as important sites involved in the modulatory effects of the noradrenergic receptors on cognitive function.

Noradrenaline functions through four different receptors: α_1, α_2, β_1 and β_2, each of which has further subtypes [96]. Both α_2 and β-adrenergic receptors have been suggested to be involved in cognitive dysfunction of schizophrenia, Alzheimer's disease and attention deficit hyper activity disorder [97]. Furthermore, it has been proposed that α_2-adrenergic antagonists could be used in Parkinson's disease [98]. Moreovere, some drugs such as imipramine [99], dexamethasone [100], a $GABA_B$ receptor agonist, baclofen [101] and histamine [102] induce impairment of memory through α_2-adrenoceptors. β-adrenoceptor also may be involved in the decrease in memory acquisition induced by theophylline [103]. In conclusion, however, activation of both α- and β-adrenoceptors may impair memory; α_2-adrenergic antagonists may improve memory in some neuropsycological disorders.

Serotonin and cognition

Serotonin (5-hydroxytryptamine, 5-HT) is a biogenic amine that is involved in a wide range of physiological functions including sleep, appetite, pain perception, sexual activity, and memory and mood control [104]. Neurochemical studies demonstrated loss of both cholinergic [105, 106] and serotoninergic amines in the brain of Alzheimer's patients [107, 108]. Cholinergic-serotonergic interactions have been suggested to play an important role in learning and memory (for reviews see [109]). There are also structures in mammalian brain in which cholinergic and serotonergic neuroanatomical substrates can be identified. These structures include the basal forebrain nuclei (diagonal band of Broca, septal region, nucleus basalis), the laterodorsal and pedunculopontine tegmental nuclei, the hipocampus, the striatum and at least some cortical areas. However, the general picture lacks precision, essentially because the histological or morphological observations often indicate possibilities rather than certitudes (for more details see [109]).

The effects of serotonin in the CNS are mediated through different 5-HT receptor types. These receptors have been classified into four main classes; named 5-HT_1, 5-HT_2, 5-HT_3 and 5-HT_4 [110, 111]. 5-HT_5, 5-HT_6 and 5-HT_7 receptors have also been cloned but are not yet fully characterized. Moreover, four subtypes of 5-HT_1 have been demonstrated and named 5-HT_{1A}, $5\text{-HT}_{1B/1D}$, and 5-HT_{1F}.

Some behavioral and neurobiological studies did not indicate any link between cognition and 5-HT receptor subtypes (see [112]). On the contrary, there is evidence indicating that different subtypes of the receptors potentially interact to contribute to a particular function. Riad et al. [113] demonstrated that 5-HT_{1A} receptor agonists promote the growth and branching of neurites of cholinergic cells in primary culture of fetal septal neurons. In regard to the role of 5-HT receptors in the control of acetylcholine release, it has been suggested that in enthorhinal cortex 5-HT activates 5-HT_3 receptors located on GABAergic neurons that in turn inhibit cholinergic function [114]. Studies in experimental animals have indicated that decrease in cholinergic and serotonergic activity produces a synergic decrement in learning and dementias of the Alzheimer type [115]. 5-HT_{1A} receptors can play a key role in cholinergic-serotonergic interactions, and may be potential targets for a possible pharmacotherapy of Alzheimer's disease [116]. There is evidence that 5-HT_{1A} and/or 5-HT_{1C} agonists may provoke a new approach to the treatment of learning disorders in aging or Alzheimer's disease [117]. Blockade of 5-HT_{1A} receptors may compensate the loss of cholinergic excitatory input on pyramidal cells, probably by favoring the action of other excitatory transmitters [118]. Stimulation of 5-HT_{1B} receptors [119] and 5-HT_{1A} receptors in the CA1 region of the dorsal hippocampus has proposed to impair spatial but not visual discrimination in rats [120]. Fornix transection in the marmoset produces a specific effect on memory for, and acquisition of, visuo-spatial tasks, and this cognitive deficit was alleviated by a 5-HT_{1A} antagonist [121]. It has also been demonstrated that serotonin has an important role in cognitive processes, since excessive release, but not depletion, of serotonin leads to memory impairments in rats [122]. Stancampiano et al. [123] proposed that hippocampal acetylcholine could be involvd in attentional and cognitive functions underlying motivational processes, while serotonin could be implicated in non-cognitive processes (i.e. in the control of motor and feeding behavior). Since serotonin and acetylcholine neurotransmission is stimulatory activated during the spatial memory task, this suggestes that these neurotransmitter systems regulate behavioral and cognitive functions. The authors suggested that the combined degeneration of serotonin and acetylcholine systems is relevant in the behavioral and cognitive disorders observed in Alzheimer's disease. However, most studies indicate impairments of memory by stimulation of 5-HT_{1A} receptors and also 5-HT neurotransmission may be necessary in learning in some memory establishment and regulation [124]. It seems possible that serotonin in combination with other neurotransmitter systems be involved in memory processes.

GABAergic system and cognition

γ-Aminobutyric acid (GABA) is the main inhibitory neurotransmitter in the brain. GABA acts at various pharmacological distinct receptor subtypes, GABA_A, GABA_B and GABA_C [125–127]. There is extensive evidence indicating that the administration of GABAergic agents affects memory retention and learning [128–130]. Generally, GABA receptor agonists impair while its antagonist facilitate memory

[131–134]. GABA is cleaved from the synaptic cleft by uptake *via* specific transporters. Inhibition of such transporters increases the effectiveness of physiologically released GABA. Schmitt and Hiemke [135] have demonstrated that tiagabine, a GABA transporter inhibitor impairs spatial learning of rats in the Moriss watermaze. Baclofen, a selective $GABA_B$ receptor agonist has been shown to impair spatial learning in rats [136]. The role $GABA_B$ receptors play in neural transmission and inhibitory regulation may significantly contribute to the processes of learning, information storage, and memory. Pharmacological manipulation of $GABA_B$ receptors may powerfully alter neuronal transmission and synaptic plasticity in the hippocampus [137]. Furthermore, Brucato et al. [138] demonstrated that $GABA_B$ receptor blockade suppresses the induction of long-term potentiation (LTP) in the dentate gyrus *in vivo*. Additionally, they provided evidence of a behavioral task that is dependent upon $GABA_B$ receptor function. While $GABA_B$ receptor blockade produced no change in performance on the radial maze, a deficit in the acquisition of spatial memory was observed in the water maze. The investigators indicate that $GABA_B$ receptors are important for the induction of LTP in the dentate gyrus. In addition, they suggest that $GABA_B$ may play a critical role in spatial learning tasks where stress may influence performance. The intermediate and medial hyperstriatum is a forebrain in the domestic chick that is a site of information storage for the learning process of imprinting. McCabe et al. [139] have proposed that this regional plasticity in GABAergic neurones is involved in the learning mechanisms of learning and memory, and that taurine also contributes to these mechanisms.

It has been shown that endogenous GABA causes tonic inhibition of actylcholine release in the ventral hippocampus *via* septal $GABA_A$ receptors and, to a lesser extent, *via* $GABA_B$ receptors in the medial septum and hippocampus [140] (F10). Moreover, high levels of septal GABA receptor activity might impair memory by down-regulating acetylcholine levels in the hippocampal region [141]. The interactions between cholinergic and GABAergic systems in learning and memory have been shown by several investigators [130, 142–146]. Hippocampus, amygdala and septum operate in parallel in memory consolidation in the avoidance task [147]. In amygdala, cholinergic muscarinic receptors enhance and GABA receptors inhibit memory consolidation [147]. The hippocampus, a model region for the study of learning and memory processes [148], is rich in cholinergic synapses that are under the inhibitory control of the GABAergic system [149]. Post-training intrahippocampal injection of GABAergic drugs has been shown to impair memory retention of passive avoidance learning in rats. The experiments showed that both $GABA_A$ and $GABA_B$ receptor activation may induce impairment of memory retention [100]. Both $GABA_A$, $GABA_B$ receptor agonists impaired improvement of acquisition of memory by an anticholinesterase, physostigmine in mice [146]. In a study using intracerebroventricular injection of GABA receptor agonists and antagonists in rats, it was found that $GABA_B$ receptor antagonists improved memory by itself and even $GABA_B$ receptors may be involved in imipramine-induced impairment of memory [150]. Other investigators also have shown that selective $GABA_B$ receptor antagonists can enhance cognitive performance in a variety of learning paradigms [151].

Conclusion: Both $GABA_A$ and $GABA_B$ receptors may impair memory processes, while $GABA_B$ receptor antagonists may improve memory. The effect of GABA may be elicited through interaction with acetylcholine release or interaction with muscarinic recptors. Amygdala may be an important site for GABA response. However, more studies may be required to elucidate the role of GABA in the sites involved in memory.

Histaminergic system and cognition

Accumulating evidence has established histamine as a central neurotransmitter [152–154]. The tuberomammillary nucleus consists of histamine synthesizing neurons located in the region of the posterior hypothalamus [155], with different varicose fibers in almost all parts of the brain [156, 157], including neostriatum, hippocampus, and tectum [153, 158]. The actions of histamine appear to be mediated by three different types of receptors, which differ in pharmacology, localization and intracellular response that they mediate [159]. Histamine receptors include postsynaptic histamine H_1 and H_2 and presynaptic histamine H3 receptors which control the release of neuronal histamine [160–162] and many other neurotransmitters such as noradrenaline, dopamine, serotonin and acetylcholine as auto- and heteroreceptors, respectively [163–165]. A role of histamine and its receptors in some specific brain processes such as cognition and novel environment-motivated exploration has also been characterized [166]. Futhermore, on the basis of lesion studies, histamine has been implicated in the processes underlying the functional recovery from brain damage, in the learning, memory and reinforcement (for review see [167]).

In behavioral experiments, histaminergic modulation of cholinergic activity is suggested by results of experiments showing that several histamine receptor ligands can antagonize spatial learning deficits caused by scopolamine [168, 169]. Furthermore, performance in several learning tasks thought to depend on cholinergic transmission [170, 171] is enhanced by lesions of tuberomammillary nucleus of the hypothalamus that produce a striking loss of histamine markers in the tuberomammillary nucleus [167, 172]. In a passive avoidance task, we have shown that histamine reduced, but the histamine H_1 receptor antagonist, pyrilamine and the histamine H_2 receptor cimetine increased memory [173], which has also been shown previously [174]. It has been reported that activation of histamine H_1 receptors attenuated histamine-induced memory impairment [172] or increased memory recall [175], whereas activation of the histamine H_2receptor was ineffective [176]. In contrast, there is a report indicating that the histamine H_2 receptors appear to exert some type of modulating effect on the inhibitory action of the histamine H_1 receptor activity [166]. Moreover, different versions of avoidance learning (active, passive and inhibitory avoidance) have been used to study the associations between histamine and memory and reinforcement. The conclusions of these studies are contradictory [167, 172, 177, 178], although they offer a wide range of data that in recent years has supported an inhibitory effect of histamine.

In our experiments, the interaction of the histaminergic system with the cholinergic system has been tested [173]. In these experiments, post-training intracerebroventricular (i.c.v.) injections of cholinergic agonists, acetylcholine or nicotine improved memory retention, while the anticholinergic drug scopolamine reduced memory retention. The histamine receptor antagonists potentiated, while histamine decreased the response induced by actylcholine or nicotine. Thus, the histaminergic system may interact with the cholinergic system on memory retention. In support of these findings, it has been shown that stimulation of muscarinic receptors by muscarinic agonists decreases the release of histamine in rat brain [179], and that histamine modulates the activity of cultures of cholineric cells *via* histamine H_1 and H_2 receptors [180].

Opioids-histamine interactions and cognition

Histamine release in target regions is under the control of inhibitory M_1 muscarinic [181] and opioid μ-receptors [182], as well as facilitatory μ-opioid receptors [183]. Post-training i.c.v. administration of morphine also reduced, while the opioid receptor antagonist, naloxone or the partial agonist mixed agonist/antagonist pentazocine increased memory retention. Thus the opioid may elicit an inhibitory role in memory retention.

Stimulation of κ-opioid receptors has been shown to attenuate memory dysfunctions resulting from the blockade of muscarinic M_1 receptors [184], while treatment with morphine and other opioid receptor agonists may disrupt memory [185]. However, there is also a report indicating that endogenous opioid systems do not play a major role in modulating neural mechanisms that maintain accurate spatial memory [186].

Morphine has been shown to elicit an increase in histamine release [187]. Post-training administration of different doses of histamine attenuated memory retention [188, 102]. Histamine H_1 receptor antagonist pyrilamine, or the histamine H_2 receptor antagonist cimetidine, increased memory retention [102]. Both the antagonists decreased the histamine response, which has also been shown previously [174].

Interactions of opioid and histaminergic systems on memory retention have also been studied [102]. Histamine reduced memory retention and showed potentiation of morphine-induced impairment of memory retention, while histamine receptor antagonists increased memory and reduced the morphine response, indicating that two histamine and opioid systems have a close interaction. This is supported by reports that high doses of morphine can induce increased histamine release in the rat central grey [187], and opioids enhance brain turnover histamine that can be blocked by naloxone [183]. It should be also considered that histamine H_2 receptor antagonist but not histamine H_1 receptor blocked morphine-induced locomotor hyperactivity in mice [189]. Moreover, several H_1 antagonists have been shown to have potentiating effects when administered both alone [190, 191] and in combination with other opioids and even tend to augment the pleasurable effects of the latter [192, 193]. Furthermore, there is evidence that opioids may modulate neural processes that are essential to memory consolidation. It can be concluded that morphine and histamine may influence memory through a common pathway.

Morphine state-dependent learning

Learning and memory in laboratory animals are known to be affected by opioids and their antagonists [194]. A method based on the measurement of step-down latency in passive avoidance task has been developed for the study of learning and memory in the laboratory animals. The latency is reduced by pre-training beta-endorphin treatment [195–197] and enhanced by the same dose of the drug when administered 24 h later in the pre-test session [198]. This is known as the state-dependent learning (St-D). Kameyama et al. [199] obtained the same results after the administration of moderate doses of morphine (5–10 mg/kg) in mice. The exact mechanism of this action of morphine is not fully elucidated. However, it has been demonstrated that μ-opioid receptors are directly involved [200, 201]. Different hypotheses have been proposed to explain the memory enhancement effect of morphine when the drug is used in the pre-test session [202]. According to Introini-Collison and Baratti [203], and Ragozzino and Gold [204] the memory enhancement of morphine response is mediated by the cholinergic system.

We have shown that the administration of a central and peripheral anticholinesterase drug (physostigmine) not only mimicked the effect of morphine administration on the test day, but also when co-administered with morphine it increased memory recall [146]. The results demonstrated also that the administration of the peripheral anticholinesterase drug (neostigmine) failed to show an intrinsic activity or to change the memory impairment of morphine. In agreement with the above results, both atropine (as a central and peripheral antimuscarinic agent) and mecamylamine (as a central and peripheral nicotinic receptor antagonist) prevented the memory recall by morphine on the test day. The above results are suggestive of the part played by the cholinergic system in the effects of morphine on memory.

Opioids not only modulate memory processes but also produce analgesia by actions at several sites. Tyce and Yaksh [205] have demonstrated that systemic administration of opioids increases norepinephrine (NE) concentrations in lumbar cerebrospinal fluid (CSF). They have suggested that spinally released NE produces analgesia in part by activating spinal cholinergic interneurons to release acetylcholine (ACh). In summary, opioid analgesia seems to be partly a result of the cascade of norepinephrine release followed by the release of acetylcholine. Moreover, acetylcholine produces analgesia when administered spinally, an effect blocked by muscarinic receptor antagonists [206]. Analgesia from central opioid injection is also partially reversed by spinal injection of muscarinic receptor antagonists [207]. Furthermore, centrally administered opioids increase ACh concentrations in the CSF [208]. Many investigators have studied the interaction between opioids and the cholinergic system in memory performance. Introini-Collison and Baratti [203] have demonstrated that muscarinic agonists antagonize β-endorphine-induced memory impairment. Baratti et al. [209] have shown that the memory-enhancing effect of naloxone can be blocked by the muscarinic antagonist atropine. These observations may suggest that activation of opioid receptors (when administered on the pre-training day) inhibits the activity of the cholinergic system and consequently impairs memory function. Some studies have demonstrated that opioid receptor agonists such as morphine and endorphin, possessing higher affinity for μ-opioid receptors, inhibit cholinergic activity in the hippocampus [142]. Moreover, it has been reported that mu and delta opioid receptors locate on cholinergic terminals, which are normally under tonic inhibition by the opiate system [210]. Ragozzino and Gold [211] have demonstrated that morphine injected into the medial septum of rats, at a dose that impairs memory, decreases hippocampal acetylcholine output. They have suggested that learning and memory impairment caused by acute administration of morphine may be, at least partially, related to a decrease in hippocampal acetylcholine release. Other studies by using *in vivo* microdialysis have revealed that acute morphine administration significantly decreased release of acetylcholine in some brain regions [212, 213].

Although the above evidence has implicated a close correlation between the opioid and cholinergic systems, the exact mechanism of their interactions is still not clear. However, it is also possible that morphine-modulating memory processes are mediated in ways other than through direct opiate-cholinergic interaction [214]. For example, it has been reported that glucocorticoids and their receptors are involved in memory improvement [215, 216].

As mentioned in the method section, the effects of morphine and cholinergic modulator drugs either alone or in combination, were also studied on the locomotor activity of the animals. According to the results of the present experiment, although hexamethonium showed no effect on the memory recall, it increased locomotor activity of the animals. On the other hand, both atropine and mecamylamine inhibited memory recall by morphine without showing any effect on the locomotor activity. Furthermore, both neostigmine and physostigmine decreased locomotor activity, but only physostigmine, which enters into the CNS, increased memory recall. The above results are in agreement with the results reported by other investigators. Sanberg and Fibiger [217] have demonstrated that oral administration of taurine resulted in the impairment in retention of a step-down passive avoidance task in rats without changes on spontaneous locomotor activity. McNamara et al. [218] have reported that treatment with (+/−)3,4-methylenedioxymethamphetamine (MDMA) increased locomotor activity without a significant change in step-down passive avoidance behavior in rats. Barros et al. [219] have studied the effects of bupropion and sertraline on memory retrieval and found it unrelated to locomotor activity as well. These findings suggest that the locomotor activity and memory recall of the step-down passive avoidance task are not inter-related.

In conclusion, considering the effect of physostigmine (enhancement of the memory recall), and atropine or mecamylamine (prevention of the memory recall) by acute administration of morphine, on the test day, one may suggest that morphine-induced memory impairment is closely related to its inhibition of the central cholinergic activity.

Administration of glucose on the test day did not affect memory recall, but increased the enhancement of memory induced by morphine [220]. Furthermore, insulin alone did not alter memory, but co-administered with morphine on the test day it significantly reduced the enhancing effect of morphine on memory.

The fact that blood glucose level changes were parallel to that of memory recall performance suggest that the latter is a direct effect of blood glucose changes after the administration of glucose or insulin, in the presence of exogenous morphine [220]. This confirmed the previous reports on the beneficial effect of glucose administration on memory enhancement using a variety of behavioral tests [221, 222]. Three possibilities have been coined to explain how glucose might enhance memory. One hypothesis proposes that the circulating glucose levels modulate brain processes involved in memory through the activation of the cholinergic system by increasing the synthesis of acetylcholine, as discussed in the introduction [204]. The second hypothesis suggests that glucose interacts directly with the opioidergic system and reverses several actions of the opioid drugs including the induction of memory impairment [223]. The effects of glucose on the cholinergic and opioidergic systems might be interrelated. One explanation is that the memory deficit by morphine is due to a decrease in hippocampal acetylcholine release which glucose attenuates by increasing the activity of the cholinergic system. The second hypothesis may be summarized as follows: when glucose was co-administered with morphine, it enhanced acetyl-CoA production, which in combination with choline increases acetylcholine synthesis, followed by enhanced stimulation of mu receptors by morphine [211]. The third hypothesis suggests that glucose administration increases its metabolism followed by an increase in the intraneuronal ATP levels resulting in a blockade of ATP dependent potassium channels [224]. The channel blockade depolarizes the neuron and increases neurotransmitter release [225]. According to the latter hypothesis, glucose may modulate memory-dependent behavior by regulating the ATP-dependent potassium channels.

Whatever the mechanism of this effect of glucose might be, its memory enhancing action can be demonstrated only in the presence of morphine and not when glucose was administered alone. A similar observation was made with insulin as well.

The administration of glucose or insulin alone, although not statistically significant, tended to increase and decrease, respectively, the locomotor activity without significant changes in memory retrieval. When glucose was co-administered with morphine, the effects of this combination on locomotor activity and memory recall were dependent on the doses of glucose. At the dose of 50 mg/kg, glucose co-administered with morphine showed no effect on locomotor activity or memory retrieval. At a higher dose (100 mg/kg), glucose administration showed no effect on locomotor activity but increased significantly the memory retrieval when compared with morphine alone. At the highest dose, (200 mg/kg), glucose administration increased significantly the locomotor activity without a significant change in the memory recall. The results obtained after the administration of morphine + three different doses of insulin were dependent on the parameter studied. Insulin at the doses of (5, 10 and 20 IU/kg) when co-administered with morphine decreased significantly the locomotor activity. Insulin only at the dose of 20 IU/kg, when co-administered with morphine, increased memory recall.

In summary, the above results show that increased locomotor activity, observed in the present experiment, was not concomitant with a decrease in the memory recall. Moreover, when locomotor activity was decreased the memory recall was not increased. This suggests that the locomotor activity and memory recall of the step-down passive avoidance task are not inter-related. This hypothesis confirms the results reported by other investigators. Sanberg and Fibiger [217] have demonstrated that oral administration of taurine resulted the impairment in retention of a step-down passive avoidance task in rats without changes on spontaneous locomotor activity. McNamara et al. [218] have reported that treatment with (+/−)3,4-methylenedioxymethamphetamine (MDMA) increased locomotor activity without a significant change in step-down passive avoidance behavior in rats. Vianna et al. [226] have studied the involvement of protein kinase C isoforms on memory retrieval and found it unrelated to locomotor activity or anxiety level of rats. Barros et al. [219] have studied the effects of bupropion and sertraline on memory retrieval and found it unrelated to locomotor activity as well.

In conclusion, the co-administration of glucose and morphine increased the effects of morphine on memory enhancement on the test day. Three mechanisms have been proposed to explain the effects of glucose on memory in the present experiments: increased activity of the cholinergic system, direct effect of glucose on opioidergic system and the modulation of the ATP-dependent potassium channels.

It has been shown that central K_{ATP} channel openers produce an antinociceptive effect similar to that of morphine [227]. Moreover, the K_{ATP} channel blockers antagonize opioid analgesia [228, 229], suggesting involvement of K_{ATP} channels in the analgesic effect of opioids. Stimulation of opioid receptors may also open potassium channels [230, 231]. In a study showed that the pre-test administration of the K_{ATP} channel blocker, glibenclamide, and not of diazoxide, restored the morphine-induced impairment of acquisition and showed retrieval. However, the pre-test administration of the K_{ATP} channel opener, diazoxide, did not retrieve the morphine-induced memory impairment, but when used with morphine, the drug decreased morphine state dependence. The response induced by glibenclamide was antagonized by diazoxide pretreatment. This suggests the involvement of K_{ATP} channels in the memory retrieval, but not interaction of K_{ATP} channel modulators with the action of morphine on the test day [232]. One may conclude that the observed effect of glibenclamide in the present experiment was not exerted through the activation of μ-opioid receptors. In accordance with this hypothesis, other investigators have also reported that glibenclamide has no significant affinity for opioid receptors. Therefore, the possibility exists that blockade of K_{ATP} channels facilitates memory recall after pre-test administration of morphine by a mechanism which is not dependent on opioid receptors.

Introini-Collison and Baratti [203] reported that the impairment of memory retention induced by post-training β-endorphin was reversed by physostigmine. Furthermore, it has been demonstrated that intraseptal morphine administration, at a dose that impairs performance of memory tasks, reduces acetylcholine output in the hippocampal formation, which suggests the involvement of the cholinergic system in some morphine actions [204]. On the other hand, Stefani and Gold [233] have demonstrated that K_{ATP} channel modulators increase acetylcholine levels in the hippocampus, which is suggestive of the involvement of the cholinergic system in the effects of K_{ATP} channel modulators as well. In this experiment the administration of scopolamine significantly prevented the effect of glibenclamide on memory retrieval on the test day.

In conclusion, the effect of glibenclamide on the test day, observed in the present experiments, is most likely exerted through an antagonistic effect on K_{ATP} channels and is less likely to be through its effects on the μ-opioid receptors. The retrieval of memory on the test day by glibenclamide may be exerted through its effect on the cholinergic system.

Adenosine systems and cognition

Adenosine is a key modulator of neuronal excitability and synaptic transmission [234]. Four types of receptors, named A_1, A_{2A}, A_{2B}, and A_3, mediate adenosine actions, to which G proteins are coupled [235]. Adenosine A_1 receptors are most prevalent and have the highest affinity among the adenosine receptors in the CNS. A_1 receptors inhibit neurotransmitter release [236]. A_2 receptors tend to enhance neuronal excitability and neurotransmitter release *via* high affinity, subtype (A_{2A}), or lower affinity subtype (A_{2B}) receptors [237]. The hippocampal formation is highly enriched with A_1 receptors [238] and low levels of A_{2A} [239, 240]. There is considerable evidence that endogenous adenosine modulates the excitability of hippocampal neurons *via* A_1 [241, 242] and A_2 [243] receptor mediated mechanisms. A_1 receptors affect activity-dependent synaptic plasticity in the hippocampus, attenuating long-term depression and inhibiting long-term potentiation [244]. Several adenosine A_1 receptor agonists and antagonists have been suggested to alter inhibition of avoidance learning. A_1 receptor mechanism has been shown to be involved in amnesia induced by post-training administration of pentylentetrazole [245]. Activation of these receptors decreases the acquisition of passive avoidance learning in mice [246], while blockade of adenosine A_1 and A_2 receptors facilitate memory acquisition and retention [235, 246]. It has been proposed that adenosine A_1 receptors in the posterior cingulate cortex inhibit memory consolidation in a way that their blockade facilitates memory for inhibitory avoidance in rats [247]. However, aminophylline, an adenosine receptor antagonist, has been suggeted to exacerbate status epilepticus induced by neuronal damage in rats [248], while a selective A_1 receptor agonist reduced postischemic brain damage and memory deficits in gerbils [249]. Furthermore, an interaction between angiotensin IV and adenosine A_1 receptors in passive avoidance task in rats has been shown. The adenosine antagonist theophylline increased, while the selective adenosine A_1 receptor agonist attenuated memory [250]. Moreover, adenosine A_{2A} but not A_1 receptors have been proposed to be involved in memory retention and consolidation [251].

Conclusion: Adenosine receptor subtypes are involved in memory processes, but clarification of the exact role of each receptor subtype in memory may need extensive experiments.

Cholecystokinin and cognition

Cholecystokinin (CCK) is one of the most abundant neurotransmitter peptides in the brain [252]. The sulphated octapeptide cholecystokinin (CCK8) exerts its effects through two G-protein-coupled receptors [253, 254]: CCK-B receptors (type B "brain") are found essentially in the CNS and CCK-A receptors (type A "alimentary") are highly concentrated in the gastrointestinal tract, but are also found in particular brain structures [255, 256]. In the brain, CCK-A receptors are only present in certain regions including the hippocampus, nucleus tractus solitarius, posterior nucleus accumbens, ventral tegmental area, and substantia nigra, whereas CCK-B receptors are widely distributed throughout the central nervous system (CNS) [255, 257]. There is also substantial evidence that CCK acts as a neurotransmitter and that it exerts a modulatory influence on several classic neurotransmitters including dopamine, serotonin, norepinephrine, GABA, glutamate and endogenous opioids [258, 259]. High concentrations of CCK are present in the hippocampus and frontal cortex, where they are involved in learning and memory processing [260]. Several behavioral studies have reported an involvement of CCK-related peptides in the modulation of learning and memory processes. In most of them active and passive avoidance tests have been used [261–263]. It is reported that nonselective agonists of CCK receptors such as CCK-8 and cerulein (ceruletide) prolong extinction of already learned tasks [264, 265], accelerate habituation to a novel environment [258], and prevent experimental amnesia in rodents (for review see [262]). It has been also suggested that CCK-A and CCK-B receptor agonists may have different roles in memory functions [266]. In particular, a balance between CCK-A mediated facilitates effects and CCK-B mediated inhibitory effects on memory retention have been proposed [267]. Impaired learning and memory in OLETF rats, which are without CCK-A receptors because of genetic abnormality, has also been shown [268]. However, the data describing the effects of CCK-B receptor agonists on memory function in laboratory animals has been variable.

It has been suggested that the CCK system in the hippocampus is involved in stress-induced impairment of spatial recognition memory [269]. There are reports indicating that selective CK-B receptor agonists (i.e. CCK-4, Bc 264) impair memory function in rodents [264, 267, 270]. Other reports suggest that intravenous administration of CCK-4 may adversely affect short-term memory consolidation and retrieval in young healthy individuals without decreasing psychomotor performance [271]. Furthermore, other reports indicated that systemic administration of selective CCK-B agonists improved the cognitive performances of rats measured in the spontanous alteration test and a spatial two-trial memory task [272–275]. These effects have been suggested to be dependent on the dopaminergic system in the anterior part of the nucleus accumbens [272]. Since, opposite behavioral and biochemical responses were observed when a CCK-B agonist was injected in the anterior nucleus accumbens [276, 277], the systemic effect of CCK-B agonists could not be from direct interaction between CCK-B receptors and dopaminergic terminals in the nucleus accumbens. There is data suggesting physiological involvement of the CCK system through its interaction with CCK-B receptors in the hippocampus to improve performance of rodents in the spatial recognition memory [278]. Consistent with this observation, the facilitatory effect of CCK-B receptor activation on memory processes in conditions of affective motivation is mediated by dopaminergic projections in the central amygdala [279] and involves the hippocampus [280].

Nitric oxide and cognition

It has been accepted that the free radical gas nitric oxide (NO) is an intracellular messenger in the CNS [70]. This messenger is a soluble, short-lived and freely diffusable gas which is produced from L-arginine by the enzyme NO synthase (NOS), which is found in various regions in the brain, including the hippocampus [281]. The enzyme activity can be inhibited by nitro analogues of L-arginine (for review see [282]). Activation of N-methyl-D-aspartate (NMDA) receptors in cerebellar [283, 284] and hippocampal [285] slices induces NO synthesis *via* this enzymatic pathway. Once produced, NO rapidly diffuses through membranes and activates guanylate cyclase, thereby increasing intracellular levels of cGMP and modulating neuronal activity [283, 285].

The role of NO in learning and memory formation has been the subject of a number of studies. Both NO release and NMDA receptor activation are necessary for induction of long-term potentiation (LTP), which is considered to involve the electrophysiological events related to synaptic plasticity and learning [286–288], and can be inhibited by nitro analogues of L-arginine [289]. Although, several behavioral investigations carried out in different rodent models have demostrated that compounds that block NO synthase inhibit learning, other studies have not supported this (see [290]). There is evidence concerning changes of NO-producing neurons during learning and memory. A memory-related up-regulation of NOS neurons in rat brain has been demonstrated, which provided further support for the involvement of NO in spatial learning and memory [291]. The results of a study demonstrated that, under the experimental circumstances used, nitric oxide is involved only in the facilitated learning and memory processes caused by pharmaceutical effect of L-arginine, and not involved in normal learning processes [292]. Moreover, it has been indicated that NO is involved in different stages of memory and a possible role for the NO donors in human memory disorders has been suggested (see [290]).

Glutamate receptor and cognition

Glutamate is the dominant excitatory neurotransmitter in the mammalian brain [293–296]. The receptors for glutamate are either ionotropic-ligand-gated ion channels for sodium and calcium, or metabotropic, with the signal transducted to other intracellular messengers like inositol triphosphate or cyclic AMP. Fast transmission is mediated by ionotropic glutamate receptors which are further classified according to their interaction with non-physiological glutamate analog, to NMDA (N-methyl-D-aspartate), AMPA (amino-3-hydroxy-5-methyl-4-isoxazolepropionic acid) and KA (Kainate) receptors, the latter two often being denominated together as non-NMDA receptors (for review see [297, 298]).

Both ionotropic (iGluR) and metabotrophic (mGluR) receptors are differentially distributed on pre- and postsynaptic sites to contribute to neuronal communication and signal processing, functions that determine learning and memory formation [299, 300]. Glutamate receptors have been implicated in several forms of diseases, including dissociative thought disorder, schizophrenia or various other forms of dementia [301, 302], and also in long-term potentiation (LTP) and long-term depression (LTD) [288, 303, 304]. NMDA receptor antagonists impair acquisition and retention in various learning tasks, suggesting involvement of NMDA receptors in synaptic plasticity of the central nervous system [288]. Furthermore, some studies indicate that metabotropic glutamate receptors are critically involved in synaptic plasticity of various brain structures, and seem to have an essential role in some learning and memory processes [305]. However, behavioral effects of the substances interacting with NMDA or metabotropic receptors differ widely in their amplitude and time- course according to the learning task used [305, 306]. Intracerebroventricular administration of a competetive NMDA receptor antagonist (D-2-amino-5-phosphonovalerate; D-AP5) blocks the induction of long-term potentiation in the hippocampus and impairs acquisition of a spatial learning task in a water maze [307]. However, the antagonist does not affect acquisition of a visual discrimination task in the water maze [307], and several others indicated that NMDA receptor antagonists may impair visual learning tasks either in water maze or in a radial maze [308–310]. It has also been suggested that NMDA receptor antagonists impair both working and reference memory in rats not pretrained to the tasks before treatment, or in pretrained rats tested in a novel environment; in contrast, the same substances did not affect working or reference memory in rats pretrained and tested in the familiar environment [311]. These data may indicate that NMDA receptor activation is involved in coding spatial representaions, but this role appears to depend on various factors, mainly the experience of the subject and contextual factors. Ungerer et al. [312] showed that NMDA receptor antagonists did not affect acquisition, retrieval or forgetting processes, and did not impair working or short-term memory. They stated that most impairment in learning and short-term memory processes consecutive to the administration of NMDA antagonists has been obtained following pretraining administration of NMDA antagonists at doses known to induce nonspecific effects, such as anxiolytic-like effects, motor or sensorial disturbances or antinociceptive effects. These authors, by using posttraining administration of the drugs, indicated that NMDA antagonists were not able to affect retention performance. In contrast, the antagonists induced significant deficits in long-term retention in the Y-maze avoidance and bar-press learning tasks. Ungerer et al. [312] suggested that mechanisms underlying the post-training performance increment require the activation of NMDA receptors. They also proposed that both NMDA and mGluRs are involved in spontaneous improvement of performance in the bar-press learning task. There are also reports suggesting that systemic administration of NMDA antagonist MK-801 impaired memory. However, nucleus accumbens may be involved in learning and memory [313–319], lesions of this structure by Ibotenic-acid which is known to induce loss of cell bodies [320], did not impair displaced-object discrimination or any other parameter measured. Therefore, it has been suggested that systemic administration of MK-801 exerts its effects upon structure other than nucleus accumbens. It has also been concluded that further studies are needed for better understanding of the glutamate transmission in the nucleus accumbens and associated structures in modulating memory and information processing [321].

References

1. Dringenberg HC (2000) Alzheimer's disease: more than a "cholinergic disorder" – evidence that cholinergic-monoaminergic interactions contribute to EEG slowing and dementia. *Behav Brain Res* 115: 235–249
2. Bymaster FP, Heath I, Hendrix JC, Shannon HE (1993) Cooperative behavioral and neurochemical activities of cholinergic antagonists in rats. *J Pharmacol Exp Ther* 267: 16–24
3. Terry AV, Buccafusco JJ, Jackson WJ (1993) Scopolamine reversal of nicotine enhanced delayed matching-to-sample performane in mokeys. *Pharmacol Biochem Behav* 45: 925–929
4. Moran PM (1993) Differential effects of scopolamine and mecamylamine on working and reference memory in the rat. *Pharmacl Biochem Behav* 45: 533–538
5. Kopelman MD (1986) The cholinergic neurotransmitter system in human memory and dementia: a review. *Q J Exp Psychol* 38: 535–573
6. Freo U, Pizzolato G, Dam M, Ori C, Battistin L (2002) Art review of cognitive and functional neuroimaging studies of cholinergic drugs: implications for therapeutic potentials. *J Neural Transm* 109: 857–870
7. Gallagher M, Nagahara AH, Burwell RD (1995) Cognition and hippocampal systems in aging: animal models. In: McGauph JL, Weinberger N, Lynch G (eds): *Brain and memory; modulation and mediation of neuroplasticity*. Oxford University Press, New York, 103–126
8. Rapp PR, Amaral DG (1992) Individual differences in the behavioral and neurological consequences of normal aging. *Trends Neurosci* 16: 104–110
9. Muir JC (1997) Acetylcholie, aging, and Alzheimer's disease. *Pharmacol Biochem Behav* 56: 687–696
10. Dunnett SB, Everitt BJ, Robbins TW (1991) The basal forebrain-cortical cholinergic system: interpreting the functional consequences of excitotoxic lesions. *Trends Neurosci* 14: 494–501
11. Fibiger HC (1991) Cholinergic mechanisms in learning, memory and dementia: a review of evidence. *Trends Neurosci* 14: 220–223
12. Mesulam MM, Geula C (1988) Nucleus basalis and cortical cholinergic innervation in the human brain: observation based on the distribution of AchE and ChAT. *J Comp Neurol* 275: 216–240
13. Mesulam MM, Geula C, Botwell MA, Hersch I (1989) Human reticular formation; cholinergic neurons of the peduncolopontine and lateral tegmental nuclei and some cytochemical comparison to forebrain cholinergic neurons. *J Comp Neurol* 281: 611–633
14. Gotti C, Fornasari D, Clementi F (1997) Human neuronal nicotinic receptors. *Prog Neurobiol* 53: 199–237
15. Diamond B, Deluca J, Kelley SM (1997) Memory and executive functions in amnestic and non-amnestic patients with aneurysms of the anterior communicating artery. *Brain* 120: 1015–1025
16. Abe K, Inokawa M, Kashiwagi A, Yamagihara T (1998) Amnesia after a discrete basal forebrain lesion. *J Neurol Neurosurg Psychiat* 65: 126–130
17. Damasio AR, Graff-Radford NR, Eslinger PJ, Damasio H, Kassell N (1985) Amnesia following basal forebrain lesions. *Arch Neurol* 42: 263–271
18. Chrobak JJ, Hanin I, Schmechel DE Walsh TJ (1988) AF64A-induced working memory impairment behavioral neurochemical and histological correlates. *Brain Res* 463: 107–117

19. Jarrard LE, Okaichi H, Steward O Goldschmidt RB (1984) On the role of hippocampal connections in the performance of place cue tasks: comparisons with damage to hippocampus. *Behav Neurosci* 98: 946–954
20. Markowska AL, Olton DS, Givens B (1995) Cholinergic manipulations in the medial septal area: age-related effects on the working memory and hippocampal electrophysiology. *J Neurosci* 15: 2063–2073
21. Stackman RW, Walsh TJ (1995) Distinct profile of working memory errors following acute or chronic disruption of the cholinergic septohippocampal pathway. *Neurobiol Learn Mem* 64: 226–236
22. Coyle JT, Price DL, Delong MR (1983) Alzheimer's disease: a disorder of cortical cholinergic innervation. *Science* 219: 1184–1190
23. Procter AW, Lowe S, Palmer AM, Francis PT, Esiri MM, Stratmann GC, Najlerahim A, Patel AJ, Hunt A Bowen DM (1988) Topographical distribution of neurochemical changes in Alzheimer's disease. *J Neurol Sci* 84: 125–140
24. Dunnett SB Fibiger HC (1993) Role of forebrain cholinergic systems in learning and memory relevance to the cognitive deficits of aging and Alzheimer's dementia. In: Cuello AC (ed): *Cholinergic function and dysfunction. Prog Brain Res* 98: 413–420
25. Le Novere N Changeux JP (1995) Molecular evolution of the nicotinic acetylcholine receptor: an example of multigene family in excitable cells. *J Mol Evol* 40: 155–172
26. Balfour DJ (1982) The effects of nicotine on brain neurotransmitter systems. *Pharmacol Ther* 16: 269–282
27. Samini M, Shayegan Y, Zarrindast MR (1995) Nicotine-induced purposeless chewing in rats: possible dopamine receptor mediation. *J Psychopharmacol* 9: 16–19
28. Zarrindast M., Sedaghati F Borzouyeh M (1998) Nicotine-induced grooming: a possible dopaminergic and/or cholinergic mechanism. *J Psychopharmacol* 12: 407–411
29. Zarrindast MR, Zarghi A Amiri A (1995) Nicotine-induced hypothermia through an indirect dopaminergic mechanism. *J Psychopharmacol* 9: 20–24
30. Zarrindast MR, Homayoun H, Babaie A, Etminani A Gharib B (2000) Involvement of adrenergic and cholinergic system in nicotine-induced anxiogenesis in mice. *Eur J Pharmacol* 407: 145–158
31. Zarrindast MR, Shekarchi M Rezayat M (1999) Effect of nicotine on apomorphine-induced licking behaviour in rats. *Eur Neuropsychopharmacol* 9: 235–238
32. Zarrindast MR, Mohadess G, Rezvani-pour M (2000) Effect of nicotine on sniffing induced by dopaminergic receptor stimulation. *Eur Neuropsychopharmacol* 10: 397–400
33. Zarrindast MR, Farzin D (1996) Nicotine attenuates naloxone-induced jumping behaviour in morphine-dependent mice. *Eur J Pharmacol* 298: 1–6
34. Zarrindast MR, Babaei-Nami A Farzin D (1996) Nicotine potentiates morphine antinociception: a possible cholinergic mechanism. *Eur Neuropsychopharmacol* 6: 127–133
35. Zarrindast MR, Pazouki M Nassiri-Rad Sh (1997) Involvement of cholinergic and opioid receptor mechanisms in nicotine-induced antinociception. *Pharmacol Toxicol* 81: 209–213
36. Zarrindast MR, Haeri-Zadeh F, Zarghi A Lahiji P (1998) Nicotine potentiates sulpiride-induced catalepsy in mice. *J Psychopharmacol* 12: 279–282
37. Zarrindast MR, Barghi-Lashkari S, Shafizadeh M (2001) The possible cross-tolerance between morphine- and nicotine-induced hypothermia in mice. *Pharmacol Biochem Behav* 68: 283–289
38. Picciotto MR Zoli M (2002) Nicotinic receptors in aging and dementia. *J Neurobiol* 53: 641–655
39. Peeke SC, Peeke HV (1984) Attention, memory, and cigarette smoking. *Psychopharmacology* 84: 205–216

40. Gilliam DM, Schlessinger K (1985) Nicotine-produced relearning deficit in C57BL/6J and DBA/2J mice, *Psychopharmacology* 86: 291–295
41. Chiou CY, Long JP, Potrepka R, Spratt JL (1970) The ability of various nicotinic agents to release acetylcholine from synaptic vesicles. *Arch Int Pharmacodyn Ther* 187: 88–96
42. Zarrindast MR, Sadegh M, Shafaghi B (1996) Effects of nicotine on memory retrieval in mice. *Eur J Pharmacol* 295: 1–6
43. Ichihara K, Nabeshima T, Kameyama T (1988) Effects of haloperidol, sulpiride and SCH 23390 on passive avoidance learning in mice. *Eur J Pharmacol* 151: 435–442
44. Haratounian V, Barnes E, Davis KL (1985) Cholinergic modulation of memory in rats. *Psychopharmacol (Berl)* 87: 266–271
45. Quartermain D, Judge ME, Leo P (1988) Attenuation of forgetting by pharmacological stimulation of aminergic neurotransmitter systems. *Pharmacol Biochem Behav* 30: 77–81
46. Gozzani JL, Izquierdo I (1976) Possible peripheral adrenergic and central dopaminergic influences in memory consolidation. *Psychopharmacol* 49: 109–111
47. Martin BR, Onaivi ES, Martin TJ (1989) What is the nature of mecamylamine's antagonism of the central effects of nicotine. *Biochem Pharmacol* 38: 3391–3397
48. Clarke PB (1990) Dopaminergic mechanisms in the locomotor stimulant effects of nicotine. *Biochem Pharmacol* 40: 1427–1432
49. Zarrindast MR, Hajian-Heydari A, Hoseini-Nia T (1992) Characterization of dopamine receptors involved in apomorphine-induced pecking in pigeons. *Gen Pharmacol* 23: 427–430
50. Bracs PU, Gregory P, Jackson DM (1980) Passive avoidance in rats: disruption by dopamine applied to the nucleus accumbens. *Psychopharmacol* 83: 70–75
51. Hyttel J (1984) Functional evidence for selective dopamine D-1 receptor blockade by SCH 23390. *Neuropharmacol* 23:1395–401
52. Stoof JC, Kebabian JW (1984) Two dopamine receptors: biochemistry, physiology and pharmacology. *Life Sci* 35:2281–2296
53. Bischoff S, Heinrich M, Sonntag JM, Krauss J (1986) The D-1 dopamine receptor antagonist SCH 23390 also interacts potently with brain serotonin (5-HT2) receptors. *Eur J Pharmacol* 129: 367–70
54. Bijak M, Smialowski A (1989) Serotonin receptor blocking effect of SCH 23390. *Neuropharmacol* 23: 1395–1401
55. Hicks PE, Schoemaker H, Langer SZ (1984) 5HT-receptor antagonist properties of SCH 23390 in vascular smooth muscle and brain. *Eur J Pharmacol* 105: 339–342
56. McGaugh JL (1988) Modulation of memory storage processes. In: Salomon PR, Goethals PRGR, Kelley CM, Stephenes BR (eds): *Perspectives of memory research.* Springer Press, New York, 33–64
57. Beatty WW, Butters N, Janowsky D (1986) Memory failure after scopolamine treatment: implications for cholinergic hypothesis of dementia. *Behav Neural Biol* 45: 196–211
58. Eagger SA, Levy R, Sahakian BJ (1991) Tacrine in Alzheimer's disease. *Lancet* 337: 989–992
59. Jones GMM, Sahakian BJ, Levy R, Warburton DM, Gray JA (1992) Effects of acute subcutaneous nicotine on attention, information processing and short-term memory in Alzheimer's disease. *Psychopharmacol* 108: 485–494
60. Dunnett SB, Toniolo G, Fine A, Ryan CN, Björklund A, Iversen SD (1985) Transplantation of embryonic ventral forebrain neurons to the nucleus basalis magnocellularis-II. Sensorimotor and learning impairments. *Neuroscience* 16: 787–797
61. Jay TM (2003) Dopamine: a potential substrate for synaptic plasticity and memory mechanisms. *Prog Neurobiol* 69: 375–390

62. Grecksch G, Matties H (1981) The role of dopaminergic mechanisms in the rat hippocampus for the consolidation in a brightness discrimination. *Psychopharmacol* 75: 165–168
63. Sara SJ (1986) Haloperidol facilitates memory retrieval in the rat. *Psychopharmacol* 89: 307–310
64. Pakard MG, White NM (1989) Memory facilitation produced by dopamine agonists: role of receptor subtypes and mnemonic requirements. *Pharmacol Biochem Behav* 33: 511–518
65. Nail-Boucherie K, Dourmap N, Jaffard R Costentin J (1998) The specific dopamine uptake inhibitor GBR 12783 improves learning of inhibitory avoidance and increases hippocampal acetylcholine release. *Cognitive Brain Res* 7: 203–205
66. Schwartz JC, Giros B, Martres MP Sokoloff P (1992) The dopamine receptor family: molecular biology and pharmacology. *Semin Neurosci* 4: 99–108
67. Gingrich JA, Caron MG (1993) Recent advances in the molecular biology of dopamine receptors. *Ann Rev Neurosci* 16: 299–321
68. Seeman P, Van Tol HHM (1994) Dopamine-receptor pharmacology. *Trends Pharmacol Sci* 15: 264–270
69. Kebabian JW, Calne DB (1979) Multiple receptors for dopamine. *Nature* 277: 93–96
70. Snyder SH (1992) Nitric oxide and neurons. *Curr Opin Neurobiol* 2: 323–327
71. Sokoloff P, Giros B, Martres MP, Bouthenet ML, Schwartz JC (1990) Molecular cloning and characterization of a novel dopamine receptor (D3) as a target for neuroleptics. *Nature* 347: 146–151
72. Bernabeun R, Bevilaqua L, Ardenghi P, Bromberg E, Schmitz, P, Bianchin M, Izquierdo I, Medina JH (1997) Involvement of hippocampal cAMP/cAMP-dependent protein kinase signaling pathways in a late memory consolidation phase of aversively motivated learning in rats. *Proc Natl Acad Sci USA* 94: 7041–7046
73. Izquierdo I, Medina JH, Izquierdo LA, Barros DM, de Souza MM, Mello e Souza T (1998) Short- and long-term memory are differentially regulated by monoaminergic systems in the rat brain. *Neurobiol Learn Mem* 69: 219–224
74. Wilkerson A, Levin ED (1999) Ventral hippocampal dopamine D1 and D2 systems and spatial working memory in rats. *Neuroscience* 89: 743–749
75. Hersi A, Rowe W, Gaudreau P, Quirion R(1995) Dopamine D1 receptor ligands modulate cognitive and hippocampal acetylcholine release in memory-impaired aged rats. *Neuroscience* 69: 1067–1074
76. Setler PE, Sarau HM, Zirkle CL ,Saunders HL(1978)The central effects of a novel dopamine agonist. *Eur J Pharmacol* 50: 419–430
77. El-Ghundi M, Fletcher PJ, Drago J, Silbey DR, O'Dowd BF, George SR (1999) Spatial learning deficit in dopamine D1 receptor knockout mice. *Eur J Pharmacol* 383: 95–106
78. Seeman P(1980) Brain dopamine receptors. *Pharmacol Rev* 32: 229–313
79. Jackson DM, Ross SB, Hashizume (1988) Dopamine-mediated behaviours produced in naive mice by bromocriptine plus SKF 38393. *J Pharm Pharmacol* 40: 221–223
80. Arnt J, Hyttel J, Meier E (1988) Inactivation of dopamine D-1 or D-2 receptors differentially inhibits stereotypies induced by dopamine agonists in rats. *Eur J Pharmacol* 155: 37–47
81. Umegaki H, Munoz J, Meyer RC, Spangler EL, Yoshimura J, Ikari H, Iguchi A, Ingram DK (2001) Involvement of dopamine D2 receptors in complex maze learning and acetylcholine release in ventral hippocampus of rats. *Neuroscience* 103: 27–33
82. Hitchcott PK, Bonardi CMT, Phillips GD (1997) Enhanced stimulus-reward learning by intra-amygdala administration of a D3 dopamine receptor agonist. *Psychopharmacol* 133: 240–248

83. Obserztyn M, Kostowski W (1983) Noradrenergic agonists and antagonists: effects on avoidance behaviour in rats. *Acta Physiol Pol* 34: 401–407
84. Introini-Collison IB, To S, McGaugh JL (1992) Fluoxetine effects on retention of inhibitory avoidance: enhancement by systemic but not intra-amygdala injections. *Psychobiology* 20: 28–32
85. Sirvio J, MacDonald E (1999) Central alpha1-adrenoceptors: their role in the modulation of attention and memory formation. *Pharmacol Ther* 83: 49–65
86. Gold PE, Zornetzer SF (1983) The mnemon and its juices: neuromodulation of memory processes. *Behav Neurol Biol* 38: 151–189
87. Introini-Collison IB, Saghafi D, Novack GD, McGaugh GL (1992) Memory-enhancing effects of post-training dipivefrin and epinephrine: involvement of peripheral and central adrenergic receptors. *Brain Res* 572: 81–86
88. Devauges V, Sara SJ (1991) Memory retrieval enhancement by locus coeruleus stimulation: evidence for mediation by beta-receptors. *Behav Brain Res* 43: 93–97
89. Chen MF, Chiu TH, Lee EHY (1992) Noradrenergic mediation of the memory-enhancing effect of corticotropin-releasing factor in the locus coeruleus of rats. *Psychopharmacology* 17: 113–124
90. Liang KC, Juler R, McGaugh JL (1986) Modulating effect of posttraining epinephrine on memory: Involvement of the amygdala noradrenergicsystem. *Brain Res* 368: 125–133
91. Introini-Collison IB, Nagahara AH, McGaugh JL (1989) Memory-enhancement with intra-amygdala naloxone in blocked by concurrent administration of propranolol. *Brain Res* 476: 94–101
92. Quirarte GL, Roozendaal B, McGaugh JL(1997) Glucocorticoid enhancement of memory storage involves noradrenergic activation in the basolateral amygdala. *Proc Natl Acad Sci USA* 94:14048–14053
93. Ayyagari V, Harrell LE, Parsons DS (1991) Interaction of neurotransmitter systems in the hippocampus: A study of behavioral effects of hippocampus systematic ingrowth. *J Neurosci* 11: 2848–2854
94. Mongeau R, Blier P, de Montigny C (1997) The serotonergic and noradrenergic systems of the hippocampus their interactions and the effects of antidepressant treatments. *Brain Res Rev* 23: 145–195
95. Watabe AM, Zaki PA, O'Dell TJ (2000) Coactivation of beta-adrenergic and cholinergic receptors enhances the induction of long-term potentiation and synergistically activates mitogen-activated protein kinase in the hippocampal CA1 region. *J Neurosci* 20: 5924–5931
96. Goldman-Pakic PS, Lidow MS, Gallager DW (1990) Overlap of dopaminergic, adrenergic and serotonergic receptors and complementarity of their subtypes in primate prefrontal cortex. *J Neurosci* 10: 2125–2138
97. Friedman PI, Adler DN, Davis KL (1999) The role of norepinephrine in the pathophysiology of cognitive disorders: Potential applications to the treatment of cognitive dysfunction in schizophrenia and Alzheimer's disease. *Biol Psychiatr* 46: 1243–1252
98. Haapalinna A, Sirviö J, MacDonald E, Virtanen R, Heinonen E (2000) The effects of a specific α_2-adrenoceptor antagonist, atipamezole, on cognitive performance and brain neurochemistry in aged Fisher 344 rats. *Eur J Pharmacol* 387: 141–150
99. Zarrindast MR, Ghiasvand M, Homayoun H, Rostami P, Shafaghi B, Khavandgar S (2003) Adrenoceptor mechanisms underlying imipramine-induced memory deficits in rats. *J Psychopharmacol* 17: 83–88
100. Zarrindast MR, Bakhsha A, Rostami P, Shafaghi B (2002) Effects of intrahippocampqal injection of GABAergic drugs on memory retention of passive avoidance learning in rats. *J Psychopharmacol* 16: 313–319

101. Zarrindast MR, Khodjastefar E, Oryan Sh, Torkaman-Boutorabi A (2001) Baclofen-impairment of memory retention in rats: possible interaction with adrenoceptor mechanisms. *Eur J Pharmacol* 411: 283–288
102. Zarrindast MR, Eidi M, Eidi A, Oryan Sh (2002) Effects of histamine and opioid systems on memory retention of passive avoidance learning in rats. *Eur J Pharmacol* 452: 193–197
103. Zarrindast MR, Jamali-Raeufy N, Shafaghi B (1995) Effects of high doses of theophylline on memory acquisition. *Psychopharmacol* 122: 307–311
104. Roth BL (1994) Multiple serotonin receptors: clinical and experimental aspects. *Ann Clin Psychiatry* 6: 67–78
105. Davies P, Maloney AJ (1976) Selective loss of central cholinergic neurons in Alzheimer's disease. *Lancet* II: 1043
106. Perry EK, Gibson PH, Blessed G, Perry RH, Tomlinson BE (1997) Neurotransmitter enzyme abnormalities in senile dementia. *J Neurol Sci* 34: 247–265
107. Mann DMA, Yates PO (1983) Serotonin nerve cells in Alzheimer's disease. *J Neurol Neurosurg Psychiat* 46: 96–98
108. Yamamoto T, Hirano A (1985) Nucleus raphe dorsalis in Alzheimer's disease: Neurofibrillary tangles and loss of large neurons. *Ann Neurol* 17: 573–577
109. Cassel JC, Jeltsch H (1995) Serotonergic modulation of cholinergic function in the central nervous system: cognitive implications. *Neuroscience* 69: 1–41
110. Hoyer D, Clarke DE, Fozard JR, Hartig PR, Martin GR, Mylecharane EJ, Saxena PR, Humphrey PPA (1994) VII. International union of pharmacology classification of receptors for 5-hydroxytryptamine (serotonin). *Pharmacol Rev* 46: 157–203
111. Martin GR, Humphrey PPA (1994) Receptors for 5-hydroxytryptamine: Current perspectives on classification and nomenclature. *Neuropharmacol* 33: 261–273
112. Buhot MC (1997) Serotonin receptors in cognitive behaviors. *Curr Opin Neurobiol* 7: 243–254
113. Riad M, Emerit MB, Hamon M (1994) Neurotrophic effects of ipsapirone and other 5-HT_{1A} receptor agonists on septal cholinergic neurons in culture. *Devl Brain Res* 82: 245–258
114. Ramirez MJ, Cenarruzabeitia E, Lasheras B, Del Rio J (1996) Involvement of GABA systems in acetylcholine release induced by 5-HT3 receptor blockade in slices from rat entorhinal cortex. *Brain Res* 712: 274–280
115. Ricaturte GA, Markowska AL, Wenk GL, Hatzidimitriou G, Wlos J, Olton DS (1993) 3,4-Methylendioxymethamphetamine, serotonin and memory. *J Pharmacol Exp Ther* 266: 1097–1105
116. Kia HK, Brisorgueil MJ, Daval G, Langlois X, Hamon M, Verge D (1996) $Serotonin_{1A}$ receptors are expressed by a subpopulation of cholinergic neurons in the rat medial septum and diagonal band of broca-A double immunocytochemical study. *Neuroscience* 74: 143–154
117. Harvey JA (1996) Serotonergic regulation of associative learning. *Behav Brain Res* 73: 47–50
118. Carli M, Luschi R, Samanin R (1995) (S)-WAY 100135, a 5-HT_{1A} receptor antagonist, prevents the impairment of spatial learning caused by intrahippocampal scopolamine. *Eur J Pharmacol* 283: 133–139
119. Buhot M-C, Patra SK, Naili S (1995) Spatial memory deficits following stimulation of hippocampal 5-HT_{1B} receptors in the rat. *Eur J Pharmacol* 285: 221–228
120. Carli M, Luschi R, Garofalo P, Samanin R (1995) 8-OH-DPAT impairs spatial but not visual learning in a water maze by stimulating 5-HT_{1A} receptors in the hippocampus. *Behav Brain Res* 67: 67–74

121. Harder JA, Maclean CJ, Alder JT, Drancis PT, Ridley RM (1996) The 5-HT_{1A} antagonist, WAY 100635 ameliorates the cognitive impairment induced by fornix transection in the marmoset. *Psychopharmacol* 127: 245–254
122. Santucci AC, Knott PJ, Haroutunian V (1996) Excessive serotonin release, not depletion, leads to memory impairments in rats. *Eur J Pharmacol* 295: 7–17
123. Stancampiano R, Cocco S, Cugusi C, Sarais L, Fadda F (1999) Serotonin and acetylcholine release response in the rat hippocampus during a spatial memory task. *Neuroscience* 89: 1135–1143
124. Olvera-Cortes E, Barajas-Perez M, Morales-Villagrán A, González-Burgos I (2001) Central serotonin depletion induces egocentric learning improvement in developing rats. *Neuroscience Lett* 313: 29–32
125. Hill DR, Bowery NG (1981) 3H-baclofen and 3H-GABA bind to bicuculline-insensitive $GABA_B$ sites in rat brain. *Nature* 290: 149–152
126. Matsumoto RR (1989) GABA receptors: are cellular differences reflected in function? *Brain Res Rev* 14: 203–225
127. Malcangio M, Bowery N (1996) GABA and its receptors in the spinal cord. *Trends Pharmacol Sci* 17: 457–462
128. Castellano C, McGaugh GH (1990) Effects of post-training bicuculline and muscimol on retention: lack of state dependency. *Behav Neural Biol* 54: 156–164
129. Brioni JD, Nagahara AH, McGaugh JL (1989) Involvement of the amygdala GABAergic system in the modulation of memory storage. *Brain Res* 487: 105–112
130. Nakagawa Y, Ishibashi Y, Yoshii T, Tagashira E (1995) Involvement of cholinergic systems in the deficit of place learning in Morris water maze task induced by baclofen in rats. *Brain Res* 683: 209–214
131. Brioni JD, McGaugh JL (1988) Post-training administration of GABAergic antagonists enhance retention of aversively motivated tasks. *Psychopharmacol* 96: 505–510
132. Castellano C, McGaugh GH (1989) Retention enhancement with post-training picrotoxin: lack of state dependency. *Behav Neural Biol* 51: 165–164
133. Castellano C, Brioni JD, Nagahara AH, McGaugh JL (1989) Post-training systemic and intra-amygdala administration of the GABA-B agonist baclofen impair retention. *Behav Neural Biol* 52: 170–179
134. McGaugh JL, Castellano C, Brionin (1990) Picrotoxin enhances latent extinction of conditioned fear. *Behav Neurosci* 104: 262–265
135. Schmitt U, Hiemke C (2002) Tiagabine, a γ-amino-butyric acid transporter inhibitor impairs spatial learning of rats in the Morris water-maze. *Behav Brain Res* 133: 391–394
136. McNamara RK, Skelton RW (1996) Baclofen, a selective $GABA_B$ receptor agonist, dose-dependently impairs spatial learning in rats. *Pharmacol Biochem Behav* 53: 303–308
137. Brucato FH, Mott DD, Lewis DV, Swartzwelder HS (1995) $GABA_B$ receptors modulate synaptically-evoked responses in the rat dendate gyrus *in vivo*. *Brain Res* 677: 326–332
138. Brucato FH, Levin ED, Mott DD, Lewis DV, Wilson WA, Swartzwelder HS (1996) Hippocampal long-term potentiation and spatial learning in the rat: effects of $GABA_B$ receptors blockade. *Neuroscience* 74: 331–339
139. Mccabe BJ, Horn G, Kendrick KM (2001) GABA, taurine and learning: release of amino acids from slices of chick brain following filial imprinting. *Neuroscience* 105: 317–324
140. Moor E, DeBoer P, Westerink BHC (1998) GABA receptors and benzodiazepine binding sites modulate hippocampal acetylcholine release *in vivo*. *Eur J Pharmacol* 359: 119–126

141. Degroot A, Parent MB (2001) Infusion of physostigmine into the hippocampus or the entorhinal cortex attenuate avoidance retention deficits produced by intra-septal infusions of the GABA agonist muscimol. *Brain Res* 920: 10–18
142. Decker MW, McGaugh JL (1991) The role of interactions between the cholinergic and other neuromodulatory systems in learning and memory. *Synapse* 7: 151–168
143. Dudchenko P, Sarter M (1991) GABAergic control of basal forebrain cholinergic neurons and memory. *Behav Brain Res* 42: 33–41
144. Konopaki J, Golebiewski H (1993) Theta-like activity in hippocampal formation slices: cholinergic-GABAergic interaction. *Neuroreport* 4: 963–966
145. Stackman RW, Walsh TJ (1994) Baclofen produced dose-related working memory impairments after intraseptal injection. *Behav Neural Biol* 61: 181–185
146. Zarrindast MR, Lahiji P, Shafaghi B, Sadegh M (1998) Effects of GABAergic drugs on physostigmine-induced improvement in memory acquisition of passive avoidance learning in mice. *Gen Pharmacol* 31: 81–86
147. Izquidero I, Da Cunha C, Rosat R, Jerusalinsky D, Ferreira MBC, Medina JH (1992) Neurotransmitter receptors involved in post-training memory processing by the amygdala, medial septum, and hippocampus of the rat. *Behav Neural Biol* 58: 16–26
148. Markam H, Segal M (1990) Long-lasting facilitation of excitatory postsynaptic potentials in the rat hippocampus by acetylcholine. *J Physiol* 427: 381–393
149. Izquierdo I, Medina JH (1991) $GABA_A$ receptorsmodulation of memory: the role of endogenous benzodiazepines. *Trends Pharmacol Sci* 12: 260–265
150. Zarrindast MR, Shamsi T, Azarmina P, Rostami P, Shafaghi B (2004) GABAergic system and imipramine-induced impairment of memory retention in rats. *Eur Neuropsychopharmacol* 14: 59–64
151. Getova D, Bowery NG (1998) The modulatory effects of high affinity $GABA_B$ receptor antagonists in an active avoidance learning paradigm in rats. *Psychopharmacology* 137: 369–373
152. Haas HL, Reiner PB, Greene RW (1991) Histaminergic and histaminoceptive neurons: electrophysiological studies in vertebrates. In: Wanatabe T, Wada H. (eds): *Histaminergic neurons; morphology and function*. CRC Press. Boca Ratton, 195–208
153. Schwarts JC, Arrang JM, Garbarg M, Pollard H, Ruat M (1991) Histaminergic transmission in the mammalian brain. *Physiol Rev* 71: 1–51
154. Onodera K, Yamatodani A, Watanabe T, Wada H (1994) Neuropharmacology of the histaminergic neuron system in the brain and its relationship with behavioral disorders. *Prog Neurobiol* 42: 685–702
155. Inagaki N, Yamatodani A, Ando-Yamamoto M, Tohyama M, Watanabe T, Wada H (1988) Organization of histaminergic fibers in the rat brain. *J Comp Neurol* 273: 282–300
156. Panula P, Yang HY, Costa E (1984) Histamine-containing neurons in the rat hypothalamus. *Prog Natl Acad Sci USA* 81: 2572–2576
157. Watanabe T, Taguchi Y, Shiosaka S, Tanaka J, Kubota H, Terano Y, Tohyama M, Wada H (1984) Distribution of histaminergic neuron system in the central nervous system of rats; a fluorecent immunohistochemical analysis with histidine decarboxylase as a marker. *Brain Res* 295: 13–25
158. Niigawa H, Yamatodani A, Nishimura T, Wada H, Cacabelos R (1988) Effect of neurotoxic lesions in the mammillary bodies on the distribution of brain histamine. *Brain Res* 459: 183–186
159. Leurs R, Smit MJ, Timmerman H (1995) Molecular pharmacological aspects of histamine receptors. *Pharmacol Ther* 66: 413–463
160. Prell CD, Green JP (1986) Histamine as a neuroregulator. *Annu Rev Neurosci* 9: 209–254

161. Schwarts JC, Arrang JM, Garbarg M (1986) Three classes of histamine receptor in brain. *Trends Pharmacol Sci* 7: 24–28
162. Arrang JM, Garbarg M, Schwartz JC (1985) Autoregulation of histamine release in brain by presynaptic H3-receptors. *Neuroscience* 15: 553–562
163. Endou M, Kazuhiko Y, Sakurai E, Fukudo S, Hongo M, Watanabe T (2001) Food-deprived activity stress decreased the activity of the histaminergic neuron system in rats. *Brain Res* 981: 32–41
164. Pollard H, Moreau J, Arrang J M, Schwartz J C (1993) A detailed autoradiographic mapping of histamine H_3 receptors in rat brain areas. *Neuroscience* 52: 169–189
165. Schlicker E, Malinowska B, Kathman M, Gothert M (1994) Modulation of neurotransmitter release *via* histamine H_3 heteroreceptors. *Fundam Clin Pharmacol* 8: 128–137
166. Alvarez EO, Ruarte MB, Banzan AM (2001) Histaminergic systems of the limbic complex on learning and motivation. *Behav Brain Res* 124: 195–202
167. Huston JP, Wagner U, Hasenohrl RU (1997) The tuberomammillary nucleus projections in the control of learning, memory and reinforcement process: evidence for an inhibitory role. *Behav Brain Res* 83: 97–105
168. Smith CPS, Hunter AJ, Bennet GW (1994) Effects of (R)-alpha-methylhistamine and scopolamine on spatial learning in the rat assessed using a water maze. *Psychopharmacol (Berl)* 114: 651–656
169. Miyazaki S, Imaizumi M, Onodera K (1995) Effects of thioperamide, a histamine H_3-receptor antagonist, on a scopolamine-induced learning deficit using an elevated plus-maze test in mice. *Life Sci* 57: 2137–2144
170. Fontana DJ, Inouye GT, Johnson RM (1994) Linopirdine (DuP 996) improves performance in several tests of learning and memory by modulation of cholinergic neurotransmission. *Pharmacol Biochem Behav* 49: 1075–1082
171. Quirion R, Wilson A, Rowe W, Aubert I, Richard J, Doods H, Parent A, White N, Meaney MJ (1995) Facilitation of acetylcholine release and cognitive performance by an m2-muscarinic receptor antagonist in aged memory-impaired rats. *J Neurosci* 15: 1455–1462
172. Frisch C, Hasenõhrl RU, Haas HL, Weiler HT, Steinbusch HWM, Huston JP (1998) Facilitation of learning after lesions of the tuberomammillary nucleus region in adult and aged rats. *Exp Brain Rev* 118: 447–456
173. Eidi M, Zarrindast MR, Eidi A, Oryan Sh, Parivar K (2003) Effects of histamine and cholinergic systems on memory retention of passive avoidance learning in rats. *Eur J Pharmacol* 465: 91–96
174. Tasaka K, Kamei C, Akahori H, Kitazumi K (1985) The effects of histamine and some compounds on conditioned avoidance response in rats. *Life Sci* 37: 2005–2015
175. De Almeida MAM, Izquierdo I (1986) Memory facilitation by histamine. *Arch Int Pharmacodyn Ther* 283: 193–198
176. Kamei C, Tasaka K (1991) Participation of histamine in the step-through active avoidance response and its inhibition by H_1-blockers. *Jpn J Pharmacol* 57: 473–482
177. Frisch C, Hasenõhrl RU, Krauth J, Huston JP (1998) Anxiolytic-like behavior after lesion of the tuberomammillary nucleus E_2-region. *Exp Brain Res* 119: 260–264
178. Seguro-Torres P, Wagner U, Massanes-Rotger E, Aldavert-era L, Marti-Nicolovius M, Morgado-Bernal I (1996) Tuberomammillary nucleus lesion facilitates two-way active avoidance retention in rats. *Behav Brain Res* 82: 113–117
179. Gulat-Murray C, Lafitte A, Arrang JM, Schwartz JC (1989) Regulation of histamine release and synthesis in the brain by muscarinic receptors. *J Neurochem* 52: 248–254
180. Khateb A, Fort P, Pegna A, Jones BE, Mühlethaler M (1995) Cholinergic nucleus basalis neurons are excited by histamine *in vitro*. *Neuroscience* 69: 495–506

181. Prast H, Fischer HP, Prast M, Philippu A (1994) *In vivo* modulation of histamine release by autoreceptors and muscarinic acetylcholine receptors in the rat anterior hypothalamus. *Naunyn-Schmiedeberg's Arch Pharmacol* 350: 599–604
182. Arrang JM, Gulat-Marnay C, Defontaine N, Schwartz JC (1991) Regulation of histamine release in rat hypothalamus and hippocampus by presynaptic galanin receptors. *Peptides* 12: 1113–1117
183. Itoh Y, Oishi R, Nishibori M, Saeki K (1988) Involvement of mu receptors in the opioid-induced increase in the turnover of mouse brain histamine. *J Pharmacol Exp Ther* 244: 1021–1026
184. Ukai M, Itoh J, Kobayashi T, Shinkai N, Kameyama T (1997) Effects of the κ-opioid dynorphin A(1-13) on learning and memory in mice. *Behav Brain Res* 83: 169–172
185. Izquierdo I (1980) Effect of β-endorphin and naloxone on acquisition, memory, and retrieval shuttle avoidance and habituation learning in rats. *Psychopharmacol* 69: 111–115
186. Beatty WW (1983) Opiate antagonists, morphine and spatial memory in rats. *Pharmacol Biochem Behav* 19: 397–401
187. Brake KE, Hough LB (1992) Morphine-induced increases of extracellular hiatamine levels in the periaqueductal gray *in vivo*; a microdialysis study. *Brain Res* 572: 146–153
188. Flood JF, Uezu K, Morley JE (1998) Effect of histamine H_2 and H_3 receptor modulation in the septum on post-training memory processing. *Psychopharmacol* 140: 279–284
189. Mickley GA (1986) Histamine H_2 receptors mediate morphine-induced locomotor hyperactivity of the C57BL/6J mouse. *Behav Neurosci* 100: 79–84
190. Wauquier A, Niemegeers CJE (1981) Effects of chlorpheniramine, pyrilamine and astemizole on intracranial self-stimulation in rats. *Eur J Pharmacol* 72: 245–248
191. Zimmermann P, Wagner U, Krauth J, Huston JP (1997) Unilateral lesion of dorsal hippocapmpus enhances reinforcing lateral hypothalamic stimulation in the contralateral hemisphere. *Brain Res Bull* 44: 256–271
192. Shannon HE, Su TP (1982) Effects of the combination of tripelennamina and pentazocine at the behavioral and molecular levels. *Pharmcol Biochem Behav* 17: 789–795
193. Suzuki T, Takamori K, Misawa M, Onodera K (1995) Effects of the histaminergic system on the morphine-induced conditions place preference in mice. *Brain Res* 675: 195–202
194. Izquierdo I (1979) Effect of naloxone and morphine on various forms of memory in the rat: possible role of endogenous opiate mechanisms in memory consolidation. *Psychopharmacol (Berl)* 66: 199–203
195. Izquierdo I, Dias RD (1983) Effect of ACTH, epinephrine, beta-endorphin, naloxone, and of the combination of naloxone or beta-endorphin with ACTH or epinephrine on memory consolidation. *Psychoneuroendocrinol* 8: 81–87
196. De Almeida MA, Izquierdo I (1984) Effect of the intraperitoneal and intracerebroventricular administration of ACTH, epinephrine, or beta-endorphin on retrieval of an inhibitory avoidance task in rats. *Behav Neural Biol* 40:119–122
197. Izquierdo I, De Almeida MA, Emiliano VR (1985) Unlike beta-endorphin, dynorphine 1–13 does not cause retrograde amnesia for shuttle avoidance or inhibitory avoidance learning in rats. *Psychopharmacol* 87: 216–218
198. Izquierdo I, Dias RD (1983) Endogenous state-dependency: memory regulation by post-training and pre-testing administration of ACTH, beta-endorphin, adrenaline and tyramine. *Braz J Med Biol Res* 16: 55–64
199. Kameyama T, Nabeshima T, Kozawa T (1986) Step-down-type passive avoidance- and escape-learning method. Suitability for experimental amnesia models. *J Pharmacol Methods* 16: 39–52

200. Shiigi Y, Takahashi H, Kaneto H (1990) Facilitation of memory retrieval by pretest morphine mediated by mu but not delta and kappa opioid receptors. *Psychopharmacol (Berl)* 102: 329–332
201. Bruins-Slot LA, Colpaert FC (1999) Opiate state of memory: receptor mechanisms. *J Neurosci* 19: 10520–10529
202. Khavandgar S, Homayoun H, Torkaman-Boutorabi A, Zarrindast MR (2002) The effects of adenosine receptor agonists and antagonists on morphine state-dependent memory of passive avoidance. *Neurobiol Learn Mem* 78: 390–405
203. Introini-Collison IB, Baratti CM (1984) The impairment of retention induced by beta-endorphin in mice may be a reduction of central cholinergic activity. *Behav Neural Biol* 41: 152–163
204. Ragozzino ME, Gold PE (1994) Task-dependent effects of intra-amygdala morphine injections: attenuation by intra-amygdala glucose injections. *J Neurosci* 14: 7478–7485
205. Tyce GM, Yaksh TL (1981) Monoamine release from cat spinal cord by somatic stimuli: an intrinsic modulatory system. *J Physiol* 314: 513–529
206. Yaksh TL, Dirksen R, Harty GJ (1985) Antinociceptive effects of intrathecally injected cholinomimetic drugs in the rat and cat. *Eur J Pharmacol* 117: 81–88
207. Dirksen R, Nijhuts GMM (1983) The relevance of cholinergic transmission blockade at the spinal level to opiate effectiveness. *Eur J Pharmacol* 91: 215–221
208. Bouaziz H, Tong CY, Yoon Y, Hood DD, Eisenach JC (1996) Intravenous opioids stimulate norepinephrine and acetylcholine release in spinal cord dorsal horn—systematic studies in sheep and an observa tion in a human. *Anesthesiology* 84: 143–154
209. Baratti CM, Introini IB, Huygens P (1984) Possible interaction between centralcholinergic muscarinic and opioid peptidergic systems during memory consolidation in mice. *Behav Neural Biol* 40: 155–169
210. Heijna MH, Padt M, Hogenboom F, Portoghese PS, Mulder AH, Schoffelmeer ANM (1990) Opioid receptor-mediated inhibition of dopamine andacetylcholine release from slices of rat nucleus accumbens, olfactorytubercle and frontal cortex. *Eur J Pharmacol* 181: 267–278
211. Ragozzino ME, Gold PE (1995) Glucose injections into the medial septum reverse the effects of intraseptal morphine infusions on hippocampal acetylcholine output and memory. *Neuroscience* 68: 981–988
212. Lapchak PA, Araujo DM, Collier B (1989) Regulation of endogenous acetylcholine release from mammalian brain slices by opiate receptors: hippocampus, striatum and cerebral cortex of guinea pig and rat. *Neuroscience* 31: 313–325
213. Rada P, Mark GP, Pothos E, Hoebel BG (1991) Systemic morphine simultaneously decreases extracellular acetylcholine and increases dopamine in the nucleus accumbens of freely moving rats. *Neuropharmacol* 30: 1133–1136
214. Li Z, Wu CF, Pei G, Xu NJ (2001) Reversal of morphine-induced memory impairment in mice by withdrawal in Morris water maze. Possible involvement of cholinergic system. *Pharmacol Biochem Behav* 68: 507–513
215. Roozendaal B (2000) Glucocorticoids and the regulation of memory consolidation. *Psychoneuroendocrinol* 25: 213–238
216. Yau JL, Noble J, Seckl JR (1999) Continuous blockade of brain mineralocorticoid receptors impairs spatial learning in rats. *Neurosci Lett* 277: 45–48
217. Sanberg PR, Fibiger HC (1979) Impaired acquisition and retention of a passive avoidance response after chronic ingestion of taurine. *Psychopharmacol (Berl)* 62: 97–99
218. McNamara MG, Kelly JP, Leonard BE (1995) Some behavioural and neurochemical aspects of subacute (+/−)3.4-methylenedioxyamphetamine administration in rats. *Pharmacol Biochem Behav* 52: 479–484

219. Barros DM, Izquierdo LA, Medina JH, Izquierdo I (2002) Bupropion and sertraline enhance retrieval of recent and remote long-term memory in rats. *Behav Pharmacol* 13: 215–220
220. Jafari MR, Zarrindast MR, Djahanguiri B (2004) Effects of different doses of glucose and insulin on morphine state dependent memory of passive avoidance in mice. *Psychopharmacol (Berl)* 175: 457–462
221. White NM (1991) Peripheral and central memory enhancing actions of glucose. In: Hogrefe and Huber (eds): *Peripheral signaling of the brain: role in neural-immune interactions, learning and memory.* Frederickson R.C.A. Toronto, 421–442
222. Kopf SR, Buchholzer ML, Hilgert K, Loffelholtz K, Klein J (2001) Glucose plus choline improve passive avoidance behaviour and increase hippocampal acetylcholine release in mice. *Neuroscience* 103: 365–371
223. Stone WS, Walser B, Gold SD, Gold PE (1991) Scopolamine- and morphine-induced impairments of spontaneous alteration performance in mice: reversal with glucose and with cholinergic and adrenergic agonists. *Behav Neurosci* 105: 264–271
224. Amoroso S, Schmid-Antomarchi H, Fosset M, Lazdunski M (1999) Glucose, sulfonylureas, and neurotransmitter release: role of ATP-sensitive K1 channels. *Science* 247: 852–854
225. Stefani MR, Nicholson GM, Gold P (1999) ATP-sensitive potassium channel blockade enhances spontaneous alternation performance in the rat: a potential mechanism for glucose-mediated memory enhancement. *Neuroscience* 93: 557–563
226. Vianna MR, Barros DM, Silva T, Choi H, Madche C, Rodrigues C, Medina JH, Izquierdo I (2000) Pharmacological demonstration of the differential involvement of protein kinase C isoforms in short- and long- term memory formation and retrieval of one-trial avoidance in rats. *Psychopharmacol (Berl)* 150: 77–84
227. Narita M, Takahashi Y, Suzuki T, Misawa M, Nagasa H (1993) An ATP-sensitive potassium channel blocker abolishes the potentiating effect of morphine on the bicuculline-induced convulsion in mice. *Psychopharmacol* 110: 500–502
228. Ocana M, Pozo EP, Barrios M, Robles LI, Baeyens JM (1990) An ATP-dependent potassium channel blocker antagonizes morphine analgesia. *Eur J Pharmacol* 86: 77–78
229. Raffa BR, Martinez P (1995) The glibenclamide-shift of centrally-acting antinociceptive agents in mice. *Brain Res* 677: 277–282
230. Werz MA, MacDonald RL (1983) Opioid peptides with different affinity for mu and delta receptors decrease sensory neuron calcium–dependent action potentials. *J Pharmacol Exp Ther* 227: 394–402
231. North RA (1989) Twelfth Gaddum memorial lecture. Drug receptors and the inhibition of nerve cells. *Br J Pharmacol* 98: 13–28
232. Zarrindast MR, Jafari MR, Ahmadi S, Djahanguiri B (2004) Influence of central administration ATP-dependent K+ channel on morphine state-dependent memory of passive avoidance. *Eur J Pharmacol* 487: 143–148
233. Stefani MR, Gold PE (2001) Intrahyppocampal infusion of K_{ATP} channel modulators influence spontaneous alteration performance: relationships to acethylcholine release in the hippocampus. *J Neurosci* 15: 609–614
234. Sebastião AM, Ribeiro JA (2000) Fine-tuning neuromodulation by adenosine. *TIPS* 21: 341–346
235. Hauber W, Bareib A (2001) Facilitative effects of an adenosine A_1/A_2 receptor blockade on spatial memory performance of rats: selective enhancement of reference retention during the light period. *Behav Brain Res* 118: 43–52

236. Cunha RA (2001) Adenosine as a neuromodulator and as a homeostatic regulator in the nervous system: different roles, different sources and different receptors. *Neurochem Int* 38: 107–125
237. Sebastião AM, Ribeiro JA (1996) Adenosine A_2 receptors-mediated excitability actions on the nervous system. *Prog Neurobiol* 48: 167–189
238. Murphy KM, Snyder SH (1982) Hetrogenity of adenosine A_1 receptor binding in brain tissue. *Mol Pharmacol* 17: 139–179
239. Cunha RA, Johnsson B, Constantino MD, Sebastiao AM, Fredholm BB (1996) Evidence for high-affinity binding sites for the adenosine A_{2A} receptor agonist [3H] CGS21680 in the rat hippocampus and cerebral cortex that are different from striatal A_{2A} receptors. *Naunyn-Schmiedeberg's Arch Pharmacol* 353: 261–271
240. Rosin DI, Robeva A, Woodard RI, Guyenet PG, Linden J (1998) Immunohistochemical localization of adenosine A_{2A} receptors in the rat central nervous system. *J Comp Neurol* 401: 163–186
241. Cunha RA, Sebastiao AM, Ribeiro JA (1998) Inhibition by ATP of hippocampal synaptic transmission requires localized extracellular catabolism by ecto-nucleotidases into adenosine and channeling to adenosine A_1 receptors. *J Neurosci* 18: 1987–1995
242. De Mendonça A, Sebastiao AM, Ribeiro JA (1995) Inhibition of NMDA receptor-mediated currents in isolated rat hippocampal neurones by adenosine A_1 receptor activation. *Neuro Report* 6: 1097–1100
243. Cunha RA, Constantino MD, Riberiro JA (1997) ZM241385 is an antagonist of the facilitatory responses produced by the A_{2A} adenosine receptor agonists CGS21680 and HENECA in the rat hippocampus. *Br J Pharmacol* 122: 1279–1284
244. De Mendonça A, Riberiro JA (1997) Adenosine and neuronal plasticity. *Life Sci* 60: 241–245
245. Homayoun H, Khavandgar S, Zarrindast MR (2001) Effects of adenosine receptor agonists and antagonists on pentylentetrazole-induced amnesia. *Eur J Pharmacol* 430: 289–294
246. Zarrindast MR, Shafaghi B (1994) Effects of adenosine receptor agonists and antagonists on acquisition of passive avoidance learning. *Eur J Pharmacol* 256: 233–239
247. Pereira GS, Mello e Souza T, Vinade ERC, Choi H, Rodrigues C, Battastini AMO, Izquierdo I, Sarkis JJF, Bonan CD (2002) Blockade of adenosine A1 receptors in the posterior cingulate cortex facilitates memory in rats. *Eur J Pharmacol* 437: 151–154
248. Hung PL, Lai MC, Yang SN, Wang CL, Liou CW, Wu MC, Wang TJ, Huang LT (2002) Aminophylline exacerbates status epilepticus-induced neuronal damages in immature rats: a morphological, motor and behavioral study. *Epilepsy Res* 49: 218–225
249. Von Lubitz DKJE, Beenhakker M, Lin RCS, Carter MF, Paul SIA, Bischofberger N, Jacobson KA (1996) Reduction of postischemic brain damage and memory deficits following treatment with the selective adenosine A1 receptor agonist. *Eur J Pharmacol* 302: 43–48
250. Tchekalarova J, Kambourova T, Georgiev V (2001) Interaction between angiotensin IV and adenosine A_1 receptor related drugs in passive avoidance conditioning in rats. *Behav Brain Res* 123: 113–116
251. Kopf SR, Melani A, Pedata F, Pepeu G (1999) Adenosine and memory storage: effect of A_1 and A_2 receptor antagonists. *Psychopharmacol* 146: 214–219
252. Dockray GJ (1976) Immunochemical evidence of cholecystokinin like peptide in brain. Nature 264: 568–570
253. Wank SA, Pisegna JR, De Weerth A (1992) Brain and gastrointestinal cholecystokinin receptor family: structure and functional expression. *Proc Natl Acad Sci USA* 89: 8691–8695

254. Lee YM, Beinborn M, McBride EW, Lu M, Kolakowski Jr LF, Kopin AS (1993) The human brain cholecystokinin-B/gastrin receptor. Cloning and characterization. *J Biol Chem* 268: 8164–8169
255. Hill DR, Camphell NJ, Shaw TM, Woodruff GM (1987) Autoradiographic localization and biochemical characterization of peripheral type CCK receptors in rat CNS using highly selective nonpeptide CCK antagonists. *J Neureosci* 7: 2967–2976
256. Mercer LD, Beart PM (1997) Histochemistry in rat brain and spinal cord with an antibody directed at the cholecystokinin A receptor. *Neurosci Lett* 225: 97–100
257. Hill DR, Shaw TN, Graham W, Woodruff GN (1990) Autoradiographical detection of CCK-A receptors in primate brain using ^{125}I-Bolton Hunter CCK-8 and ^{3}H-MK 329. *J Neurosci* 10: 1070–1081
258. Crawley JN (1984) Cholecystokinin accelerates the rate of habituation to a novel environment. *Pharmacol Biochem Behav* 20: 23–27
259. Daugé V, Roques BP (1995) Opioid and CCK systems in anxiety and reward. In: Bradwejn J, Vasar E (eds) *Cholecystokinin and anxiety: from neuron to behavior*. R.G. Landes Company, Austin, 151–171
260. Ding XZ, Bayer BM (1993) Increases in CCK mRNA and peptide in different brain area following acute and chronic administration of morphine. *Brain Res* 625: 139–144
261. Katsuura G, Itoh S (1986) Preventive effect of cholecystokinin octapeptide on experimental amnesia in rats. *Peptides* 7: 105–110
262. Itoh S, Lal H (1990) Influences of cholecystokinin and analogues on memory processes. *Drug Rev Res* 21: 257–276
263. Daugé V, Léna I (1998) CCK in anxiety and cognitive processes. *Neurosc Biochem Rev* 22: 815–825
264. Derrien M, Daugé V, Blommaert A, Roques BP (1994) The selective CCK-B agonist, BC 264, impairs socially reinforced memory in the three runaway test in rats. *Behav Brain Res* 65: 139–146
265. Kadar T, Fekete M, Telegdy G (1981) Modulation of passive avoidance behavior of rats by intracerebroventricular administration of cholecystokinin sulfate ester and nonsulfated cholecystokinin octapeptide. *Acta Physiol Acad Sci Hung* 58: 269–274
266. Harro J, Orland L (1993) Cholecystokinin receptors and memory: a radial maze study. *Pharmacol Biochem Behav* 44: 509–517
267. Lemaire M, Bohme GA, Piot O, Roques BP, Blanchard JC (1994) CCK-A and CCK-B selective receptor agonists and antagonists modulate olfactory recognition in male rats. *Psychopharmacol* 115: 435–440
268. Nomoto S, Miyake M, Ohta M, Funakoshi A, Miyaksaka K (1999) Impaired learning and memory OLETF rats without cholecystokinin (CCK)-A receptor. *Physiol Behav* 66: 869–872
269. Daugé V, Pophillat M, Creté D, Melik-Parsadaniantz S, Roques P (2003) Involvement of brain endogenous cholecystokinin in stress-induced impairment of spatial recognition memory. *Neuroscience* 118: 19–23
270. Katsuura G, Itoh S (1986) Passive avoidance deficit following intracerebroventricular administration of cholecystokinin tetrapeptide amide in rats. *Peptides* 7: 809–814
271. Shlik J, Koszycki D, Bradwejn J (1998) Decrease in short-term memory function induced by CCK-4 in healthy volunteers. *Peptides* 19: 969–975
272. Ladurelle N, Keller G, Blommaert A, Roques BP, Daugé V (1997) The CCK-B agonist, BC 264, increases dopamine in the nucleus accumbens and facilitates motivation and attention after intraperitoneal injection in rats. *Eur J Neurosci* 9: 1804–1814

273. Millon ME, Léna I, DaNascrimento S, Noble F, Daugé V, Garbay C, Roques BP (1997) Development of new potent agonists able to interact with two postulated subsites of the cholecystokinin CCK-B receptor. *Lett Peptide Sci* 4: 407–410
274. Léna I, Simon H, Roques BP, Daugé V (1999) Opposing effects of two selective CCK-B agonists, on the retrieval phase of a two-trial memory task after systemic injection in the rat. *Neuropharmacology* 38: 543–553
275. Taghzouti K, Léna I, Dellu F, Roques BP, Daugé V, Simon H (1999) Cognitive enhancing effects in young and old rats of pBC 264, a selective CCK-B receptor agonist. *Psychpharmacol* 143: 141–149
276. Daugé V, Derrien M, Blanchard JC, Roques BP (1992) The selective CCK-B agonist BC 264 injected in the anteralateral part of the nucleus accumbens, reduced the spontaneous alteration behaviour of rats. *Neuropsychopharmacol* 31: 67–75
277. Ladurelle N, Keller G, Roques BP, Daugé V (1993) Effects of CCK8 and of the CCK-B selective agonist BC 264 on extracellular dopamine content in the anterior nucleus accumbens: a microdialysis study in freely moving rats. *Brain Res* 628: 254–262
278. Sebret A, Léna I, Crété D, Matsui T, Roques BP, Daugé V (1999) Rat hippocampal neurons are critically involved in physiological improvement of memory processes induced by cholecystokinin-B receptor stimulation. *J Neurosci* 19: 7230–7237
279. Winnicka MM, Wisniewski K (1999) Dopaminergic projection to the central amygdala mediates the facilitatory effect of CCK-8US and caerulein on memory in rats. *Pharmacol Res* 39: 445–450
280. Winnicka MM, Wisniewski K (2000) Bilateral 6-OHDA lesions to the hippocampus attenuate the facilitatory effect of CCK-8US and caerulein on memory in rats. *Pharmacol Res* 41: 347–353
281. Vincent SR (1994) Nitric oxide: A radical neurotransmitter in the central nervous system. *Prog Neurobiol* 42: 129–160
282. Moncada S (1992) The L-arginine: nitric-oxide pathway. *Acta Physiol Scand* 145: 201–227
283. Garthwaite J, Charles SL, Chess-Williams R (1988) Endothelium-derived relaxing factor release on activation of NMDA receptors suggests role as intercellular messenger in the brain. *Nature* 336: 385–388
284. Garthwaite J, Garthwaite G, Palmer RMJ, Moncada S (1989) NMDA receptor activation induces nitric oxide synthesis from arginine in rat brain slices. *Eur J Pharm* 172: 413–416
285. East SJ, Garthwaite J (1991) NMDA receptor activation in rat hippocampus induces cyclic GMP formation through the L-arginine-nitric oxide pathway. *Neurosci Lett* 123: 17–19
286. Haley JE, Wilcox GL, Chapman PF (1992) The role of nitric oxide in hippocampal long-term potentiation. *Neuron* 8: 211–216
287. Haley JE, Schuman EM (1994) Involvement of nitric oxide in synaptic plasticity and learning. *Neurosciences* 6: 11–20
288. Bliss TVP, Collingridge GL (1993) A synaptic model of memory: long-term potentiation in the hippocampus. *Nature* 361: 31–39
289. Böhme GA, Bon C, Stutzman JM, Doble A, Blanchard JC (1991) Possible involvement of nitric oxide in long-term potentiation. *Eur J Pharmacol* 199: 379–381
290. Pitsikas N, Rigamonti AE, Cella SG, Muller EE (2002) Effects of the nitric oxide donor molsidomine on different memory components as assessed in the object-recognition task in the rat. *Psychopharmacol* 162: 239–245
291. Zhang S, Chen J, Wang S (1998) Spatial learning and memory induce up-regulation of nitric oxide-producing neurons in rat brain. *Brain Res* 801: 101–106

292. Telegdy G, Kokavszky R (1997) The role of nitric oxide in passive avoidance learning. *Neuropharmacol* 36: 1583–1587
293. Colingridge GL, Lester RAJ (1989) Excitatory amino acid receptors in the vertebrate central nervous system. *Pharmacol Rev* 41: 143–210
294. Hollmann M, Heinemann S (1994) Cloned glutamate receptors. *Annu Rev Neurosci* 17: 31–108
295. Ozawa S, Kamiya H, Tsuzuki K (1998) Glutamate receptors in the mammalian central nervous system. *Prog Neurobiol* 54: 581–618
296. Michaelis EK (1998) Molecular biology of glutamate receptors in the central nervous system and their role in excitotoxicity, oxidative stress and aging. *Prog Neurobiol* 54: 369–415
297. Pláteník J, Kuramoto N, Yoneda Y (2000) Molecular mechanisms associated with long-term consolidation of the NMDA signals. *Life Sci* 67: 335–364
298. Riedel G, Platt B, micheau J (2003) Glutamate receptor function in learning and memory. *Behav Brain Res* 140: 1–47
299. Storm-Mathisen J, Leknes AK, Bore A, Vaaland JL, Edminson P, Haug FMS, Ottersen OP (1983) First visualisation of glutamate and GABA in neurones by immunocytochemistry. *Nature* 301: 517–520
300. Storm-Mathisen J, Danbolt NC, Ottersen OP (1995) Localization of glutamate and its membrane transport proteins. In: Stone TW (eds): *CNS neurotransmitters and neuro-modulators*. CRC Press, Boca Raton, 1–18
301. Ellison G (1995) The N-methyl-D-aspartate antagonists phencyclidine, ketamine and dizocilpine as both behavioral and anatomical models of the dementias. *Brain Res Rev* 20: 250–267
302. Pellicciari R, Costantino G (1999) Metabotropic G-protein coupled glutamate receptors as therapeutic targets. *Curr Opp Chem Biol* 3: 433–440
303. Bear MF, Abraham WC (1996) Long-term depression in hippocampus. *Ann Rev Neurosci* 19: 437–462
304. Riedel G, Wetzel W, Reymann KG (1996) Comparing the role of metabotropic glutamate receptors in long-term potentiation and in learning and memory. *Progr Neuropharmacol Biol Psychiatr* 20: 761–789
305. Riedel G, Reymann KG (1996) Metabotropic glutamate receptors in hippocampal long-term potentiation and learning and memory. *Acta Physiol Scand* 157: 1–19
306. Danysz W, Zajaczkowski W, Parsons CG (1995) Modulation of learning processes by ionotropic glutamate receptor ligangds. *Behav Pharmacol* 6: 455–474
307. Morris RGM, Anderson E, Lynch GS, Baudry M (1986) Selective impairment of learning and blockade of long-term potentiation by an N-methyl-D-aspartate receptor antagonist, AP5. *Nature* 319: 774–776
308. Upchurch M, Wehner JM (1990) Effects of N-methyl-D-aspartate antagonism on spatial learning in mice. *Psychopharmacol* 100: 209–214
309. Saucier D, Cain DP (1995) Spatial learning without NMDA receptor-dependent long-term potentiation. *Nature* 378: 186–189
310. Lyfold GL, Jarrard LE (1991) Effects of the competitive NMDA antagonist CPP on performance of a place and cue radial maze task. *Psychobiol* 19: 157–160
311. Caramanos Z Shapiro ML (1994) Spatial memory and N-methyl-D-aspartate receptor antagonists APV and MK-801: memory impairments depend on familiarity with the environment, drug dose, and training duration. *Behav Neurosci* 108: 30–43
312. Ungerer A, Mathis C, Mélan C (1998) Are glutamate receptors specially implicated in some forms of memory processes? *Exp Brain Res* 123: 45–51

313. Annett LE, McGregor A, Robbins TW (1989) The effects of ibotenic acid lesion of the nucleus accumbens on spatial learning and extinction in the rat. *Behav Brain Res* 31: 321–242
314. Floresco SB, Seamans JK, Phillips AG (1996) Differential effects of lidocaine infusions into the ventral CA1/subiculum or the nucleus accumbens on acquisition and retention of spatial information. *Behav Brain Res* 81: 163–171
315. Maldonado-Irizarry CS, Kelley AE (1994) Differential behavioral effects following microinjection of an NMDA receptor antagonist into nucleus accumbens subregions. *Psychopharmacol* 116: 65–72
316. Maldonado-Irizarry CS, Kelley AE (1995) Excitatory amino acid receptors within nucleus accumbens subregions differentially mediate spatial learning in rat. *Behav Pharmacol* 6: 527–539
317. Ploeger GE, Spuijit BM, Cools AR (1994) Spatial localization in the Morris water maze in rats: acquisition is affected by intra-accumbens injections of dopamine antagonist haloperidol. *Behav Neurosci* 108: 927–934
318. Schacter GB, Yang CR, Innis NK, Mogenson GJ (1989) The role of the hippocampus-nucleus accumbens pathway in radial-arm maze performance. *Brain Res* 494: 339–349
319. Seaman JK, Phillips AG (1994) Selective memory impairments produced by transient lidocaine-induced lesions of the nucleus accumbens in rats. *Behav Neurosci* 108: 456–468
320. Schwarez R, Hockfeld T, Fuxe K, Jonsson G, Goldtein M, Terenius L (1979) Ibotenic acid-induced neuronal degeneration: a morphological and neurochemical study. *Exp Brain Res* 37: 199–216
321. Adriani W, Felici A, Sargolini F, Roullet P, Usiello A, Oliverio A, Mele A (1998) N-methyl-D-aspartate and dopamine receptor involvement in the modulation of locomotor activity and memory processes. *Exp Brain Res* 123: 52–59

Neurotransmitter Interactions and Cognitive Function
Edited by Edward D. Levin
© 2006 Birkhäuser Verlag/Switzerland

Interactions between CRF and acetylcholine in the modulation of cognitive behaviour

Geoff Warnock, Jos Prickaerts and Thomas Steckler

Dept. Psychiatry, RED Europe, Johnson & Johnson Pharmaceutical Research and Development, Turnhoutseweg 30, 2340 Beerse, Belgium

Introduction

Corticotropin-releasing factor

The 41-amino acid polypeptide corticotropin-releasing factor (CRF), also named corticotropin-releasing hormone (CRH), is well known as a hypothalamic hormone which controls the hypothalamic-pituitary-adrenocortical (HPA) axis during basal activity and stress [1, 2].

Besides being the most dominant trigger of HPA axis activation, CRF also serves a neurotransmitter function in the brain, where it modulates, for example, anxiety-related behaviour, food intake, reproductive behaviour, motor function and sleep, and coordinates the behavioural and autonomic changes during stress. Moreover, CRF and CRF-related peptides such as the Urocortins (Urocortin 1, 2 and 3) seem to play an important role in the modulation of cognitive processes [3–6].

Two CRF receptor subtypes have been identified in the brain, the CRF_1 and the CRF_2 receptors (Fig. 1c) [7–17]. Three splice variants of the CRF_2 receptor have been described: $CRF_{2(a)}$ [12], $CRF_{2(b)}$ [12] and $CRF_{2(c)}$ [10]. The $CRF_{2(a)}$ receptor shares approximately 71% sequence identity with the CRF_1 receptor [12], and is the dominant CRF_2 splice variant located at neuronal membranes, whereas the $CRF_{2(b)}$ receptor is predominantly found in non-neuronal elements, such as choroid plexus, arterioles, heart and skeletal muscle [18]. The $CRF_{2(c)}$ splice variant has only been detected in limbic regions of the human central nervous system [10] and its functional role is not known. Thus, of the CRF_2 splice variants, $CRF_{2(a)}$ is the subtype of most interest in the context of this chapter and for simplicity will henceforth be referred to as CRF_2.

Expression of the CRF_1 receptor has been observed in frontal cortical areas, the cholinergic basal forebrain, the brainstem cholinergic nuclei, the ventral tegmental area, the superior colliculus, the basolateral nucleus of the amygdala (BLA), the cerebellum, the red nucleus, the trigeminal nuclei, the anterior pituitary, the hippocampus, substantia nigra pars compacta and pars reticularis, the locus coeruleus and at the level of the substantia innominata (SI) (Fig. 1c) ([19]; see [5], for review).

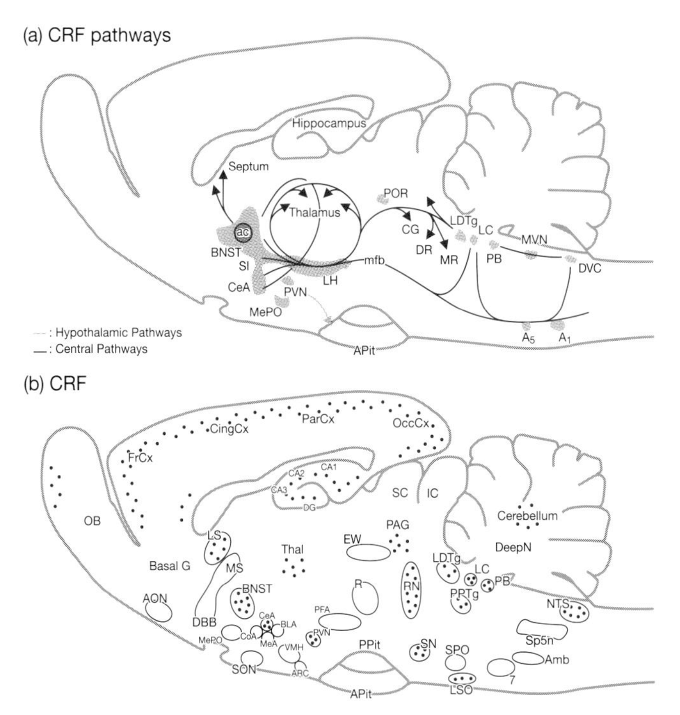

Figure 1. (a) CRF pathways, (b) CRF peptide, (c) CRF receptor mRNA, and (d) Urocortin 1, 2, 3 mRNA distributions in the rodent brain; adapted from [145]; Abbreviations: 7, facial nucleus; 12, hypoglossal nucleus; A1, A5, noradrenaline-containing cell groups; ac, anterior commissure; Amb, ambiguus nucleus; AON, anterior olfactory nucleus, APit, anterior pituitary, Arc, arcuate nucleus, Basal G, basal ganglia, BLA, basolateral amygdala, BNST, bed nucleus of the stria terminalis, CA1, 2, 3, fields CA1, 2, 3, of the hippocampus, cc, corpus callosum, CeA, central nucleus of the amygdala, CG, central grey matter, CingCx, cingulate cortex, CoA, cortical nucleus of the amygdala, DBB, diagonal band of broca, DeepN, deep nuclei, DG, dentate gyrus, DR, dorsal raphe, DVC, dorsal motor nucleus of the vagus, EW, edinger westphal nucleus, FrCx, frontal cortex, Hipp, hippocampus, IC, inferior colliculi, LC, locus coeruleus, LDTg, laterodorsal tegemental nucleus, LH, lateral hypothalamus, LS, lateral septum, LSO, lateral superior olive, MeA, medial nucleus of the amygdala, MePO, median preoptic nucleus, mfb, median forebrain bundle, MPO, median preoptic area, MR, median raphe, MS, medial septum, MVN, medial vestibular nucleus, NTS, nucleus of the solitary tract, OB, olfactory bulb, OccCx, occipital cortex, PAG, periaqueductal grey, ParCx, parietal cortex, PB, parabrachial nucleus, PFA, perifornical area, POR, perioculomotor nucleus, PPit, posterior pituitary, PPTg, pendunculopontine tegmental nucleus, PVN, paraventicular nucleus of the hypothalamus, R, red nucleus, RN, raphe nuclei, SC, superior colliculi, SI, substantia inominata, SN, substantia nigra, SON, supraoptic nucleus, Sp5n, spinal trigeminus nucleus, SPO, superior paraolivary nucleus, Thal, thalamus, VMH, ventromedial hypothalamus.

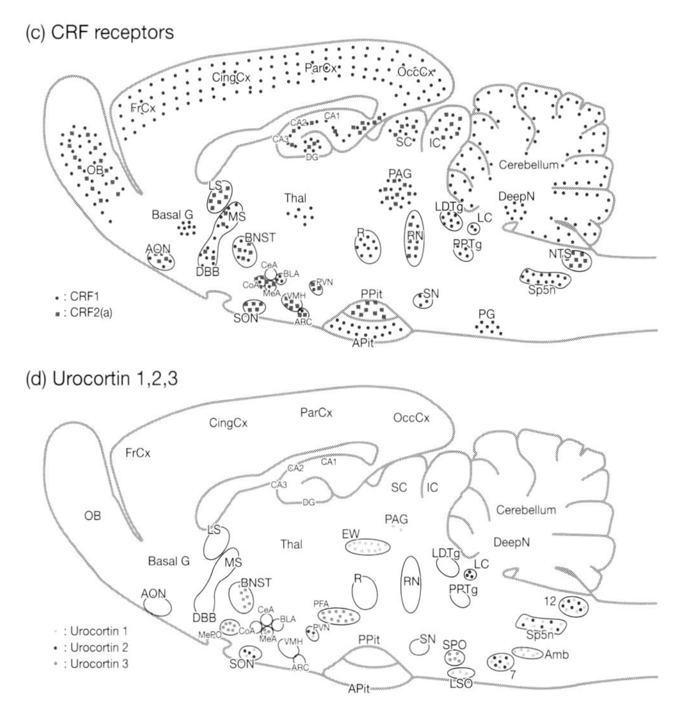

Figure 1. Continued.

The CRF_2 receptor is primarily localised in subcortical regions, including the lateral septum, the paraventricular hypothalamic nucleus (PVN), the ventromedial nucleus of the hypothalamus, the cortical and medial nuclei of the amygdala, and the serotonergic raphe nuclei (Fig. 1c) [5, 12]. Both the CRF_1 and CRF_2 receptors are moderately to strongly expressed in the olfactory bulbs, the hippocampus, the entorhinal cortex, the bed nucleus of the stria terminalis and the periaqueductal grey [12]. Many of these brain regions are strongly implicated in the mediation of cognitive processes, such as arousal, attention, learning and memory, which also raises the possibility that activation of the two CRF receptor subtypes will affect these types of behaviour.

CRF-positive neurons and their projections are also found in various brain areas linked to cognition, such as the hippocampus and cerebral cortex [20–25], with particularly high densities of CRF-positive neurons being found in the prefrontal and cingulate cortices, and throughout the neocortex [24]. The locus coeruleus, which is strongly implicated in arousal [26], also receives dense CRF projections from the PVN [27], the bed nucleus of the stria terminalis [28] and central nucleus of the amygdala (CeA) [29, 30] (Fig. 1a).

Over recent years, a number of additional CRF-related peptides have been discovered in both rodents and humans, called Urocortin 1, 2 and 3. Urocortin 1 [31] shares approximately 45% sequence identity with r/hCRF, has a high and approximately equal affinity for both CRF_1 and CRF_2 receptors. Urocortin 2 and Urocortin

3 [32–34] share approximately 34% amino acid identity with CRF and are highly selective for the CRF_2 receptor.

Urocortin 1, 2 and 3-positive neurons are sparsely distributed in subcortical regions in the rodent brain (Fig. 1d). Urocortin 1 is most strongly expressed in the Edinger-Westphal nucleus [35], with other areas of expression partially overlapping CRF_2 expressing regions. Of note, Urocortin 1 expression has also been reported in one of the cholinergic brainstem nuclei, the laterodorsal tegmental nucleus (LDTg) [36], which could represent one point of interaction with the cholinergic system. Urocortin 2 is expressed, amongst other sites, in the noradrenergic locus coeruleus [34]. Both the cholinergic LDTg [37] and the noradrenergic locus coeruleus [38] are strongly implicated in the mediation of arousal, which would suggest that both Urocortin 1 and Urocortin 2 might play a role in the modulation of arousal. Conversely, Urocortin 3 expression has neither been reported in any regions implicated in cognitive processes, or in the cholinergic system [33].

Based on the central distribution of the Urocortins and CRF_2 receptor expressing neurons, it has been suggested that Urocortin 1 may serve as the major CRF_2 ligand in the hindbrain, whereas Urocortin 3 may serve as the major CRF_2 ligand in the forebrain [39]. Urocortin 2 may signal at CRF_2 receptors expressed in regions lacking Urocortin 1 or Urocortin 3 innervations, for example, in the hippocampus and certain regions of the cerebral cortex [39].

Neuroanatomical and neurochemical evidence that the cholinergic system is an important site for CRF action

The cholinergic system

The major cholinergic pathways in the brain arise from the two major clusters of cholinergic nuclei: the cholinergic basal forebrain and the cholinergic brainstem nuclei. The basal forebrain cholinergic system includes the medial septum (MS), the ventral and the horizontal limbs of the diagonal band of Broca (vDBB and hDBB), and the nucleus basalis magnocellularis (NBM)/substantia innominata (SI) complex. The MS and vDBB both project to the hippocampal formation [37–40] (Fig. 2), and the vDBB provides the major cholinergic innervation to the olfactory bulbs. The NBM projects primarily to the frontal and parietal cortices, as well as to the BLA [40, 41].

The two major cholinergic nuclei located in the brainstem are the pedunculopontine tegmental nucleus (PPTg) and the LDTg. These nuclei project to the dopaminergic substantia nigra pars compacta, the noradrenergic locus coeruleus, the serotonergic raphe nuclei, thalamus, hypothalamus, basal forebrain and medial prefrontal cortex [42, 43] (Fig. 2).

Cholinergic receptors can be divided into nicotinic and muscarinic receptors. The nicotinic receptors are composed of α and β subunits (for a review see [44]). The predominant high-affinity nicotinic receptor in the central nervous system is composed of the $\alpha4$ and $\beta2$ subunits. This receptor is found throughout the brain in rodents, monkeys and humans [44–46], mainly associated with presynaptic cholinergic nerve terminals [47, 48].

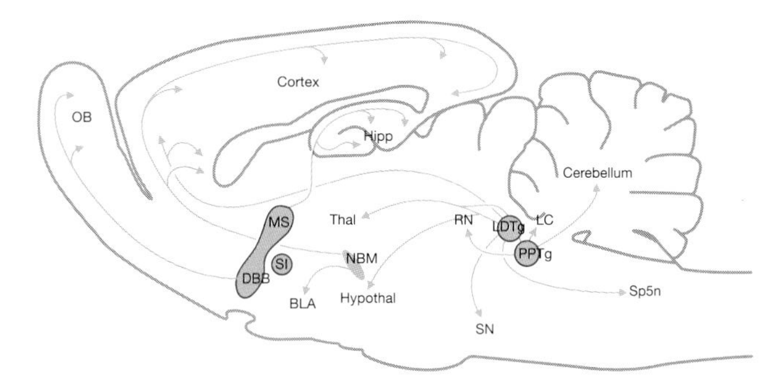

Figure 2. Major cholinergic nuclei and their projections in the rodent brain. All projections shown are to/from CRF/CRF receptor expressing regions, with the exception of the NBM. Areas outlined in red indicate those regions in which co-expression has been documented. For full cholinergic projections irrespective of CRF see [146]. Abbreviations: see Fig. 1, Hypothal, hypothalamus; NBM, nucleus basalis magnocellularis.

Five subtypes of muscarinic receptors have been defined, M_1–M_5, of which M_1, M_2 and M_4 receptor subtypes are the predominant muscarinic receptors in the central nervous system [49]. The M_1 receptor is thought to be responsible for the postsynaptic action of acetylcholine [50], while M_2 and M_4 receptors are localized both post- and presynaptically and are thought to regulate acetylcholine release by functioning as inhibitory autoreceptors on cholinergic terminals [50–53]. The M_1 receptor is found in the cerebral cortex, hippocampus, medial and BLA, nucleus accumbens and caudate putamen [50]. The M_2 receptor is found throughout the brain with high densities in the colliculi, thalamus and (cholinergic) brain stem nuclei [50, 54]. The M_4 receptor predominantly acts as an inhibitory autoreceptor in the striatum and modulates dopamine activity in motor tracts [55].

Interactions between CRF and the cholinergic system

The high abundance of CRF_1 receptors in frontal cortical areas, i.e., in one of the prime targets for cholinergic projections, but also in the cholinergic basal forebrain nuclei (MS, DBB and SI, but not in the NBM) and the brainstem cholinergic nuclei (LDTg and PPTg) (Fig. 1c) lends further support for important interactions between CRF and acetylcholine and suggests that the CRF_1 receptor may play a role in the mediation of attentional and executive functions, as these areas are implicated in the control of these types of behaviour [43, 56–62]. Indeed co-expression of CRF_1 receptors and choline acetyltransferase has been found in the cholinergic forebrain nuclei, except the NBM, and the cholinergic brainstem nuclei [19] (Table 1), and co-expression of CRF and acetylcholinesterase has been reported in the LDTg, projecting to the medial frontal cortex, septum and thalamus in the rat [63]. Furthermore, CRF-immunoreactive and CRF mRNA containing neurons were also found in the PPTg in humans [64], another area implicated in attentive processes (see [43]). Thus, there are a number of regions where CRF might interact with the cholinergic system

Table 1. Percentage of cholinergic neurons (choline acetyltransferase immunoreactive) expressing the CRF_1 receptor (CRF_1-immunoreactive) in the cholinergic forebrain and brainstem nuclei. Taken from [19]. Data represent mean percentage per section ± S.E.M.

Region	% Co-expressing neurons
Medial septum	80 ± 2.12
Substantia inominata	58 ± 4.18
Diagonal band of Broca	
Vertical limb	93 ± 3.18
Horizontal limb	90 ± 3.89
Nucleus basalis magnocellularis	—
Pendunculopontine tegmental nucleus	75 ± 6.52
Laterodorsal tegmental nucleus	92 ± 3.52

via activation of the CRF_1 receptor, and a number of these regions are implicated in cognitive function.

The locus coeruleus

Another neurotransmitter strongly implicated in arousal and attentional processes is noradrenaline, and in particular the noradrenergic locus coeruleus seems to play an important role in the modulation of these types of behaviour [65–67]. The locus coeruleus receives both CRF and cholinergic projections [27–30] (and see [37]) (Figs. 1a and 2) and CRF_1 receptors are expressed in this region [19]. Thus, the noradrenergic locus coeruleus might represent an area of convergence for both the cholinergic and the CRF systems through which both systems could affect arousal and attention. Indeed, intra-coeruleal CRF administration activates neurons of the locus coeruleus [30, 68], an effect that can be blocked with CRF_1 antagonists [69, 70]. Furthermore, intra-coeruleal administration of CRF induces behavioural activation (expressed as an increase in non-ambulatory spontaneous motor activity and reduced immobility in the modified Porsolt swim test), which could be indicative of increased arousal [68].

The hippocampus

Intracerebroventricular (ICV) administration of CRF has been reported to increase hippocampal acetylcholine release [71, 72] through CRF_1 receptor activation [73, 74]. This is believed to represent release of acetylcholine from the terminals of neurons projecting from the MS (which provides the major source of cholinergic input to the hippocampus) [72] (Fig. 2). In support of this, ICV CRF stimulates Fos expression within the basal forebrain and brainstem nuclei, including the MS [75]. Moreover, CRF_1 blockade with the selective CRF_1 antagonists antalarmin or SSR125543A also partially antagonised the CRF-induced release of acetylcholine in the hippocampus [74], indicating that this effect is CRF_1 receptor mediated. However,

ICV CRF has also been reported to decrease high-affinity choline uptake in the rat hippocampus [76], which might represent an alternative mechanism by which hippocampal acetylcholine levels could be increased in response to CRF.

The hippocampal cholinergic system has been suggested to be involved in the mechanisms underlying the arousal that is associated with fear and anxiety-provoking stimuli [77, 78] and one of the possible roles of the cholinergic septohippocampal system could be to ensure that the animal is appropriately responsive to its environment, being able to monitor and amend its behaviour in an appropriate manner when exposed to a fearful or anxiety-provoking stimulus [79]. Consequently, it might be speculated that stress- or CRF-induced increases in hippocampal acetylcholine release facilitate information processing by hippocampal circuits and hence induce a bias towards affectively negative information. This is in line with the fact that an increase in hippocampal acetylcholine release can be observed following exposure to a variety of stressors [71, 80–82]. Naturally, such a bias might also interfere with accurate cognitive processing. Thus, both an underactive and an overactive cholinergic system might be detrimental to proper cognitive functioning.

The frontal cortex

Synergism between the CRF and cholinergic systems has also been reported in the rat frontal cortex *in vitro*, where stimulation of adenylate cyclase was seen following co-activation of M_1 and CRF_1 receptors in frontal cortical membrane preparations [83]. An interaction at this level is further supported by the demonstration that ICV CRF decreases high-affinity choline uptake in the rat frontal cortex [76], which should lead to increased frontal cortex acetylcholine levels in response to CRF.

Thus, it is conceivable that CRF might influence cognitive function *via* activation of parts of the cholinergic basal forebrain and brainstem nuclei, as well as by co-activation of intracellular signalling pathways at higher brain areas such as the frontal cortex.

Interactions between the cholinergic and CRF systems

There is evidence that the interactions between the CRF and the cholinergic system are reciprocal, i.e. not only is CRF capable of modulating cholinergic activity, but acetylcholine is also capable of modulating CRF function.

Thus, it has been reported that chronic treatment with the non-selective muscarinic antagonist atropine produced a significant and selective increase in CRF_1 receptors in frontoparietal cortex in rats [84], which might represent a compensatory mechanism. Furthermore, acetylcholine induces CRF release from the amygdala (another brain area closely involved in the processing of stress-related information and in the modulation of an individual's responses to stress) *in vitro* [85], an effect that could be antagonised with both atropine and mecamylamine, suggesting the involvement of both muscarinic and nicotinic receptors. It would be interesting to investigate the exact amygdaloid nuclei involved in this response, as there are discreet differences in the distribution of the components of the two systerms at this level: CRF is only expressed in the CeA (Fig. 1b), which is the major output relay in the amygdala [86], and thus it is possible that acetylcholine-induced release of CRF originates from this nucleus. This could, for example, be mediated *via* activation of M_1 receptors, which are expressed in the BLA [50]. The BLA represents the major input area of the amygdala [86]. Alternatively, it could be mediated *via* activation of nicotinic receptors, which are expressed throughout the amygdala [44–46]. The NBM provides the major cholinergic input into the BLA (Fig. 2), i.e., even though no CRF receptors were observed at the level of the NBM [19], it is possible that important interactions between this cholinergic nucleus and the CRF system exist at the level of the amygdala.

The HPA axis

However, interactions between CRF and acetylcholine are not confined to higher brain areas, but can also be observed at the level of the HPA axis, partly involving noradrenergic neurons innervating the PVN (likely to project from the locus coeruleus). Thus, ICV administration of nicotine has been reported to elevate plasma corticosterone in rats through effects on noradrenaline release in the PVN [87]. This effect can be abolished by the specific CRF_1 antagonist CP-154,526, and appears to be dependent on nicotinic receptors [87]. Furthermore, injection of acetylcholine directly into the PVN increase CRF mRNA levels in the hypothalamus as a whole, as well as plasma ACTH levels [88]. In addition, the septohippocampal system might be of relevance for an appropriate termination of the stress response, as an appropriate septohippocampal cholinergic input seems to be essential for the hippocampus to convey its inhibitory effect on the HPA axis [89]. Thus, the cholinergic system modulates HPA axis activity, both indirectly *via* the hippocampus [89] and *via* noradrenergic modulation [87], as well as directly at the level of the PVN.

CRF and cognitive function

Arousal and attention

From the previous section it is evident that there is both strong anatomical and neurochemical evidence for important CRF-cholinergic interactions in the brain. But how is this interaction translated at the behavioural level? ICV infusions of CRF have been reported to decrease slow-wave sleep [70] and to shorten pentobarbital-induced sleeping time in rats [90]. This could suggest an arousing function of CRF, possibly in preparation to stressful stimuli. Conversely, stress-induced shortening of pentobarbital-induced sleeping time in rats can be reversed by the CRF_1 antagonist CRA1000, suggesting that this stress-induced increase in arousal is CRF_1-mediated [91]. The non-selective CRF antagonists α-helical CRF_{9-41} or astressin [92], as well as ICV administration of CRF antisense [93] have also been shown to reduce spontaneous waking, consistent with a role of CRF in arousal. Furthermore, stimulation of the PVN with glutamate has been shown to increase c-Fos expression in PVN CRF neurons, indicating activation, accompanied by an arousal shift measured by electrocorticogram [94].

Of note, it is important for a subject to be at the right level of arousal to perform in an optimal way. Neither a state of underarousal nor of overarousal would be helpful to deal with a threatening situation. In line with this view, it has been shown that transgenic mice overexpressing CRF, which are characterized by hightened anxiety [95, 96], also show general impairments in operant five choice serial reaction time performance, i.e., in a task taxing attentional processes [97]. This task is known to be sensitive to lesions of the PPTg, medial prefrontal cortex, and to damage of the NBM in rats [43, 58–60, 62, 98]. Since the NBM is lacking in CRF_1 receptors [19], it is conceivable that the PPTg and/or the prefrontal cortex may be involved in the effects of CRF on attention and arousal.

In further support of a role of CRF-related peptides in arousal and attention, it has been reported that ICV CRF and Urocortin 1 both facilitated initial acquisition but not consolidation in a spatial water maze task, but only under relatively easy learning conditions (inter-trial interval of 30 s), while impaired performance was seen under more difficult learning conditions (inter-trial interval > 2 min; [99]). Enhanced consolidation was seen in a passive avoidance task, and these effects were reversible by ICV treatment with the CRF_1 antagonist antalarmin [99]. Such a pattern of cognitive effects would be consistent with an increase in arousal, rather than with a true mnemonic effect. As has been pointed out [6], this profile of cognitive enhancement is consistent with increased activity of the medial septal area (vertical limb of the DBB and the MS), which has been implicated in arousal and attention, and is a site of CRF_1 receptor expression (Fig. 1c).

Learning and memory

Since both CRF_1 and CRF_2 receptors are also strongly expressed in the hippocampus (Fig. 1c), it might be further suggested that activation of these receptors by CRF has implications for spatial/contextual memory.

Indeed, intra-hippocampal infusions of CRF induce a long lasting enhancement of synaptic efficiency in the hippocampus, as measured by an increase in amplitude and slope of population excitatory postsynaptic potentials (pEPSPs) and an increase in the level of cAMP [100, 101]. Moreover, CRF has also been shown to facilitate long-term potentiation (LTP) in the mouse hippocampus *in vitro* [102, 103], which might provide indirect evidence for a role of CRF in the modulation of memory.

It has also been shown that injections of CRF or Urocortin 1 into the dorsal hippocampus before training in a context- and tone-dependent fear conditioning paradigm enhanced learning [103, 104], and similar effects were seen with post-training administration of CRF, suggesting a role of CRF-related peptides in memory consolidation. This effect was mediated through the CRF_1 receptor, but not *via* the CRF_2 receptor, as blockade of the hippocampal CRF_2 receptor with the selective peptidergic CRF_2 antagonist anti-sauvagine-30 did not alter the effects of pre-training intrahippocampal administration of CRF in fear conditioning [104]. Moreover, CRF administration into the dentate gyrus also improved retention performance in a passive avoidance task in rats [105–107], although it was suggested that this is due to facilitation of noradrenaline release in the DG [107]. On the other hand, direct intrahippocampal infusion of an antisense oligonucleotide directed against CRF mRNA has been reported to impair performance in a passive avoidance task [108]. As a caveat, it should be mentioned that it is difficult to clearly disentangle an effect on learning from effects on other types of behaviour, such as anxiety-related behaviour, motivational factors and altered pain threshold, particularly in passive avoidance, which might confound results.

In further support of a positive modulatory role for CRF_1 in cognitive behaviour, it has been shown that CRF_1 knockout mice display deficits in spatial recognition memory in a two-trial spatial memory task [109] and that intraperitoneal injection of the specific CRF_1 antagonist CP-154,526 prior to training impaired the induction of contextual fear conditioning, while injection prior to testing reduced the expression of conditioned fear [110, 111], and CP154,526 has also been shown to antagonise stress-induced learning deficits in a fear conditioning paradigm when administered prior to stress [103], suggesting that blockade of CRF_1 may interfere with acquisition and retrieval processes. However, the latter is difficult to distinguish from a possible anxiolytic effect, and indeed in recent studies the CRF_1 antagonists DMP-904 and DMP-696 were found to have little or no effect in the water maze or delayed non-matching to position test [112].

In contrast to the cognition-enhancing properties of CRF in the hippocampus, CRF or Urocortin 1 injection into the lateral septum impaired learning in context- and tone-dependent fear conditioning [104]. This effect appears to be CRF_2 mediated, as it was blocked by the peptidergic CRF_2 antagonist anti-sauvagine-30, and CRF_1 receptors are lacking in the rat lateral septum. Furthermore, intra-lateral-septal injection of anti-sauvagine-30 enhanced learning in these tasks when injected alone, suggesting a tonic control of learning by CRF_2. However, ICV infusion of antisense oligonucleotides against CRF_2 mRNA failed to affect performance in the spatial water maze or affect social recognition memory [113], suggesting no or at least no major role for CRF_2 in these hippocampus-dependent tasks [114, 115], although CRF_2 antisense treatment enhanced context-dependent fear conditioning in another study [116]. This latter effect may be mediated via knockdown of CRF_2 receptors in the lateral septum, producing an effect similar to pharmacological blockade of septal CRF_2 receptors [104]. Interestingly, this lateral septal CRF_2-mediated cognitive impairment appears to be dependent on dopamine D2 receptors, as it could be blocked by the specific D2 antagonist sulpiride [117].

CRF_1-mediated enhancement and CRF_2-mediated impairment of learning and memory is supported by non-localised administration of CRF_1 and CRF_2 ligands. ICV administration of CRF has been reported to mildly enhance conditioned auditory fear [118], while ICV administration of Urocortin 1 facilitated acquisition, consolidation and retrieval of the passive avoidance response [119]. In contrast, intraperitoneal administration of the CRF_2-specific agonist Stresscopin [32] (human Urocortin 3) has been shown to impair performance in a passive avoidance task, although the involvement of central CRF receptors in this effect is questionable [120].

There is also behavioural evidence for a role of CRF in the modulation of amygdala-dependent mnemonic processes, as post-training bilateral infusions of the non-selective CRF antagonist α-helical $CRF_{9\text{-}41}$ directly into the BLA of rats impaired passive avoidance retention performance [121]. In the same study it was found that the training stimulus, a brief foot shock, increased CRF levels in the CeA. This could suggest that CRF receptor activation in the BLA, likely through training-induced release of CRF from the CeA, participates in mediating stress-induced effects on memory consolidation [121]. Infusion of the CRF_1 antagonist antalarmin into the amygdala after social defeat has been shown to reduce subsequent conditioned defeat, implicating CRF_1 receptors in the amygdala-related effects of CRF on memory consolidation [122].

Enhancing the availability of free CRF by displacing it from its binding protein (CRF-BP) with the CRF-BP inhibitor CRF(6-33) has also been reported to produce cognition-enhancing effects in animal tests of learning and memory, such as in spatial water maze navigation, Y-maze visual discrimination, passive avoidance, one-way active avoidance and context- and tone-dependent fear conditioning, but without the characteristic stress effects seen after direct ICV CRF administration [104, 123–125]. One of the reasons why CRF-BP inhibitors could have less effects on stress responsivity could be the distinct distribution pattern of CRF-BP, as CRF-BP is distributed in discrete regions of the rat brain [126], including the hippocampus and frontal cortex, but has a low expression in subcortical brain areas also linked to anxiety-related behaviour, such as the septum. As such, CRF-BP inhibitor administration may be mimicking the effects of raising frontal cortical and hippocampal CRF concentration through direct CRF administration, but lacking the subcortical effects of CRF. Furthermore, CRF(6-33) reversed impairments in a social memory test in adult female rats induced by the non-specific CRF antagonist D-Phe CRF(12-41) [127]. However, it remains debatable whether the effects of CRF-BP inhibition are a direct action on mnemonic processes or indirect effects *via* action on other types of cognition. For example, in one of these studies the improvement in performance was only seen during early but not late acquisition sessions in a cued visual discrimination task [124]. Drug-induced effects on learning would be expected in late acquisition, with groups starting at comparable accuracy levels at the beginning of acquisition. This may suggest that CRF(6-33) does not directly affect learning, but enhances performance through altered arousal or attentional processes [5].

Lastly, there is evidence that stabilisation of the CRF system in early life is beneficial to later cognitive development, as administration of the CRF_1 antagonist NBI30775 to rats from postnatal day 10-17 improved performance in the water maze and object recognition test at postnatal days 50-70, to levels similar to that in rats with extra maternal care stimulated by handling during postnatal days 2-9 [128]. Reduced CRF mRNA expression the CeA, BNST and PVN and increased glucocorticoid receptor levels in the hippocampus during the postnatal handling period suggests the involvement of these regions in the response [129, 130]. The involvement of CRF in the BNST in conditioned responses in adult animals is supported by the finding that the non-selective CRF antagonist D-Phe $CRF_{(12\text{-}41)}$ reduced conditioned defeat when infused into the BNST, but not the CeA [131], and BNST CRF_2 is implicated as the CRF_2-specific antagonist anti-sauvagine-30 [132], but not the CRF_1-specific antagonist CP-154,526 also reduced conditioned defeat when infused into this region [133].

Interim summary

To summarise, CRF appears to exert a modulatory effect on attention and arousal, mediated at least in part by CRF_1 receptors, and possibly involving the PPTg, prefrontal cortex, medial septal area and/or the noradrenergic locus coeruleus. CRF_1 in the hippocampus appears to exert a positive modulatory role on learning and memory, while CRF_2 in the lateral septum has a negative effect on these processes, possibly through interaction with the dopaminergic system at this level. However, a direct effect of CRF_2 receptor activation on cognitive function at hippocampal level cannot be ruled out, as it has been demonstrated that stress-enhancement of context-dependent fear conditioning can be prevented by the specific CRF_2 receptor antagonist anti-sauvagine-30 [134], but only when administered 3h following stress, suggesting a delayed CRF_2-mediated effect. Furthermore, at the level of the amygdala, cognitive function can also be modulated by CRF, possibly involving CRF receptors in the BLA. This might be mediated *via* CRF_1 receptors, as CRF_2 receptors have not been reported to be expressed in this region.

Interactions between the CRF and the cholinergic systems in the modulation of cognitive behaviour

Based on the points discussed above, it may be argued that there is a central interaction between CRF and the cholinergic system in order to maintain appropriate processing of environmental information, in particular under stressful conditions [135]. This could take place at various levels of the basal forebrain, brainstem cholinergic nuclei, or their respective projection areas.

Given that CRF administration increases acetylcholine release at hippocampal level [71, 72], it can be expected that there might also be important interactions between CRF and acetylcholine in relation to hippocampus-mediated cognitive function.

Intrahippocampal infusion of CRF enhances context- and tone-dependent fear conditioning [104, 117], but was unable to reverse the cognitive impairment induced by the muscarinic antagonist scopolamine [117]. This would suggest that it is unlikely that the memory enhancing effects of direct intrahippocampal CRF administration are mediated through actions on the septohippocampal pathway at hippocampal level. It would further argue against an important role of a decrease in high-affinity choline uptake at the level of the hippocampus [76]. Next, it would of course be interesting to investigate the effects of direct injections of CRF into the medial septum on a scopolamine-induced fear-conditioning deficit. An interaction at the level of the lateral septum is likely, as CRF injected into this area reversed the cognitive deficit induced by scopolamine injected into this same region [117].

Despite the effects of CRF on hippocampal acetylcholine release, it has been reported that ICV CRF in combination with either scopolamine or the nicotinic antagonist mecamylamine appears to act synergistically in impairing spatial water maze discrimination learning [135]. This finding provides further support for partially independent actions of CRF and cholinergic blockade on cognitive function, also at hippocampal level, as CRF would have been expected to attenuate a scopolamine-induced impairment in this hippocampus-dependent paradigm *via* activation of cholinergic activity at this level. It is of course possible that CRF primarily activated the CRF_2 receptor at the level of the lateral septum in this study, as this region and receptor have been implicated in CRF-induced cognitive impairment [104]. Non-localised peptide delivery may indeed be responsible for mixed reports in the literature, as ICV administration of the non-specific CRF receptor antagonist α-helical-$CRF_{9\text{-}41}$ [136] blocks nicotine-induced conditioned anxiety in the social interaction paradigm, but not acute nicotine-induced anxiety [137], suggesting that the effect is not purely anxiety-related. However, both the dorsal hippocampus and lateral septum have been implicated in the anxiety-inducing effects of nicotine in the social interaction paradigm [138], providing no further clarity on the region involved. Further confusing the issue, ICV Urocortin 1-induced facilitation of acquisition, consolidation and retrieval in a passive avoidance task can be blocked with both α-helical-$CRF_{9\text{-}41}$ and atropine, although the response was also blocked by antagonists for other neurotransmitter systems [119]. As an agonist at both CRF_1 and CRF_2 receptors, it is not possible to attribute Urocortin 1-induced cognitive enhancement to either receptor, and as mentioned earlier, it can be difficult to dissociate cognitive effects from, for example, effects on anxiety-related behaviour, motivational factors and altered pain threshold, in passive avoidance.

Another very interesting brain region where CRF and cholinergic systems might interact is the BLA. Although direct evidence for important interactions between ACh and CRF in the modulation of cognitive function is lacking at the level of the amygdala, it is of note that both systems might interact at this level in modulating stress-induced changes in blood pressure [139]. Given that CRF potently modulates amygdala-dependent types of learning and memory *via* activation of CRF receptors in the CeA or BLA (see above), it would be very interesting to elucidate the possibility that the CRF and the cholinergic systems indeed interact at this level to modulate cognitive function.

Although there is evidence for important interactions between the septohippocampal cholinergic projection and CRF as well as for independent and indirect actions of the two systems at the same brain target areas, another possibility is that some of the effects of CRF are mediated through enhanced glucocorticoid levels induced by CRF activation of the HPA axis. Glucocorticoids themselves affect cognitive processes, and it is possible that some of the effects of CRF on cognition are due to their subsequent release from the adrenal gland (for a review see [140]). However, this pathway is relatively slow, and there is evidence for direct effects of CRF, such as the priming of LTP in the hippocampus [102], suggesting that these effects are not primarily glucocorticoid mediated.

Conclusions

In conclusion, there is a limited amount of evidence that CRF interacts with the cholinergic system in the modulation of cognitive behaviour, although some of the cognitive effects of CRF are clearly independent of cholinergic activity. It has been suggested that CRF may function in a parallel processing model with other neurotransmitter systems, modulating the signal in order to attribute a stressful characteristic [141]. In such a model, CRF may modulate cognitive processes through interactions with the cholinergic system leading to storage of memories with a stressful component. Of note, overactivity of such a system, possibly resulting in hyperattention (see [142]), or enhanced storage of stressful memories, might be as detrimental as an underactivity of such a system.

Indeed, the existence of two apparently opposing systems for the modulation of cognition by CRF, possibly involving CRF_1 in the medial septum and hippocampus, and CRF_2 in the lateral septum, and evidence that long-term disruption of CRF function (i.e. CRF overexpression) impairs cognitive function (such as attention, learning and memory) may suggest that a delicate balance exists, which further supports the above concept. As such, any severe (e.g. chronic) disruption of this balance (such as altered HPA axis activity in psychiatric disorders such as depression, or anxiety disorders) might result in cognitive impairment, and hence contribute to the cognitive deficits common to many psychiatric disorders. Supporting this hypothesis, it has been reported that improvements in working memory during antidepressant treatment of patients with major depression, were correlated with normalisation of the HPA axis [143]. Interestingly, a cholinergic hyperactivity has been found in depression [144]. This would further suggest that novel drugs that normalize an overactive HPA axis, such as CRF_1 antagonists, might be beneficial in attenuating cognitive dysfunction (e.g., reduce a bias towards affectively negative information) in these disorders in parallel with an attenuation of enhanced cholinergic activity.

References

1. Rivier CL, Plotsky PM (1986) Mediation by corticotropin releasing factor (CRF) of adenohypophysial hormone secretion. *Annu Rev Physiol* 48: 475–494
2. Vale W, Spiess J, Rivier C, Rivier J (1981) Characterization of a 41-residue ovine hypothalamic peptide that stimulates secretion of corticotropin and beta-endorphin. *Science* 213: 1394–1397
3. De Souza EB (1995) Corticotropin-releasing factor receptors: physiology, pharmacology, biochemistry and role in central nervous system and immune disorders. *Psychoneuroendocrinology* 20: 789–819
4. Dunn AJ, Berridge CW (1990) Physiological and behavioral responses to corticotropin-releasing factor administration: is CRF a mediator of anxiety or stress responses? *Brain Res Brain Res Rev* 15: 71–100
5. Steckler T, Holsboer F (1999) Corticotropin-releasing hormone receptor subtypes and emotion. *Biol Psychiatry* 46: 1480–1508
6. Zorrilla EP, Koob GF (2005) The roles of Urocortins 1, 2 and 3 in the brain. In: Steckler T, Kalin NH, Reul JMHM (eds): *Handbook of stress and the rain*. Elsevier, Amsterdam, 179–203
7. Chang CP, Pearse RV, O'Connell S, Rosenfeld MG (1993) Identification of a seven transmembrane helix receptor for corticotropin-releasing factor and sauvagine in mammalian brain. *Neuron* 11: 1187–1195
8. Chen R, Lewis KA, Perrin MH, Vale WW (1993) Expression cloning of a human corticotropin-releasing-factor receptor. *Proc Natl Acad Sci USA* 90: 8967–8971

9. Kishimoto T, Pearse RV, Lin CR, Rosenfeld MG (1995) A sauvagine/corticotropin-releasing factor receptor expressed in heart and skeletal muscle. *Proc Natl Acad Sci USA* 92: 1108–1112
10. Kostich WA, Chen A, Sperle K, Largent BL (1998) Molecular identification and analysis of a novel human corticotropin-releasing factor (CRF) receptor: the CRF2gamma receptor. *Mol Endocrinol* 12: 1077–1085
11. Liaw CW, Lovenberg TW, Barry G, Oltersdorf T, Grigoriadis DE, De Souza EB (1996) Cloning and characterization of the human corticotropin-releasing factor-2 receptor complementary deoxyribonucleic acid. *Endocrinology* 137: 72–77
12. Lovenberg TW, Liaw CW, Grigoriadis DE, Clevenger W, Chalmers DT, De Souza E B, Oltersdorf T (1995) Cloning and characterization of a functionally distinct corticotropin-releasing factor receptor subtype from rat brain. *Proc Natl Acad Sci USA* 92: 836–840
13. Perrin MH, Donaldson CJ, Chen R, Lewis KA, Vale WW (1993) Cloning and functional expression of a rat brain corticotropin releasing factor (CRF) receptor. *Endocrinology* 133: 3058–3061
14. Perrin M, Donaldson C, Chen R, Blount A, Berggren T, Bilezikjian L, Sawchenko P, Vale W (1995) Identification of a second corticotropin-releasing factor receptor gene and characterization of a cDNA expressed in heart. *Proc Natl Acad Sci USA* 92: 2969–2973
15. Stenzel P, Kesterson R, Yeung W, Cone RD, Rittenberg MB, Stenzel-Poore MP (1995) Identification of a novel murine receptor for corticotropin-releasing hormone expressed in the heart. *Mol Endocrinol* 9: 637–645
16. Valdenaire O, Giller T, Breu V, Gottowik J, Kilpatrick G (1997) A new functional isoform of the human CRF2 receptor for corticotropin-releasing factor. *Biochim Biophys Acta* 1352: 129–132
17. Vita N, Laurent P, Lefort S, Chalon P, Lelias JM, Kaghad M, Le Fur G, Caput D, Ferrara P (1993) Primary structure and functional expression of mouse pituitary and human brain corticotrophin releasing factor receptors. *FEBS Lett* 335: 1–5
18. Lovenberg TW, Chalmers DT, Liu, C, De Souza EB (1995) CRF2 alpha and CRF2 beta receptor mRNAs are differentially distributed between the rat central nervous system and peripheral tissues. *Endocrinology* 136: 4139–4142
19. Sauvage M, Steckler T (2001) Detection of corticotropin-releasing hormone receptor 1 immunoreactivity in cholinergic, dopaminergic and noradrenergic neurons of the murine basal forebrain and brainstem nuclei–potential implication for arousal and attention. *Neuroscience* 104: 643–652
20. Merchenthaler I (1984) Corticotropin releasing factor (CRF)-like immunoreactivity in the rat central nervous system. Extrahypothalamic distribution. *Peptides* 5: Suppl-69
21. Merchenthaler I, Vigh S, Schally AV, Stumpf WE, Arimura A (1984) Immunocytochemical localization of corticotropin releasing factor (CRF)-like immunoreactivity in the thalamus of the rat. *Brain Res* 323: 119–122
22. Morin SM, Ling N, Liu XJ, Kahl SD, Gehlert DR (1999) Differential distribution of urocortin- and corticotropin-releasing factor-like immunoreactivities in the rat brain. *Neuroscience* 92: 281–291
23. Sakanaka M, Shibasaki T, Lederis K (1987) Corticotropin releasing factor-like immunoreactivity in the rat brain as revealed by a modified cobalt-glucose oxidase-diaminobenzidine method. *J Comp Neurol* 260: 256–298
24. Swanson LW, Sawchenko PE, Rivier J, Vale WW (1983) Organization of ovine corticotropin-releasing factor immunoreactive cells and fibers in the rat brain: an immunohistochemical study. *Neuroendocrinology* 36: 165–186

25. Van Bockstaele EJ, Colago EE, Valentino RJ (1996) Corticotropin-releasing factor-containing axon terminals synapse onto catecholamine dendrites and may presynaptically modulate other afferents in the rostral pole of the nucleus locus coeruleus in the rat brain. *J Comp Neurol* 364: 523–534
26. Bremner JD, Krystal JH, Southwick SM, Charney DS (1996) Noradrenergic mechanisms in stress and anxiety: I. Preclinical studies. *Synapse* 23: 28–38
27. Valentino RJ, Page M, Van Bockstaele E, Aston-Jones G (1992) Corticotropin-releasing factor innervation of the locus coeruleus region: distribution of fibers and sources of input. *Neuroscience* 48: 689–705
28. Van Bockstaele EJ, Colago EE, Valentino RJ (1998) Amygdaloid corticotropin-releasing factor targets locus coeruleus dendrites: substrate for the co-ordination of emotional and cognitive limbs of the stress response. *J Neuroendocrinol* 10: 743–757
29. Koegler-Muly SM, Owens MJ, Ervin GN, Kilts CD, Nemeroff CB (1993) Potential corticotropin-releasing factor pathways in the rat brain as determined by bilateral electrolytic lesions of the central amygdaloid nucleus and the paraventricular nucleus of the hypothalamus. *J Neuroendocrinol* 5: 95–98
30. Valentino RJ, Foote SL, Page ME (1993) The locus coeruleus as a site for integrating corticotropin-releasing factor and noradrenergic mediation of stress responses. *Ann NY Acad Sci* 697: 173–188
31. Vaughan J, Donaldson C, Bittencourt J, Perrin MH, Lewis K, Sutton S, Chan R, Turnbull AV, Lovejoy D, Rivier C (1995) Urocortin, a mammalian neuropeptide related to fish urotensin I and to corticotropin-releasing factor. *Nature* 378: 287–292
32. Hsu SY, Hsueh AJ (2001) Human stresscopin and stresscopin-related peptide are selective ligands for the type 2 corticotropin-releasing hormone receptor. *Nat Med* 7: 605–611
33. Lewis K, Li C, Perrin MH, Blount A, Kunitake K, Donaldson C, Vaughan J, Reyes TM, Gulyas J, Fischer W et al. (2001) Identification of urocortin III, an additional member of the corticotropin-releasing factor (CRF) family with high affinity for the CRF2 receptor. *Proc Natl Acad Sci USA* 98: 7570–7575
34. Reyes TM, Lewis K, Perrin MH, Kunitake KS, Vaughan J, Arias CA, Hogenesch JB, Gulyas J, Rivier J, Vale WW, Sawchenko PE (2001) Urocortin II: a member of the corticotropin-releasing factor (CRF) neuropeptide family that is selectively bound by type 2 CRF receptors. *Proc Natl Acad Sci USA* 98: 2843–2848
35. Bittencourt JC, Vaughan J, Arias C, Rissman RA, Vale WW, Sawchenko PE (1999) Urocortin expression in rat brain: evidence against a pervasive relationship of urocortin-containing projections with targets bearing type 2 CRF receptors. *J Comp Neurol* 415: 285–312
36. Kozicz T, Yanaihara H, Arimura A (1998) Distribution of urocortin-like immunoreactivity in the central nervous system of the rat. *J Comp Neurol* 391: 1–10
37. Butcher LL (1995) Cholinergic neurons and networks. In: Paxinos G (ed) *The rat nervous system*. Academic Press, San Diego, 1003–1015
38. Aston-Jones G, Shipley MT, Grzanna R (1995) The Locus Coeruleus, A5 and A7 Noradrenergic Cell Groups. In: Paxinos G (ed) *The rat central nervous system*, Academic Press, San Diego, 183–213
39. Hauger RL, Grigoriadis DE, Dallman MF, Plotsky PM, Vale WW, Dautzenberg FM (2003) International Union of Pharmacology. XXXVI. Current Status of the Nomenclature for Receptors for Corticotropin-Releasing Factor and Their Ligands. *Pharmacol Rev* 55: 21–26
40. Mesulam MM, Mufson EJ, Wainer BH, Levey AI (1983) Central cholinergic pathways in the rat: an overview based on an alternative nomenclature (Ch1-Ch6). *Neuroscience* 10: 1185–1201

41. Hellendall RP, Godfrey DA, Ross CD, Armstrong DM, Price JL (1986) The distribution of choline acetyltransferase in the rat amygdaloid complex and adjacent cortical areas, as determined by quantitative micro-assay and immunohistochemistry. *J Comp Neurol* 249: 486–498
42. Satoh K, Fibiger HC (1986) Cholinergic neurons of the laterodorsal tegmental nucleus: efferent and afferent connections. *J Comp Neurol* 253: 277–302
43. Steckler T, Inglis W, Winn P, Sahgal A (1994) The pedunculopontine tegmental nucleus: a role in cognitive processes? *Brain Res Brain Res Rev* 19: 298–318
44. Picciotto MR, Caldarone BJ, King SL, Zachariou V (2000) Nicotinic receptors in the brain. Links between molecular biology and behavior. *Neuropsychopharmacology* 22: 451–465
45. Han ZY, Zoli M, Cardona A, Bourgeois JP, Changeux JP, Le Novere N (2003) Localization of [3H]nicotine, [3H]cytisine, [3H]epibatidine, and [125I]alpha-bungarotoxin binding sites in the brain of Macaca mulatta. *J Comp Neurol* 461: 49–60
46. Happe HK, Peters JL, Bergman DA, Murrin LC (1994) Localization of nicotinic cholinergic receptors in rat brain: autoradiographic studies with [3H]cytisine. *Neuroscience* 62: 929–944
47. McGehee DS, Role LW (1995) Physiological diversity of nicotinic acetylcholine receptors expressed by vertebrate neurons. *Annu Rev Physiol* 57: 521–546
48. Rowell PP, Winkler DL (1984) Nicotinic stimulation of [3H]acetylcholine release from mouse cerebral cortical synaptosomes. *J Neurochem* 43: 1593–1598
49. Volpicelli LA, Levey AI (2004) Muscarinic acetylcholine receptor subtypes in cerebral cortex and hippocampus. *Prog Brain Res* 145: 59–66
50. Spencer DG Jr., Horvath E, Traber J (1986) Direct autoradiographic determination of M1 and M2 muscarinic acetylcholine receptor distribution in the rat brain: relation to cholinergic nuclei and projections. *Brain Res* 380: 59–68
51. Carey GJ, Billard W, Binch H, III, Cohen-Williams M, Crosby G, Grzelak M, Guzik H, Kozlowski JA, Lowe DB, Pond AJ et al. (2001) SCH 57790, a selective muscarinic M(2) receptor antagonist, releases acetylcholine and produces cognitive enhancement in laboratory animals. *Eur J Pharmacol* 431: 189–200
52. Stillman MJ, Shukitt-Hale B, Galli RL, Levy A, Lieberman HR (1996) Effects of M2 antagonists on *in vivo* hippocampal acetylcholine levels. *Brain Res Bull* 41: 221–226
53. Tzavara ET, Bymaster FP, Felder CC, Wade M, Gomeza J, Wess J, McKinzie DL, Nomikos GG (2003) Dysregulated hippocampal acetylcholine neurotransmission and impaired cognition in M2, M4 and M2/M4 muscarinic receptor knockout mice. *Mol Psychiatry* 8: 673–679
54. Regenold W, Araujo DM, Quirion R (1989) Quantitative autoradiographic distribution of [3H]AF-DX 116 muscarinic-M2 receptor binding sites in rat brain. *Synapse* 4: 115–125
55. Bymaster FP, Heath I, Hendrix JC, Shannon HE (1993) Comparative behavioral and neurochemical activities of cholinergic antagonists in rats. *J Pharmacol Exp Ther* 267: 16–24
56. Gallagher M, Holland PC (1994) The amygdala complex: multiple roles in associative learning and attention. *Proc Natl Acad Sci USA* 91: 11771–11776
57. Garcia-Rill E (1991) The Pedunculopontine Nucleus. *Prog Neurobiol* 36: 363–389
58. Muir JL, Dunnett SB, Robbins TW, Everitt BJ (1992) Attentional functions of the forebrain cholinergic systems: effects of intraventricular hemicholinium, physostigmine, basal forebrain lesions and intracortical grafts on a multiple-choice serial reaction time task. *Exp Brain Res* 89: 611–622

59. Muir JL, Everitt BJ, Robbins TW (1994) AMPA-induced excitotoxic lesions of the basal forebrain: a significant role for the cortical cholinergic system in attentional function. *J Neurosci* 14: 2313–2326
60. Muir JL, Everitt BJ, Robbins TW (1996) The cerebral cortex of the rat and visual attentional function: dissociable effects of mediofrontal, cingulate, anterior dorsolateral, and parietal cortex lesions on a five-choice serial reaction time task. *Cereb Cortex* 6: 470–481
61. Overton P, Dean P (1988) Detection of visual stimuli after lesions of the superior colliculus in the rat; deficit not confined to the far periphery. *Behav Brain Res* 31: 1–15
62. Robbins TW, Everitt BJ, Marston HM, Wilkinson J, Jones GH, Page KJ (1989) Comparative effects of ibotenic acid- and quisqualic acid-induced lesions of the substantia innominata on attentional function in the rat: further implications for the role of the cholinergic neurons of the nucleus basalis in cognitive processes. *Behav Brain Res* 35: 221–240
63. Crawley JN, Olschowka JA, Diz DI, Jacobowitz DM (1985) Behavioral investigation of the coexistence of substance P, corticotropin releasing factor, and acetylcholinesterase in lateral dorsal tegmental neurons projecting to the medial frontal cortex of the rat. *Peptides* 6: 891–901
64. Austin MC, Rice PM, Mann JJ, Arango V (1995) Localization of corticotropin-releasing hormone in the human locus coeruleus and pedunculopontine tegmental nucleus: an immunocytochemical and *in situ* hybridization study. *Neuroscience* 64: 713–727
65. Carli M, Robbins TW, Evenden JL, Everitt BJ (1983) Effects of lesions to ascending noradrenergic neurons on performance of a 5-choice serial reaction task in rats; implications for theories of dorsal noradrenergic bundle function based on selective attention and arousal. *Behav Brain Res* 9: 361–380
66. Cole BJ, Robbins TW (1992) Forebrain norepinephrine: role in controlled information processing in the rat. *Neuropsychopharmacology* 7: 129–142
67. Usher M, Cohen JD, Servan-Schreiber D, Rajkowski J, Aston-Jones G (1999) The role of locus coeruleus in the regulation of cognitive performance. *Science* 283: 549–554
68. Butler PD, Weiss JM, Stout JC, Nemeroff CB (1990) Corticotropin-releasing factor produces fear-enhancing and behavioral activating effects following infusion into the locus coeruleus. *J Neurosci* 10: 176–183
69. Okuyama S, Chaki S, Kawashima N, Suzuki Y, Ogawa S, Nakazato A, Kumagai T, Okubo T, Tomisawa K (1999) Receptor binding, behavioral, and electrophysiological profiles of nonpeptide corticotropin-releasing factor subtype 1 receptor antagonists CRA1000 and CRA1001. *J Pharmacol Exper Ther* 289: 926–935
70. Schulz DW, Mansbach RS, Sprouse J, Braselton JP, Collins J, Corman M, Dunaiskis A, Faraci S, Schmidt AW, Seeger T et al. (1996) CP-154,526: a potent and selective nonpeptide antagonist of corticotropin releasing factor receptors. *Proc Natl Acad Sci USA* 93: 10477–10482
71. Day JC, Koehl M, Deroche V, Le Moal M, Maccari S (1998) Prenatal stress enhances stress- and corticotropin-releasing factor-induced stimulation of hippocampal acetylcholine release in adult rats. *J Neurosci* 18: 1886–1892
72. Day JC, Koehl M, Le Moal M, Maccari S (1998) Corticotropin-releasing factor administered centrally, but not peripherally, stimulates hippocampal acetylcholine release. *J Neurochem* 71: 622–629
73. Desvignes C, Rouquier L, Souilhac J, Mons G, Rodier D, Soubrie P, Steinberg R (2003) Control by tachykinin NK(2) receptors of CRF(1) receptor-mediated activation of hippocampal acetylcholine release in the rat and guinea-pig. *Neuropeptides* 37: 89–97

74. Gully D, Geslin M, Serva L, Fontaine E, Roger P, Lair C, Darre V, Marcy C, Rouby P E, Simiand J et al. (2002) 4-(2-Chloro-4-methoxy-5-methylphenyl)-N-[(1S)-2-cyclopropyl-1-(3-fluoro-4- methylphenyl)ethyl]5-methyl-N-(2-propynyl)-1,3-thiazol-2-amine hydrochloride (SSR125543A): a potent and selective corticotrophin-releasing factor(1) receptor antagonist. I. Biochemical ad pharmacological characterization. *J Pharmacol Exp Ther* 301: 322–332
75. Bittencourt JC, Sawchenko PE (2000) Do centrally administered neuropeptides access cognate receptors?: an analysis in the central corticotropin-releasing factor system. *J Neurosci* 20: 1142–1156
76. Lai H, Carino MA (1990) Effects of noise on high-affinity choline uptake in the frontal cortex and hippocampus of the rat are blocked by intracerebroventricular injection of corticotropin-releasing factor antagonist. *Brain Res* 527: 354–358
77. Hess C, Blozovski D (1987) Hippocampal muscarinic cholinergic mediation of spontaneous alternation and fear in the developing rat. *Behav Brain Res* 24: 203–214
78. Smythe JW, Colom LV, Bland BH (1992) The extrinsic modulation of hippocampal theta depends on the coactivation of cholinergic and GABA-ergic medial septal inputs. *Neurosci Biobehav Rev* 16: 289–308
79. Bhatnagar S, Costall B, Smythe JW (1997) Hippocampal cholinergic blockade enhances hypothalamic-pituitary-adrenal responses to stress. *Brain Res* 766: 244–248
80. Acquas E, Wilson C, Fibiger HC (1996) Conditioned and unconditioned stimuli increase frontal cortical and hippocampal acetylcholine release: effects of novelty, habituation, and fear. *J Neurosci* 16: 3089–3096
81. Gilad GM, Mahon BD, Finkelstein Y, Koffler B, Gilad VH (1985) Stress-induced activation of the hippocampal cholinergic system and the pituitary-adrenocortical axis. *Brain Res* 347: 404–408
82. Imperato A, Puglisi-Allegra S, Casolini P, Angelucci L (1991) Changes in brain dopamine and acetylcholine release during and following stress are independent of the pituitary-adrenocortical axis. *Brain Res* 538: 111–117
83. Onali P, Olianas MC (1998) Identification and characterization of muscarinic receptors potentiating the stimulation of adenylyl cyclase activity by corticotropin-releasing hormone in membranes of rat frontal cortex. *J Pharmacol Exp Ther* 286: 753–759
84. De Souza EB, Battaglia G (1986) Increased corticotropin-releasing factor receptors in rat cerebral cortex following chronic atropine treatment. *Brain Res* 397: 401–404
85. Raber J, Koob GF, Bloom FE (1995) Interleukin-2 (IL-2) induces corticotropin-releasing factor (CRF) release from the amygdala and involves a nitric oxide-mediated signaling; comparison with the hypothalamic response. *J Pharmacol Exp Ther* 272: 815–824
86. Pitkanen A, Savander V, LeDoux JE (1997) Organization of intra-amygdaloid circuitries in the rat: an emerging framework for understanding functions of the amygdala. *Trends Neurosci* 20: 517–523
87. Okada S, Shimizu T, Yokotani K (2003) Extrahypothalamic corticotropin-releasing hormone mediates (-)-nicotine-induced elevation of plasma corticosterone in rats. *Eur J Pharmacol* 473: 217–223
88. Ohmori N, Itoi K, Tozawa F, Sakai Y, Sakai K, Horiba N, Demura H, Suda T (1995) Effect of acetylcholine on corticotropin-releasing factor gene expression in the hypothalamic paraventricular nucleus of conscious rats. *Endocrinology* 136: 4858–4863
89. Han JS, Bizon JL, Chun HJ, Maus CE, Gallagher M (2002) Decreased glucocorticoid receptor mRNA and dysfunction of HPA axis in rats after removal of the cholinergic innervation to hippocampus. *Eur J Neurosci* 16: 1399–1404

90. Ehlers CL, Reed TK, Henriksen SJ (1986) Effects of corticotropin-releasing factor and growth hormone-releasing factor on sleep and activity in rats. *Neuroendocrinology* 42: 467–474
91. Arai K, Ohata H, Shibasaki T (1998) Non-peptidic corticotropin-releasing hormone receptor type 1 antagonist reverses restraint stress-induced shortening of sodium pentobarbital-induced sleeping time of rats: evidence that an increase in arousal induced by stress is mediated through CRH receptor type 1. *Neurosci Lett* 255: 103–106
92. Chang FC, Opp MR (1998) Blockade of corticotropin-releasing hormone receptors reduces spontaneous waking in the rat. *Am J Physiol* 275: R793–R802
93. Chang FC, Opp MR (2004) A corticotropin-releasing hormone antisense oligodeoxynucleotide reduces spontaneous waking in the rat. *Regul Pept* 117: 43–52
94. Kita I, Seki Y, Nakatani Y, Fumoto M, Oguri M, Sato-Suzuki I, Arita H (2006) Corticotropin-releasing factor neurons in the hypothalamic paraventricular nucleus are involved in arousal/yawning response of rats. *Behav Brain Res* 169: 48–56
95. Stenzel-Poore MP, Heinrichs SC, Rivest S, Koob GF, Vale WW (1994) Overproduction of corticotropin-releasing factor in transgenic mice: a genetic model of anxiogenic behavior. *J Neurosci* 14: t–84
96. van Gaalen MM, Reul JH, Gesing A, Stenzel-Poore MP, Holsboer F, Steckler T (2002) Mice overexpressing CRH show reduced responsiveness in plasma corticosterone after a5-HT1A receptor challenge. *Genes Brain Behav* 1: 174–177
97. van Gaalen MM, Stenzel-Poore M, Holsboer F, Steckler T (2003) Reduced attention in mice overproducing corticotropin-releasing hormone. *Behav Brain Res* 142: 69–79
98. Inglis WL, Olmstead MC, Robbins TW (2001) Selective deficits in attentional performance on the 5-choice serial reaction time task following pedunculopontine tegmental nucleus lesions. *Behav Brain Res* 123: 117–131
99. Zorrilla EP, Schulteis G, Ormsby A, Klaassen A, Ling N, McCarthy JR, Koob GF, De Souza EB (2002) Urocortin shares the memory modulating effects of corticotropin-releasing factor (CRF): mediation by CRF1 receptors. *Brain Res* 952: 200–210
100. Wang HL, Wayner MJ, Chai CY, Lee EH (1998) Corticotrophin-releasing factor produces a long-lasting enhancement of synaptic efficacy in the hippocampus. *Eur J Neurosci* 10: 3428–3437
101. Wang HL, Tsai LY, Lee EH (2000) Corticotropin-releasing factor produces a protein synthesis–dependent long-lasting potentiation in dentate gyrus neurons. *J Neurophysiol* 83: 343–349
102. Blank T, Nijholt I, Eckart K, Spiess J (2002) Priming of long-term potentiation in mouse hippocampus by corticotropin-releasing factor and acute stress: implications for hippocampus-dependent learning. *J Neurosci* 22: 3788–3794
103. Blank T, Nijholt I, Grammatopoulos DK, Randeva HS, Hillhouse EW, Spiess J (2003) Corticotropin-releasing factor receptors couple to multiple G-proteins to activate diverse intracellular signaling pathways in mouse hippocampus: role in neuronal excitability and associative learning. *J Neurosci* 23: 700–707
104. Radulovic J, Ruhmann A, Liepold T, Spiess J (1999) Modulation of learning and anxiety by corticotropin-releasing factor (CRF) and stress: differential roles of CRF receptors 1 and 2. *J Neurosci* 19: 5016–5025
105. Hung HC, Chou CK, Chiu TH, Lee EH (1992) CRF increases protein phosphorylation and enhances retention performance in rats. *Neuroreport* 3: 181–184
106. Lee EH, Hung HC, Lu KT, Chen WH, Chen HY (1992) Protein synthesis in the hippocampus associated with memory facilitation by corticotropin-releasing factor in rats. *Peptides* 13: 927–937

107. Lee EH, Lee CP, Wang HI, Lin WR (1993) Hippocampal CRF, NE, and NMDA system interactions in memory processing in the rat. *Synapse* 14: 144–153
108. Wu HC, Chen KY, Lee WY, Lee EH (1997) Antisense oligonucleotides to corticotropin-releasing factor impair memory retention and increase exploration in rats. *Neuroscience* 78: 147–153
109. Contarino A, Dellu F, Koob GF, Smith GW, Lee KF, Vale W, Gold LH (1999) Reduced anxiety-like and cognitive performance in mice lacking the corticotropin-releasing factor receptor 1. *Brain Res* 835: 1–9
110. Deak T, Nguyen KT, Ehrlich AL, Watkins LR, Spencer RL, Maier SF, Licinio J, Wong ML, Chrousos GP, Webster E, Gold PW (1999) The impact of the nonpeptide corticotropin-releasing hormone antagonist antalarmin on behavioral and endocrine responses to stress. *Endocrinology* 140: 79–86
111. Hikichi T, Akiyoshi J, Yamamoto Y, Tsutsumi T, Isogawa K, Nagayama H (2000) Suppression of conditioned fear by administration of CRF receptor antagonist CP-154,526. *Pharmacopsychiatry* 33: 189–193
112. Hogan JB, Hodges DB, Jr., Lelas S, Gilligan PJ, McElroy JF, Lindner MD (2005) Effects of CRF1 receptor antagonists and benzodiazepines in the Morris water maze and delayed non-matching to position tests. *Psychopharmacology (Berl)* 178: 410–419
113. Liebsch G, Landgraf R, Engelmann M, Lorscher P, Holsboer F (1999) Differential behavioural effects of chronic infusion of CRH 1 and CRH 2 receptor antisense oligonucleotides into the rat brain. *J Psychiatr Res* 33: 153–163
114. Maaswinkel H, Baars AM, Gispen WH, Spruijt BM (1996) Roles of the basolateral amygdala and hippocampus in social recognition in rats. *Physiol Behav* 60: 55–63
115. Morris RG, Garrud P, Rawlins JN, O'Keefe J (1982) Place navigation impaired in rats with hippocampal lesions. *Nature* 297: 681–683
116. Isogawa K, Akiyoshi J, Tsutsumi T, Kodama K, Horinouti Y, Nagayama H (2003) Anxiogenic-like effect of corticotropin-releasing factor receptor 2 antisense oligonucleotides infused into rat brain. *J Psychopharmacol* 17: 409–413
117. Radulovic J, Fischer A, Katerkamp U, Spiess J (2000) Role of regional neurotransmitter receptors in corticotropin-releasing factor (CRF)-mediated modulation of fear conditioning. *Neuropharmacology* 39: 707–710
118. Stiedl O, Meyer M, Jahn O, Ogren SO, Spiess J (2005) Corticotropin-releasing factor receptor 1 and central heart rate regulation in mice during expression of conditioned fear. *J Pharmacol Exp Ther* 312: 905–916
119. Telegdy G, Tiricz H, Adamik A (2005) Involvement of neurotransmitters in urocortin-induced passive avoidance learning in mice. *Brain Res Bull* 67: 242–247
120. Klenerova V, Kaminsky O, Sida P, Hlinak Z, Krejci I, Hynie S (2003) Impaired passive avoidance acquisition in Wistar rats after restraint/cold stress and/or stresscopin administration. *Gen Physiol Biophys* 22: 115–120
121. Roozendaal B, Brunson KL, Holloway BL, McGaugh JL, Baram TZ (2002) Involvement of stress-released corticotropin-releasing hormone in the basolateral amygdala in regulating memory consolidation. *Proc Natl Acad Sci USA* 99: 13908–13913
122. Robison CL, Meyerhoff JL, Saviolakis GA, Chen WK, Rice KC, Lumley LA (2004) A CRH1 antagonist into the amygdala of mice prevents defeat-induced defensive behavior. *Ann NY Acad Sci* 1032: 324–327
123. Behan DP, Heinrichs SC, Troncoso JC, Liu XJ, Kawas CH, Ling N, De Souza EB (1995) Displacement of corticotropin releasing factor from its binding protein as a possible treatment for Alzheimer's disease. *Nature* 378: 284–287

124. Heinrichs SC, Vale EA, Lapsansky J, Behan DP, McClure LV, Ling N, De Souza EB, Schulteis G (1997) Enhancement of performance in multiple learning tasks by corticotropin-releasing factor-binding protein ligand inhibitors. *Peptides* 18: 711–716
125. Zorrilla EP, Schulteis G, Ling N, Koob GF, De Souza EB (2001) Performance-enhancing effects of CRF-BP ligand inhibitors. *Neuroreport* 12: 1231–1234
126. Potter E, Behan DP, Linton EA, Lowry PJ, Sawchenko PE, Vale WW (1992) The central distribution of a corticotropin-releasing factor (CRF)-binding protein predicts multiple sites and modes of interaction with CRF. *Proc Natl Acad Sci USA* 89: 4192–4196
127. Heinrichs SC (2003) Modulation of social learning in rats by brain corticotropin-releasing factor. *Brain Res* 994: 107–114
128. Fenoglio KA, Brunson KL, Avishai-Eliner S, Stone BA, Kapadia BJ, Baram TZ (2005) Enduring, handling-evoked enhancement of hippocampal memory function and glucocorticoid receptor expression involves activation of the corticotropin-releasing factor type 1 receptor. *Endocrinology* 146: 4090–4096
129. Fenoglio KA, Brunson KL, Avishai-Eliner S, Chen Y, Baram TZ (2004) Region-specific onset of handling-induced changes in corticotropin-releasing factor and glucocorticoid receptor expression. *Endocrinology* 145: 2702–2706
130. Fenoglio KA, Chen Y, Baram TZ (2006) Neuroplasticity of the hypothalamic-pituitary-adrenal axis early in life requires recurrent recruitment of stress-regulating brain regions. *J Neurosci* 26: 2434–2442
131. Jasnow AM, Davis M, Huhman KL (2004) Involvement of central amygdalar and bed nucleus of the stria terminalis corticotropin-releasing factor in behavioral responses to social defeat. *Behav Neurosci* 118: 1052–1061
132. Ruhmann A, Bonk I, Lin CR, Rosenfeld MG, Spiess J (1998) Structural requirements for peptidic antagonists of the corticotropin-releasing factor receptor (CRFR): development of CRFR2beta-selective antisauvagine-30. *Proc Natl Acad Sci USA* 95: 15264–15269
133. Cooper MA, Huhman KL (2005) Corticotropin-releasing factor type II (CRF-sub-2) receptors in the bed nucleus of the stria terminalis modulate conditioned defeat in Syrian hamsters (*Mesocricetus auratus*). *Behav Neurosci* 119: 1042–1051
134. Sananbenesi F, Fischer A, Schrick C, Spiess J, Radulovic J (2003) Mitogen-activated protein kinase signaling in the hippocampus and its modulation by corticotropin-releasing factor receptor 2: a possible link between stress and fear memory. *J Neurosci* 23: 11436–11443
135. Steckler T, Holsboer F (2001) Interaction between the cholinergic system and CRH in the modulation of spatial discrimination learning in mice. *Brain Res* 906: 46–59
136. Rivier J, Rivier C, Vale W (1984) Synthetic competitive antagonists of corticotropin-releasing factor: effect on ACTH secretion in the rat. *Science* 224: 889–891
137. Tucci S, Cheeta S, Seth P, File SE (2003) Corticotropin releasing factor antagonist, alpha-helical CRF(9-41), reverses nicotine-induced conditioned, but not unconditioned, anxiety. *Psychopharmacology (Berl)* 167: 251–256
138. Cheeta S, Kenny PJ, File SE (2000) Hippocampal and septal injections of nicotine and 8-OH-DPAT distinguish among different animal tests of anxiety. *Prog Neuropsychopharmacol Biol Psychiatry* 24: 1053–1067
139. Li YH, Ku YH (2002) Involvement of rat lateral septum-acetylcholine pressor system in central amygdaloid nucleus-emotional pressor circuit. *Neurosci Lett* 323: 60–64
140. Prickaerts J, Steckler T (2005) Effects of glucocorticoids on emotion and cognitive processes in animals. In: Steckler T, Kalin NH, and Reul JMHM (eds): *Handbook of Stress and the Brain*, Elsevier, Amsterdam, 359–385

141. Ingram CD (2005) Pathways and transmitter interactions mediating an integrated stress response. In: Steckler T, Kalin NH, Reul JMHM (eds): *Handbook of Stress and the Brain*, Elsevier, Amsterdam, 609–639
142. Sarter, M. (1994) Neuronal mechanisms of the attentional dysfunctions in senile dementia and schizophrenia: two sides of the same coin? *Psychopharmacology* 114: 539–550
143. Zobel AW, Schulze-Rauschenbach S, von Widdern OC, Metten M, Freymann N, Grasmader K, Pfeiffer U, Schnell S, Wagner M, Maier W (2004) Improvement of working but not declarative memory is correlated with HPA normalization during antidepressant treatment. *J Psychiatr Res* 38: 377–383
144. Janowsky DS, Overstreet DH, Nurnberger JI, Jr. (1994) Is cholinergic sensitivity a genetic marker for the affective disorders? *Am J Med Genet* 54: 335–344
145. Holmes A, Heilig M, Rupniak NMJ, Steckler T, Griebel G (2003) Neuropeptide systems as novel therapeutic targets for depression and anxiety disorders. *Trends Pharmacol Sci* 24: 580–588
146. Butcher LL (1992) The cholinergic basal forebrain and its telencephalic targets: Interrelations and implications for cognitive function. In: Levin ED, Decker MW and Butcher LL (eds): *Neurotransmitter interactions and cognitive function*. Birkhäuser Verlag, Switzerland, pp. 15–26

Neurotransmitter Interactions and Cognitive Function
Edited by Edward D. Levin
© 2006 Birkhäuser Verlag/Switzerland

Forebrain dopaminergic-cholinergic interactions, attentional effort, psychostimulant addiction and schizophrenia

Martin Sarter[1], John P. Bruno[2], Vinay Parikh[1], Vicente Martinez[1], Rouba Kozak[1] and Jerry B. Richards[3]

[1] *Department of Psychology, University of Michigan, Ann Arbor, MI 48109, USA*
[2] *Departments of Psychology and Neuroscience, Ohio State University, Columbus, OH 43210, USA*
[3] *Research Institute on Addictions, Buffalo University, NY 14203, USA*

Cholinergic systems and attention

Cholinergic neurons innervating the cortical mantle originate from areas along the medial wall of the globus pallidus (the nucleus basalis of Meynert), the ventral globus pallidus (the substantia innominata) and the horizontal limb of the diagonal band (collectively termed basal forebrain, BF). BF cholinergic projections terminate in all cortical regions and layers, indicating that this most rostral cortical input system generally modulates cortical information processing [1–3]. The BF projections to the cortex also include GABAergic and possibly glutamatergic neurons, but little is known about their organization and function [4].

Based primarily on experiments designed to test the effects of selective lesions of the BF cholinergic projection system and on studies using microdialysis to monitor acetylcholine (ACh) efflux in task-performing animals, substantial evidence in support of the attentional functions mediated *via* the cortical cholinergic input system has accumulated [5–16]. Attention is generally defined as the subject's ability to detect rarely and unpredictably occurring stimuli or signals over extended periods of time (sustained attention), to discriminate signals from "noise" or non-target signals (selective attention), or to divide attentional resources between the processing of multiple stimuli or response rules (divided attention).

Attentional functions have been conceptualized as a set of variables that contribute to the efficacy of higher cognitive processes, including learning and memory. Although the relationships between attentional functions and learning and memory have not been extensively substantiated with respect to the involvement of the cholinergic system [17, 18], the results from several experiments may be interpreted as indicating that the contributions of the cortical cholinergic input system to learning and memory are a function of the (explicit) attentional demands of learning processes [19, 20].

Neurophysiological studies demonstrated that increases in cholinergic transmission in sensory areas enhance the cortical processing of thalamic inputs [21, 22]. We recently attempted to integrate the neurophysiological and behavioral evidence on the functions of cortical cholinergic inputs and hypothesized that the cortical cholinergic input system generally acts to optimize the processing of signals in attention-demanding contexts [23]. Such signals "recruit" *via* activation of BF cholinergic projections to the cortex, anterior and posterior cortical attention systems, thereby amplifying the processing of attention-demanding signals (termed "signal-driven cholinergic modulation of detection").

In addition to the signal-driven (bottom-up) recruitment of cortical cholinergic inputs, the prefrontal cortex influences the activity of cholinergic terminals elsewhere in the cortex [24], presumably *via* direct prefrontal projections to the BF [25] and, *via* multi-synaptic cortico-cortical projections, to cholinergic terminals elsewhere in the cortex [24]. The prefrontal regulation of the activity of cortical cholinergic inputs elsewhere in the cortex is thought to mediate top-down effects, such as the knowledge-based augmentation of detection of signals and the filtering of irrelevant information (termed "cognition-based cholinergic modulation of detection"). Depending on the quality of signals and task characteristics, cortical cholinergic activity reflects the combined effects of signal-driven and cognitive modulation of detection.

Prefrontal cholinergic inputs contribute to the activation of top-down mechanisms and mediate increases in "attentional effort"

A possibly complicating yet central component of the conceptualization described above concerns the hypothesis that cholinergic inputs to the prefrontal cortex contribute to the activation of the anterior attention system, and thus to the cholinergic modulation of the detection process in other cortical areas [23]. Several lines of evidence support such a special role of cholinergic inputs to prefrontal regions. First, lesions of the cholinergic inputs to prefrontal regions are sufficient to produce impairments in attentional performance assessed by a well-practiced task [26]. Performance in a well-practiced task entails that the type, location, and probability of stimuli are familiar to the operator and thus performance depends extensively on top-down mechanisms. Furthermore, Dalley et al. observed that such lesions produced impairments in performance over time-on-task [26]; such an effect reflects weakened top-down mechanisms and the exhaustion of such mechanisms over time in animals with loss of prefrontal cholinergic inputs.

Second, in studies in which medial prefrontal neurons were recorded in attention task-performing animals, we observed that the presentation of a distractor systematically altered the firing activity of a substantial proportion of neurons in the prelimbic cortex [27]. Furthermore, the effects of distractors on prefrontal neuronal activity were attenuated by infusions of 192 IgG-saporin into the recording region, thereby destroying the cholinergic inputs to this area. Importantly, the deafferentation indeed remained very restricted to the site of the electrode tip and thus did not cause effects on performance [27]. The presentation of distractors serves as a productive tool to test the nature and capacity of top-down mechanisms because, in order to "stay on task" and recover from the detrimental performance effects of a distractor, mechanisms designed to filter the distractor and enhance the detection of signals against a "noisy background" need to be initiated. The finding that the presence of distractors is encoded in the prefrontal cortex and that cholinergic innervation is necessary for this encoding collectively supports the hypothesis that cholinergic inputs to this region contribute to the activation of the "'anterior attention system" [28] and thus to the initiation of top-down effects designed to optimize attentional performance and to counteract the consequences of detrimental events or manipulations.

Third, evidence from a recent experiment substantiated the hypothesis that prefrontal cholinergic inputs play a special role in coping with the effects of manipulations that challenge attentional performance. Prefrontal ACh efflux was measured in attentional task performing rats (using microdialysis) before and after a neuropharmacological manipulation known to produce limited impairments in performance [29]. We had previously observed that bilateral infusions of the NMDA receptor antagonist DL-2- amino-5-phosphonovaleric acid (APV) into the BF resulted in impairments in the animals' ability to detect signals while performing an operant sustained attention task. The animals' response accuracy in non-signal trials remained unchanged [30]. Importantly, these animals did not terminate performance as a result of the infusions of APV into the BF; in fact, the number of omitted trials was not affected by the smaller dose (3 nmol) and only moderately increased by the higher dose of APV (20 nmol; Fig. 1).

As illustrated in Fig. 1, bilateral infusions of APV into the BF decreased the animals' ability to detect hits (Fig. 1 depicts the animals' hit rate to longest [500 ms] signals). Following the smaller dose of APV, animals' hit rate recovered in the second task block after the infusion (T3). Following the higher dose of APV, animals continued to perform, but their detection rates for longest signals remained impaired at about 40% (for details see [29]).

Figure 2 illustrates performance-associated changes in mPFC ACh efflux over the five blocks of trials. Prior to the infusion of APV, performance-associated ACh efflux was about 140% over baseline for all animals (see also [31]). Infusions of saline did not affect this level of ACh efflux that remained relatively stable throughout the reminder of the task [31]. Infusions of APV resulted in a further increase in ACh efflux that did not differ between the two doses, up to about 200% over baseline during the last two task blocks. Following the termination of the task, ACh release returned to baseline, and this return did not differ in slope and duration between treatments (for details see [29]).

It is important to note, as would be expected, that previous studies demonstrated that in non-performing animals, blockade of BF NMDA receptors lowers basal cortical ACh output or prevents increases in ACh efflux in response to pharmacological or behavioral manipulation [32–34]. In animals performing the sustained attention task, the opposite effect on ACh efflux was observed.

The interpretation of these data depends on the validity of the assumption that if animals terminated their performance as a result of APV infusions, ACh efflux would have returned to baseline, as it did at the end of the task. However, animals obviously were motivated to continue performing and, following infusion of the smaller dose

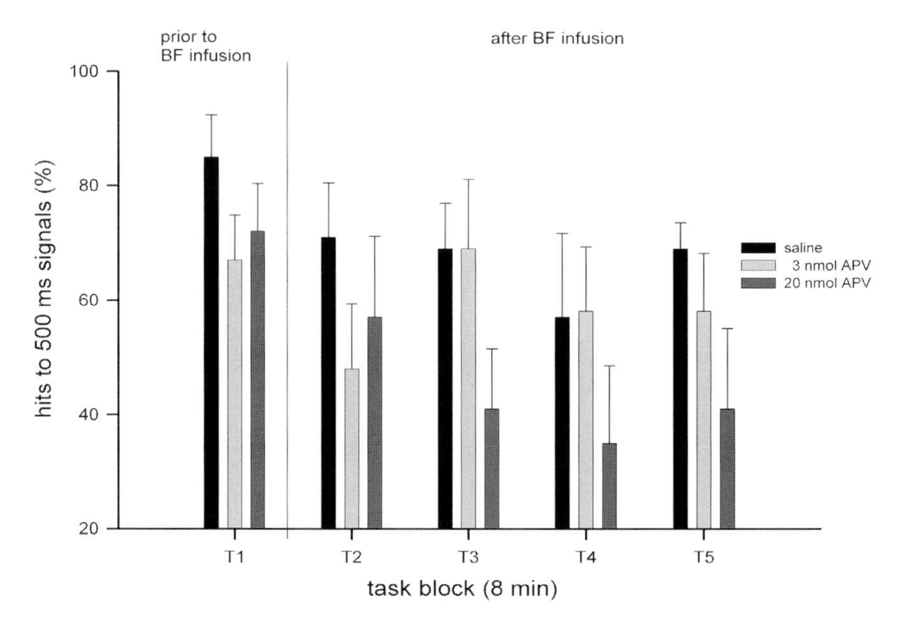

Figure 1. Impairment of attentional performance following bilateral infusion of the NMDA receptor antagonist APV into the basal forebrain (BF; modified from [29]). The figure depicts the animals' hit rate for trials presenting longest (500 ms signals). Drug was infused remotely following completion of the first block of trials (T1; 8 min) and the first collection of dialysate (see Fig. 2). Infusion of the lower concentration of APV transiently impaired performance while the hit rate remained depressed throughout the remainder of the task following infusions of the higher dose of APV.

of APV, even recovered their hit rate. Thus, we speculate that top-down mechanisms were initiated in response to the detection of impairments in performance, perhaps based on reward loss, and in order to counteract the detrimental effects of APV. This perspective suggests that "attentional effort" acts as cognitive incentive [35]. Therefore, the data shown in Fig. 2 are speculated to reflect the increased attentional effort that resulted from the APV-induced impairments in performance and the associated loss of reward. The absence of dose-response effects on performance-associated ACh efflux may reflect the possibility that the increases in attentional effort triggered by the two doses of APV were similar, that levels of ACh efflux do not predict levels of effort in accordance to a linear relationship, or that the microdialysis method lacks the sensitivity to reveal APV dose-related differences in ACh efflux.

Several neuronal routes are available to prefrontal regions to stimulate BF cholinergic systems in order to attenuate the detrimental effects of NMDA receptor blockade on cholinergic activity. Prefrontal regions directly innervate the BF [25], although details concerning the nature of this innervation remain to be explored [4]. Additionally, prefrontal regions may contact basal forebrain neurons indirectly *via* limbic regions, particularly involving the *nucleus accumbens*. Furthermore, prefrontal multi-synaptic projections to other cortical regions may contribute to the regulation

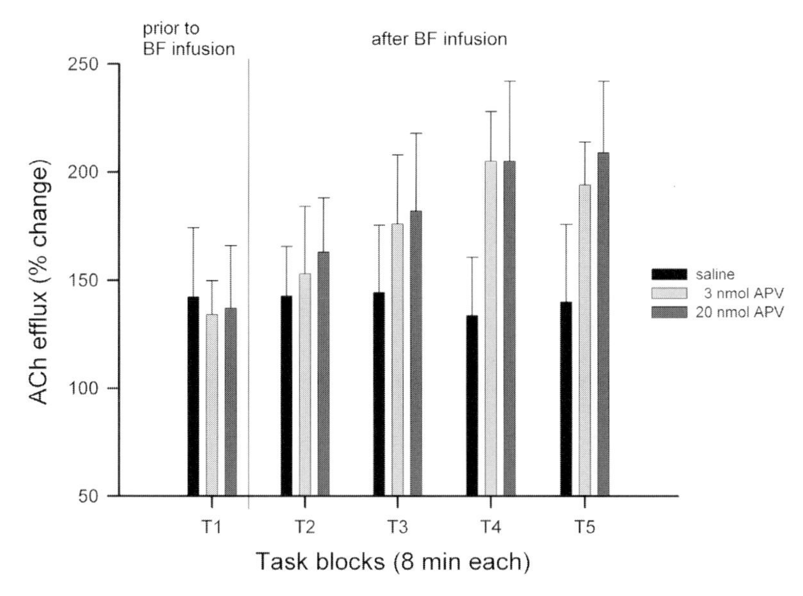

Figure 2. Prefrontal acetylcholine (ACh) efflux (% change from baseline) in animals performing the sustained attention task and following the infusions of saline or APV (modified from [29]). Following saline infusions, the performance-associated increase in ACh efflux remained unchanged during the subsequent four blocks of trials. Following infusions of APV, and while animals' performance was impaired (see Fig. 1), ACh release further increased, up to around 200% during the last two blocks of trials. These data form the basis for the hypothesis that prefrontal ACh efflux, in addition to attentional performance-associated increases, reflects increases in attentional effort (see [29] for details).

of cortical ACh efflux and thus to the cholinergic modulation of input functions by prefrontal regions [24]. Below, prefrontal-accumbens-BF circuitry will be discussed as a major neuronal system that is hypothesized to underlie the ability of subjects to increase attentional effort by recruiting the cholinergic system. Increases in the activity of cortical cholinergic inputs are thought to optimize executive functions and input processing mechanisms that, collectively, support attentional performance under challenging conditions [23]. Mesolimbic-basal forebrain, dopaminergic-cholinergic interactions are a central component of the neuronal circuitry mediating such motivated increased in attentional effort.

Accumbens dopaminergic control of cortical ACh efflux

The main GABAergic output pathway of the nucleus accumbens (NAC) reaches the basal forebrain and directly contacts the cortically projecting cholinergic neurons of this region [36, 37]. The GABAergic regulation of BF cholinergic neurons has been studied extensively [38–42]. Beginning primarily with the work by Mogenson and colleagues, the general idea evolved that increased NAC dopaminergic transmission translates into increases in BF neuronal activity, possibly via suppressing the

GABAergic inhibition of BF neurons [43]. However, the experiments conducted in the late 1980s and early 1990s generated confusing evidence, possibly because effects typically were assessed in passive animals not recruiting the circuitry of interest, and/or because experiments did not reflect the importance of assessing the effects of NAC manipulations of dopaminergic neurotransmission in interaction with activation of the converging telencephalic, glutamatergic projections to the NAC [44–51].

Because of the prediction that dopamine D2 receptor stimulation in the NAC disinhibits the activity of BF cholinergic neurons, NAC D2 receptor blockade was expected to attenuate increases in activity of these neurons. This prediction was also based on our previous studies indicating that the demonstration of effects of infusions of positive GABA modulators into the BF on cortical ACh efflux was only possible in animals that exhibited activated efflux [41]. Therefore, one of our earlier experiments on the regulation of cortical ACh efflux by accumbens DA investigated the effects of intra-NAC infusions of DA receptor antagonists on activated ACh efflux [52]. To activate ACh efflux, the negative GABA modulator FG 7142 (FG) was administered systemically [53]. Administration of FG has been suggested to represent a psychotogenic manipulation [54] in part because FG stimulates DA efflux in the medial prefrontal cortex and accumbens [55–60] and the cognitive effects of FG are attenuated by antipsychotic dopamine D2 receptor antagonists ([61–63]; see also [54, 64]).

Infusions of the D2 receptor antagonist sulpiride (Fig. 3) or haloperidol into the NAC, but not the D1 antagonist SCH 23390, significantly attenuated ACh efflux in animals treated with FG [52]. Although the interpretation of this data remains complicated by the diverse and distributed effects of FG, and although they may reflect simply the antagonism of FG-induced increases in mesolimbic DA, these results clearly indicate that NAC D2 receptors, at least under certain conditions, contribute potently to the regulation of cortical ACh efflux. The exact circuitry underlying the NAC dopaminergic regulation of cortical ACh efflux remains unsettled. In addition to the direct NAC-BF connections, multi-synaptic circuits including the amygdala and/or the ventral tegmentum (VTA) may have contributed to the mediation of the influence of the NAC on cortical ACh efflux.

Because of the interpretational complexities associated with the use of FG to increase cortical ACh efflux, a subsequent experiment was designed to assess the necessity of NAC neurotransmission in permitting behavior-associated increases in cortical ACh efflux [65]. Although this experiment was not intended to specify the behavioral or cognitive components that need to be present in order to "recruit" the NAC regulation of cortical ACh efflux, we employed a behavioral task that combined motivational with attentional variables, although at a relatively implicit level, in order to ensure the co-activation of mesolimbic dopaminergic and basal forebrain cholinergic systems. Animals were trained to perform a defined number of licks of a citric acid solution in order to gain access to a palatable, cheese-flavored food. Animals were trained to expect, and were able to time, access to the solution and, following completion of the required number of licks, the removal of a barrier to access the palatable food.

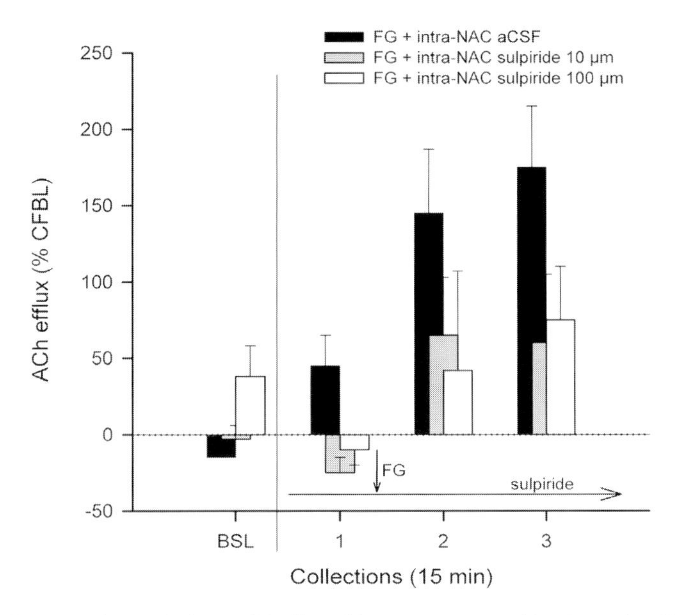

Figure 3. Increases in cortical acetylcholine efflux are attenuated by dopamine D2 receptor blockade in the nucleus accumbens (NAC; modified from [52]). Cortical ACh efflux was increased administering systemically the negative GABA modulator FG 7142 (FG; 8 mg/kg; see filled black bars). The D2 antagonist sulpiride was perfused through a dialysis probe into the NAC. Sulpiride dose-dependently attenuated the effects of FG (see [52] for details).

Animals remained for 30 min in a plastic bowl to establish baseline ACh efflux prior to the transfer into the test apparatus. Medial prefrontal ACh release was significantly increased by transferring the animals from the bowl into the test apparatus (Fig. 4) and again when the solution bottle was presented and when they were allowed to cross-over to the palatable food (see [65]). While waiting for the bottle and following the consumption of the food, ACh efflux returned to baseline. Although our collection intervals were relatively short (5 min), ACh efflux obviously could not be attributed to any specific components of these events and, as already stressed, this was not our intention in this experiment. Importantly, perfusion of tetrodotoxin (TTX), a potent blocker of voltage-regulated sodium channels, into the NAC completely attenuated any increases in cortical ACh efflux (Fig. 4).

Furthermore, TTX also decreased NAC extracellular DA levels below baseline values. Thus, these results confirm that NAC neurotransmission is necessary for the demonstration of increases in mPFC ACh efflux in a behavioral context. However, infusions of sulpiride into the NAC did not affect basal or behavior-evoked cortical ACh efflux. Compared with the data from the prior experiment that utilized FG to increase cortical ACh efflux, the lack of effect of sulpiride, a D2 antagonist, indicates that the demonstration of the necessity of dopaminergic activity in NAC depends on the manipulation used to activate cortical cholinergic output. This conclusion is also supported by the finding that increases in cortical ACh efflux that are triggered by ex-

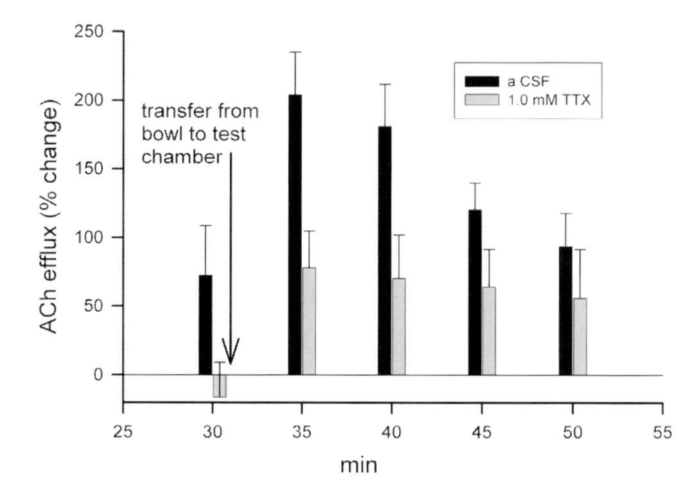

Figure 4. Blockade of neuronal transmission in the nucleus accumbens (NAC), by perfusing TTX through a dialysis probe, attenuated the increases in cortical acetylcholine (ACh) release that are normally (following the perfusion of artificial CSF (aCSF) into the NAC) observed in association with several events during a complex appetitive procedure (see [65] for details). Figure 4 depicts the effects of the transfer of the animals from a test bowl to the two-chamber apparatus (see arrow) on cortical ACh efflux. The increases in ACh efflux were largely attenuated by perfusion of TTX into the NAC (modified from [65]).

posure to a novel environment are not modulated by increases in NAC dopaminergic transmission [66].

These data indicate that NAC neurotransmission may strongly influence cortical ACh efflux, but that the conditions under which NAC dopaminergic transmission contributes to the modulation of cortical ACh efflux remain very poorly understood. We will come back to the special significance of the interactions between sulpiride and FG below, in the context of the discussion of dopaminergic-cholinergic interactions in schizophrenia.

In order to investigate the possibility that the NAC dopaminergic influence on cortical ACh efflux depends on the state of glutamatergic transmission in the NAC, we have recently begun investigating the effects of co-perfusion of glutamatergic and dopaminergic ligands into the NAC. These experiments utilize the finding that perfusion of NMDA into the NAC result in increases in cortical ACh efflux. Co-administration of the D1 receptor antagonist SCH23390 into the NAC attenuated the effects of a lower (150 μM), but not higher (250 μM), concentration of NMDA [67]. Data on D2 receptor modulation are not yet available. These data suggest that NAC D1 receptor stimulation may positively modulate the increases in cortical ACh efflux produced by NAC NMDA receptor stimulation.

The results from two experiments indicate that the NAC regulation is active while animals perform attention-taxing tasks known to reflect closely the state of prefrontal cholinergic transmission. First, antisense-induced suppression of the expression of one of the two isoforms of glutamic acid decarboxylase (GAD_{65}) by NAC neurons produced robust impairments in sustained attention performance [68]. Second, infusions of the non-selective DA receptor antagonist *cis*-flupenthixol into the NAC of task-performing animals yielded impairments in the performance of such a task [69]. Although the performance effects of these manipulations could have been due to effects on non-cholinergic systems, there is sufficient causal and correlational evidence relating performance in this task to cortical cholinergic transmission (references above) to suggest that the effects of these manipulations on performance indeed were mediated *via* effects on performance-associated increases in cortical ACh efflux.

Evidently, our understanding of the mechanisms underlying the ability of NAC DA to modulate cortical ACh efflux remains extremely premature, and the behavioral or cognitive functions that require NAC DA modulation of the activity of cortical cholinergic inputs need to be defined. However, the collective evidence is sufficient to conclude that a substantial part of the behavioral and cognitive, particularly attentional, consequences of changes in NAC function are due to the trans-synaptic modulation of basal forebrain cholinergic neurons projecting to the cortex.

Dopaminergic-cholinergic interactions in the basal forebrain

Mesolimbic dopaminergic neurons make direct contacts with the cholinergic neurons of the BF and thus may also directly influence their excitability [70–72]. Napier and colleagues extensively studied the effects of dopamine on neuronal activity in the BF and demonstrated increases in neuronal activity as a result of D1 stimulation while the majority of neurons showed decreases in firing rate following D2 stimulation [73]. In addition to suggestions about the direct dopaminergic regulation of BF neurons, there is also evidence suggesting that dopamine reduces the inhibitory, GABAergic control of these neurons *via* D1 receptor [74, 75] and also influences glutamatergic transmission in this region [76].

Relatively little is known about the dopaminergic regulation of positively identified cholinergic BF neurons, and about the significance of DA-cholinergic interactions in the BF with respect to cortical ACh efflux. Although the demonstration of systemic amphetamine-induced increases in cortical ACh efflux [77] requires an intact mesolimbic dopamine system [78], the increases in cortical ACh efflux produced by systemic amphetamine do not require BF or NAC D1 or D2 receptor stimulation ([79]; Nelson, Sarter and Bruno, *unpublished observations*).

The paucity of knowledge concerning BF DA-cholinergic interactions represents a major problem, particularly when considering the potential importance of these interactions for disease models involving abnormal dopaminergic transmission. Information about the regulation of BF DA and the direct dopaminergic regulation of cortically projecting cholinergic neurons in the basal forebrain, specifically in animals performing (attentional) tasks that recruit these neuronal systems, is urgently needed.

Dopaminergic-cholinergic interactions within the PFC

As discussed above, cholinergic inputs to the PFC are hypothesized to mediate specifically the ability to maintain or recover attentional performance during challenging conditions [29]. Dopaminergic inputs to the PFC also influence the learning and performance of attention tasks [80, 81], with D1 receptor-mediated effects appearing to mediate more robust effects when compared with drug effects acting at D2 receptors [82, 83]. As would be expected, the effects of D1 receptor manipulations in the PFC extend to other cognitive functions, particularly working memory (e.g., [84, 85]), possibly because of overlapping cognitive mechanisms assessed by attention and working memory tasks and/or the modulation of more general executive functions by PFC dopamine.

Although some data suggest rather limited or even absent functional relationships between dopaminergic and cholinergic inputs to the PFC, and that prefrontal cholinergic inputs may mediate a more restricted range of primarily attentional functions when compared with the possibility that D1 receptor stimulation exerts less specific, or broader modulation of executive functions [86], insufficient evidence renders such hypotheses to be premature (e.g., [87, 88]). Given that the working memory tasks and attention tasks tax overlapping cognitive functions, a complete dissociation between the modulatory cognitive effects of dopaminergic and cholinergic inputs would be unexpected. Rather, some degree of co-modulation of cognitive functions by dopaminergic and cholinergic inputs [89] represents a more plausible scenario. Moreover, the finding that D1 receptor stimulation in the mPFC resulted in increases in basal ACh efflux in this region in postpubertal animals with neonatal ventral hippocampus lesions, but not in sham-lesioned postpubertal animals [90], suggests that abnormally-regulated cholinergic transmission represents a likely component of the neuronal foundations of schizophrenia (below; see also [64]), and that local regulation of cholinergic activity by dopaminergic inputs may contribute directly to such cholinergic dysregulation. Research on intra-PFC interactions between dopaminergic and cholinergic inputs in relation to defined behavioral and cognitive functions is of obvious significance.

Dopaminergic regulation of the choline transporter

The synthesis of ACh is strongly influenced by the uptake of choline from the extracellular space by the high-affinity choline transporter (CHT). In fact, blockade of the CHT with hemicholinium-3 (HC) attenuates ACh synthesis and thus release. Therefore, the mechanisms that modulate the capacity of the CHT represent an important aspect of research on the regulation and function of cholinergic systems [91]. For example, we have recently demonstrated that in animals performing the attention-taxing task described above, the capacity of the CHT to transport choline is enhanced in the prefrontal cortex, and that this enhanced capacity is due to an increased translocation of CHTs from intracellular domains to plasma membrane [92].

Evidence from recent research has begun to suggest the possibility that the capacity of the CHT, including the trafficking of CHTs from intracellular domains to plasma membranes, is subject to multiple regulatory pathways. Furthermore, changes in CHT capacity may not necessarily reflect the activity of cholinergic terminals [93–97]. Alternatively, it may also be possible that the activity of cholinergic neurons represents the primary variable dictating the capacity of the CHT and that such activity-dependent regulation of the CHT can be influenced by non-cholinergic mechanisms. As will be discussed in the following, it is unclear whether the dopaminergic modulation of the CHT reflects dopaminergic effects on cholinergic neuronal activity or a regulatory mechanism unrelated to cholinergic activity, or both.

As already mentioned, interactions between dopaminergic and cholinergic neurons are central to our basic understanding of the functions of these modulators and also to the modeling of schizophrenia and psychostimulant addiction. As hyperdopaminergic mechanisms contribute essentially to the manifestation of the symptoms of these disorders, experiments addressed the question of whether the capacity of the CHT is affected in animals exhibiting abnormal increases in dopaminergic transmission. This hypothesis was studied in part by investigating the capacity of the CHT to remove choline from the extracellular pace *in vivo*, using an amperometric biosensor method for the detection of changes in extracellular choline concentrations [98] and *ex vivo* using synaptosomal choline uptake assays (e.g., [92]).

The available data support the hypothesis that hyperdopaminergic neurotransmission is associated with a decreased capacity of the CHT to remove choline from the extracellular space (Parikh et al., *unpublished data*). This conclusion is based on results from experiments in mice with a reduced expression of the dopamine transporter (DAT). DAT-knockdown (KD) mice express reduced levels of the DAT, approximately 10% of the amount seen in wild-type mice. Therefore, extracellular dopamine levels are increased by about 70% [99]. DAT-KD mice develop normally, but exhibit hyperactivity and decreased habituation to novel environments. Interestingly, locomotor activity in these mice is attenuated by the administration of psychostimulants, indicating that these animals may model aspects of attention deficit hyperactivity disorder (ADHD; [99]). Furthermore, Berridge and colleagues demonstrated that DAT-KD mice exhibited enhanced acquisition and greater incentive performance for a palatable food [100] and that they show more stereotyped grooming behavior [101]. Thus, DAT-KD animals are of interest also for the modeling of addictive processes as well as disorders characterized by highly stereotyped behaviors such as obsessive-compulsive disorder (OCD) or Tourette's syndrome [101].

We assessed choline clearance in the striatum of DAT-KD mice using the amperometric choline detection method described in detail in Parikh et al. [98]. The focus on the striatum in part was motivated by prior research that characterized the effects of DAT-KD on dopaminergic neurotransmission primarily in this region (references above), a relatively rich literature on dopaminergic-cholinergic interactions in the striatum (e.g., [102]), and by the relative ease with which recording electrodes and glass capillaries for the administration of drugs can be implanted into the striatum of mice. Furthermore, and although much of our research on the cholinergic mediation of cognitive functions has focused on cortical cholinergic inputs, striatal dopaminergic mechanisms widely influence cortical information processing (e.g., [103]) and striatal cholinergic interneurons modulate the efficacy of cortico-striatal output [104].

Thus, it is not unexpected that striatal dopaminergic mechanisms also influence attentional performance [105] and that cholinergic striatal interneurons represent a crucial component of the striatal circuitry involved in cognitive functions [106].

Mice were generated as described in Zhuang et al. [99] at Buffalo University. Animals were anesthetized with urethane (1.25–1.5 g/kg) and placed in a stereotaxic frame. Microelectrode/micropipette arrays were prepared as described in [98] and inserted into the striatum. We assessed potassium-evoked ACh release by ejecting different volumes (50–400 nL) of 70 mM KCl. To determine the capacity of CHTs, pressure ejections of choline (5 mM), with and without co-ejection of the potent and specific CHT blocker hemicholinium-3 (HC; 10, 50 μM) were conducted. Data were recorded at 10 Hz using a FAST-16 recording system (for additional details concerning calibration and other technical issues see [98]).

As shown previously [107], the clearance of choline is retarded by co-administration of the specific CHT blocker HC. The clearance of choline was quantified by calculating the uptake rate of choline for a section of the clearance curve ranging from a 40–80% decrease from peak concentration ($t_{40\text{-}80}$). In wild type mice, co-ejection of 10 μM or 50 μM HC reduced striatal choline uptake rate by 20% and 43%, respectively.

The focus on the effects of HC in these experiments is based on the fact that only a component of the clearance of choline is caused by high-affinity choline uptake, as indicated by the partial effects of HC. Furthermore, differences in choline clearance are not always readily apparent in the absence of HC, possibly because CHTs may be close to being saturated by endogenous extracellular choline concentrations [108] and changes in clearance of exogenous choline may be due to other mechanisms, such as diffusion of choline away from the surface of the electrode or the capacity of low-affinity choline transporters. As the present experiments were designed to address specifically the effects of DAT-KD on the capacity of the CHT, data on the effects of HC are necessary for conclusions about putative differences between CHT capacity in wild-type and DAT-KD mice.

DAT-KD mice exhibited a 37% reduction in uptake rate of the HC-sensitive component of choline clearance curve obtained from the difference of uptake rates of choline signal in response to pressure ejection of choline in the absence and presence of HC, as compared to the wildtypes. Moreover, the mutants displayed higher choline signals in response to potassium-induced terminal depolarization, reflecting lower clearance of choline hydrolyzed from ACh. These data were interpreted as indicating that there is a reduced capacity of CHTs in the striatum of DAT knockdown animals. *Ex vivo* synaptosomal assays measuring choline uptake confirmed this hypotheses (Parikh et al., *unpublished data*).

As mentioned earlier, the interpretation of these findings is not straight forward. It is not clear whether increased dopamine receptor stimulation can translate into a decreased activity of cholinergic interneurons and therefore a down-regulation of the CHT [102]. However, while some studies demonstrated D1 receptor stimulation-induced increases in striatal ACh release (e.g., [109–111]), the overwhelming evidence suggests that endogenous dopamine release primarily inhibits the activity of striatal cholinergic interneurons *via* D2 and D5 receptors [111–113]. Supplemental materials provided by Zhuang et al. [99] indicate that in the striatum, D1 and D2 receptor densities were similar in wild-type and DAT-KD mice. Furthermore, D2-receptor mediated autoreceptor function was unchanged in DAT-KD mice, as indicated by voltammetric data (reported in the appendix of [99]).

Additional complexities burden the understanding of dopaminergic-cholinergic interactions in the striatum as well as in the cortex. For example, repeated amphetamine exposure facilitates striatal [114] and cortical [77] ACh efflux, but such increases appear unlikely to be a direct result of local increases in dopaminergic neurotransmission [114]. Dopaminergic and cholinergic neurons may interact on the basis of complex additional neuronal mechanisms and *via* distributed neuronal circuits (e.g., [115, 116]); moreover, the mechanisms underlying these interactions may differ as a function of the state of activity, or the degree of "recruitment" of dopaminergic and cholinergic neurons.

Functional implications: dopaminergic-cholinergic modulation, schizophrenia and craving in psychostimulant addicts

Evidence in support of the hypothesis that alterations in the regulation of forebrain cholinergic systems contribute essentially to the mediation of the cognitive symptoms of schizophrenia and psychostimulant addiction has been discussed previously [64, 117]. Abnormally regulated mesolimbic dopaminergic neurons, directly or *via* distributed neuronal circuits, influence the excitability of cortical cholinergic inputs and thereby affect the modulation of cortical information processing in general and the attentional functions that are critically mediated *via* this neuronal system in particular.

With respect to schizophrenia, dopaminergic-cholinergic interactions are of central importance for two major hypotheses (see [64] for details). First, during acute psychotic periods when mesolimbic dopaminergic neurons exhibit a "sensitized" state, abnormal cortical cholinergic activity mediates the impairments in attentional functions, including the dysfunctional filtering of irrelevant stimuli, which contribute to the manifestation of positive symptoms. Second, lasting cognitive impairments in schizophrenia have been hypothesized to be associated with abnormalities in prefrontal dopaminergic neurotransmission, possibly insufficient D1 receptor function [118]. The persistent abnormalities in cholinergic neurotransmission in these patients (reviewed in [64]) in part may be a result of interactions with dopaminergic dysfunction. Such interactions form the basis for new therapeutic approaches involving muscarinic receptor modulators and drugs that modulate ACh efflux in this region in order to enhance these patients' cognitive capabilities (see also [119]).

The role of a sensitized mesolimbic dopamine system in the mediation of compulsive drug seeking has been extensively conceptualized [120–123]. To a substantial degree, compulsive drug seeking behavior can be considered a cognitive disorder that involves general impairments in cognitive flexibility (e.g., [124]) and, more specifically, a disruption of the ability to disengage from the processing of drug-related stimuli. The evidence described above indicates the potentially close relationships between mesolimbic dopaminergic and cortical cholinergic neuronal systems. Furthermore, cocaine self-administration affects the regulation of cortical cholinergic input systems [125, 126], although the exact nature and the dynamics of these effects remain to be determined by *in vivo* studies. Our recent experiments demonstrated that attentional impairments that resulted from amphetamine sensitization were associated with the induction of Fos-immunoreactivity in basal forebrain cholinergic neurons [127]. Based on these data and current hypothesis about the functions of cortical cholinergic input systems, particularly the role of cholinergic inputs to prefrontal regions in the organization of top-down mechanisms for the optimization of input functions elsewhere in the cortex [23], it is hypothesized that abnormal cholinergic transmission in this region contributes to the compulsive processing of drug-related information, general cognitive inflexibility, and thus compulsive drug seeking and relapse. Clearly, the exact nature of the dysregulation of cortical cholinergic inputs in addiction remains to be investigated. Importantly, the range of this dysregulation and its functional significance will not become apparent by studying the regulation of basal ACh efflux but rather by monitoring the effects of drug-related stimuli on attentional performance-associated activation of the cortical cholinergic input system.

The present discussion about the clinical significance of dopaminergic-cholinergic interactions ignores evidence in support of the regulation of dopamine release by cholinergic mechanisms (e.g., [128–130]) and does not address the importance of such interactions for the understanding of age-related cognitive disorders and the development of treatments for such disorders (see the chapters by Levin and Dringenberg in this volume). It will not only be important to characterize such interactions further in the context of the functions mediated *via* these neuronal systems, but the interacting behavioral and cognitive functions that are based on meso-cortical and striatal interactions between dopaminergic and cholinergic neurons need to be defined more precisely in order to develop useful perspectives about the significance of such interactions for various cognitive disorders. In other words, efforts to demonstrate behavioral and neuropharmacological "effects" of combined manipulations of multiple neurotransmitter systems need to be augmented by experiments in which such manipulations are conducted in animals performing procedures that explicitly recruit multiple neurotransmitter systems and their interactions.

Conclusions

It has been almost 15 years since the ancestor of this volume appeared. The amount of progress that has been made is remarkable. From a somewhat distanced "appreciation of the concert" [131], our views and understanding of neurotransmitter interactions have dramatically evolved. The present evidence allows us to attribute increasingly precise cognitive operations to increasingly well-defined interactions between neurotransmitter systems in defined neuronal circuits. Furthermore, 15 years ago major cognitive disorders often were discussed with respect to single neurotransmitter systems while today, core cognitive dysfunctions are hypothesized to develop and escalate as a result of the abnormal interplay between multiple systems and across multiple synapses. However, given the enormous gaps in our current knowledge, such as the regulation and function of non-cholinergic BF projections to the cortex [4], or the poorly understood modulation of critical steps in neurotransmission (such as the choline transporter) by other transmitters [91], current concepts and perspectives, including those described and discussed in this chapter, remain premature. If the progress made during the last 15 years serves as a valid harbinger, the present volume will become outdated very soon.

Acknowledgements

The authors' research was supported by PHS grants MH063114, NS37026 (MS, JPB), MH057436 (JPB, MS), and KO2 MH01072 (MS).

References

1. Semba K (2000) Multiple output pathways of the basal forebrain: organization, chemical heterogeneity, and roles in vigilance. *Behav Brain Res* 115: 117–141
2. Woolf NJ (1991) Cholinergic systems in mammalian brain and spinal cord. *Prog Neurobiol* 37: 475–524
3. Zaborszky L, Pang K, Somogyi J, Nadasdy Z, Kallo I (1999) The basal forebrain corticopetal system revisited. *Ann N Y Acad Sci* 877: 339–367
4. Sarter M, Bruno JP (2002) The neglected constituent of the basal forebrain corticopetal projection system: GABAergic projections. *Eur J Neurosci* 15: 1867–1873
5. McGaughy J, Everitt BJ, Robbins TW, Sarter M (2000) The role of cortical cholinergic afferent projections in cognition: impact of new selective immunotoxins. *Behav Brain Res* 115: 251–263
6. Sarter M, Givens B, Bruno JP (2001) The cognitive neuroscience of sustained attention: where top-down meets bottom-up. *Brain Res Rev* 35: 146–160
7. Sarter M, Bruno JP (2000) Cortical cholinergic inputs mediating arousal, attentional processing and dreaming: differential afferent regulation of the basal forebrain by telencephalic and brainstem afferents. *Neuroscience* 95: 933–952
8. Sarter M, Bruno JP (1997) Cognitive functions of cortical acetylcholine: toward a unifying hypothesis. *Brain Res Rev* 23: 28–46
9. Sarter M, Bruno JP, Turchi J (1999) Basal forebrain afferent projections modulating cortical acetylcholine, attention, and implications for neuropsychiatric disorders. *Ann N Y Acad Sci* 877: 368–382
10. Everitt BJ, Robbins TW (1997) Central cholinergic systems and cognition. *Ann Rev Psychol* 48: 649–684
11. Mesulam MM (1990) Large-scale neurocognitive networks and distributed processing for attention, language, and memory. *Ann Neurol* 28: 597–613
12. Chiba AA, Bucci DJ, Holland PC, Gallagher M (1995) Basal forebrain cholinergic lesions disrupt increments but not decrements in conditioned stimulus processing. *J Neurosci* 15: 7315–7322
13. Dalley JW, McGaughy J, O'Connell MT, Cardinal RN, Levita L, Robbins TW (2001) Distinct changes in cortical acetylcholine and noradrenaline efflux during contingent and noncontingent performance of a visual attentional task. *J Neurosci* 21: 4908–4914

14. Voytko ML (1996) Cognitive functions of the basal forebrain cholinergic system in monkeys: memory or attention? *Behav Brain Res* 75: 13–25
15. McGaughy J, Dalley JW, Morrison CH, Everitt BJ, Robbins TW (2002) Selective behavioral and neurochemical effects of cholinergic lesions produced by intrabasalis infusions of 192 IgG-saporin on attentional performance in a five-choice serial reaction time task. *J Neurosci* 22: 1905–1913
16. McGaughy J, Kaiser T, Sarter M (1996) Behavioral vigilance following infusions of 192 IgG-saporin into the basal forebrain: selectivity of the behavioral impairment and relation to cortical AChE-positive fiber density. *Behav Neurosci* 110: 247–265
17. Sarter M, Bruno JP, Givens B (2003) Attentional functions of cortical cholinergic inputs: what does it mean for memory? *Neurobiol Learn Mem* 80: 245–256
18. Baxter MG, Murg SL (2002) The basal forebrain cholinergic system and memory. Beware of dogma. In: LR Squire, DL Schachter (eds): *Neuropsychology of memory.* The Guilford Press, New York, 425–436
19. Butt AE, Bowman TD (2002) Transverse patterning reveals a dissociation of simple and configural association learning abilities in rats with 192 IgG-saporin lesions of the nucleus basalis magnocellularis. *Neurobiol Learn Mem* 77: 211–233
20. Berger-Sweeney J, Stearns NA, Frick KM, Beard B, Baxter MG (2000) Cholinergic basal forebrain is critical for social transmission of food preferences. *Hippocampus* 10: 729–738
21. Weinberger NM (2003) The nucleus basalis and memory codes: auditory cortical plasticity and the induction of specific, associative behavioral memory. *Neurobiol Learn Mem* 80: 268–284
22. Edeline JM (2003) The thalamo-cortical auditory receptive fields: regulation by the states of vigilance, learning and the neuromodulatory systems. *Exp Brain Res* 153: 554–572
23. Sarter M, Hasselmo ME, Bruno JP, Givens B (2005) Unraveling the attentional functions of cortical cholinergic inputs: interactions between signal-driven and top-down cholinergic modulation of signal detection. *Brain Res Rev* 48: 98–111
24. Nelson CL, Sarter M, Bruno JP (2005) Prefrontal cortical modulation of acetylcholine release in the posterior parietal cortex. *Neuroscience* 132:347–359.
25. Zaborszky L, Gaykema RP, Swanson DJ, Cullinan WE (1997) Cortical input to the basal forebrain. *Neuroscience* 79: 1051–1078
26. Dalley JW, Theobald DE, Bouger P, Chudasama Y, Cardinal RN, Robbins TW (2004) Cortical cholinergic function and deficits in visual attentional performance in rats Following 192 IgG-Saporin-induced lesions of the medial prefrontal cortex. *Cereb Cortex* 14: 922–932
27. Gill TM, Sarter M, Givens B (2000) Sustained visual attention performance-associated prefrontal neuronal activity: evidence for cholinergic modulation. *J Neurosci* 20: 4745–4757
28. Posner MI, Petersen SE (1990) The attention system of the human brain. *Ann Rev Neurosci* 13: 25–42
29. Kozak R, Bruno JP, Sarter M (2006) Augmented prefrontal acetylcholine release during challenged attentional performance. *Cereb Cortex* 16: 9–17
30. Turchi J, Sarter M (2001) Bidirectional modulation of basal forebrain N-methyl-D-aspartate receptor function differentially affects visual attention but not visual discrimination performance. *Neuroscience* 104: 407–417
31. Arnold HM, Burk JA, Hodgson EM, Sarter M, Bruno JP (2002) Differential cortical acetylcholine release in rats performing a sustained attention task *versus* behavioral control tasks that do not explicitly tax attention. *Neuroscience* 114: 451–460

32. Giovannini MG, Giovannelli L, Bianchi L, Kalfin R, Pepeu G (1997) Glutamatergic modulation of cortical acetylcholine release in the rat: a combined *in vivo* microdialysis, retrograde tracing and immunohistochemical study. *Eur J Neurosci* 9: 1678–1689
33. Rasmusson DD, Szerb IC, Jordan JL (1996) Differential effects of alpha-amino-3-hydroxy-5-methyl-4-isoxazole propionic acid and N-methyl-D-aspartate receptor antagonists applied to the basal forebrain on cortical acetylcholine release and electroencephalogram desynchronization. *Neuroscience* 72: 419–427
34. Fadel J, Sarter M, Bruno JP (2001) Basal forebrain glutamatergic modulation of cortical acetylcholine release. *Synapse* 39: 201–212
35. Berridge KC, Robinson TE (2003) Parsing reward. *Trends Neurosci* 26: 507–513
36. Zaborszky L (1992) Synaptic organization of basal forebrain cholinergic projection neurons. In: E Levin, MW Decker, LL Butcher (eds): *Neurotransmitter Interactions and Cognitive Function.* Birkhäuser, Boston, 27–65
37. Zaborszky L, Cullinan WE (1992) Projections from the nucleus accumbens to cholinergic neurons of the ventral pallidum: a correlated light and electron microscopic double-immunolabeling study in rat. *Brain Res* 570: 92–101
38. Dudchenko P, Sarter M (1991) GABAergic control of basal forebrain cholinergic neurons and memory. *Behav Brain Res* 42: 33–41
39. Holley LA, Turchi J, Apple C, Sarter M (1995) Dissociation between the attentional effects of infusions of a benzodiazepine receptor agonist and an inverse agonist into the basal forebrain. *Psychopharmacol* 120: 99–108
40. Moore H, Sarter M, Bruno JP (1995) Bidirectional modulation of cortical acetylcholine efflux by infusion of benzodiazepine receptor ligands into the basal forebrain. *Neurosci Lett* 189: 31–34
41. Moore H, Sarter M, Bruno JP (1993) Bidirectional modulation of stimulated cortical acetylcholine release by benzodiazepine receptor ligands. *Brain Res* 627: 267–274
42. Sarter M, Bruno JP, Dudchenko P (1990) Activating the damaged basal forebrain cholinergic system: tonic stimulation *versus* signal amplification. *Psychopharmacol* 101: 1–17
43. Yang CR, Mogenson GJ (1989) Ventral pallidal neuronal responses to dopamine receptor stimulation in the nucleus accumbens. *Brain Res* 489: 237–246
44. Nicola SM, Surmeier J, Malenka RC (2000) Dopaminergic modulation of neuronal excitability in the striatum and nucleus accumbens. *Annu Rev Neurosci* 23: 185–215
45. O'Donnell P, Grace AA (1993) Dopaminergic modulation of dye coupling between neurons in the core and shell regions of the nucleus accumbens. *J Neurosci* 13: 3456–3471
46. Floresco SB, Blaha CD, Yang CR, Phillips AG (2001) Modulation of hippocampal and amygdalar-evoked activity of nucleus accumbens neurons by dopamine: cellular mechanisms of input selection. *J Neurosci* 21: 2851–2860
47. Floresco SB, Todd CL, Grace AA (2001) Glutamatergic afferents from the hippocampus to the nucleus accumbens regulate activity of ventral tegmental area dopamine neurons. *J Neurosci* 21: 4915–4922
48. Meredith GE (1999) The synaptic framework for chemical signaling in nucleus accumbens. *Ann N Y Acad Sci* 877: 140–156
49. Mulder AB, Hodenpijl MG, Lopes da Silva FH (1998) Electrophysiology of the hippocampal and amygdaloid projections to the nucleus accumbens of the rat: convergence, segregation, and interaction of inputs. *J Neurosci* 18: 5095–5102
50. O'Donnell P (1999) Ensemble encoding in the nucleus accumbens. *Psychobiology* 27: 187–197
51. Brady AM, O'Donnell P (2004) Dopaminergic modulation of prefrontal cortical input to nucleus accumbens neurons *in vivo*. *J Neurosci* 24: 1040–1049

52. Moore H, Fadel J, Sarter M, Bruno JP (1999) Role of accumbens and cortical dopamine receptors in the regulation of cortical acetylcholine release. *Neuroscience* 88: 811–822
53. Moore H, Stuckman S, Sarter M, Bruno JP (1995) Stimulation of cortical acetylcholine efflux by FG 7142 measured with repeated microdialysis sampling. *Synapse* 21: 324–331
54. Sarter M, Bruno JP, Berntson GG (2001) Psychotogenic properties of benzodiazepine receptor inverse agonists. *Psychopharmacol* 156: 1–13
55. Murphy BL, Arnsten AF, Goldman-Rakic PS, Roth RH (1996) Increased dopamine turnover in the prefrontal cortex impairs spatial working memory performance in rats and monkeys. *Proc Natl Acad Sci USA* 93: 1325–1329
56. Bassareo V, Tanda G, Petromilli P, Giua C, Di Chiara G (1996) Non-psychostimulant drugs of abuse and anxiogenic drugs activate with differential selectivity dopamine transmission in the nucleus accumbens and in the medial prefrontal cortex of the rat. *Psychopharmacol* 124: 293–299
57. Bradberry CW, Lory JD, Roth RH (1991) The anxiogenic beta-carboline FG 7142 selectively increases dopamine release in rat prefrontal cortex as measured by microdialysis. *J Neurochem* 56: 748–752
58. Brose N, O'Neill RD, Boutelle MG, Anderson SM, Fillenz M (1987) Effects of an anxiogenic benzodiazepine receptor ligand on motor activity and dopamine release in nucleus accumbens and striatum in the rat. *J Neurosci* 7: 2917–2926
59. Tam SY, Roth RH (1985) Selective increase in dopamine metabolism in the prefrontal cortex by the anxiogenic beta-carboline FG 7142. *Biochem Pharmacol* 34: 1595–1598
60. McCullough LD, Salamone JD (1992) Anxiogenic drugs beta-CCE and FG 7142 increase extracellular dopamine levels in nucleus accumbens. *Psychopharmacol* 109: 379–382
61. Ninan I, Kulkarni SK (1999) Effect of olanzapine on behavioural changes induced by FG 7142 and dizocilpine on active avoidance and plus maze tasks. *Brain Res* 830: 337–344
62. Murphy BL, Arnsten AF, Jentsch JD, Roth RH (1996) Dopamine and spatial working memory in rats and monkeys: pharmacological reversal of stress-induced impairment. *J Neurosci* 16: 7768–7775
63. Murphy BL, Roth RH, Arnsten AF (1997) Clozapine reverses the spatial working memory deficits induced by FG7142 in monkeys. *Neuropsychopharmacol* 16: 433–437
64. Sarter M, Nelson CL, Bruno JP (2005) Cortical cholinergic transmission and cortical information processing following psychostimulant-sensitization: implications for models of schizophrenia. *Schizophren Bull* 31: 1–22
65. Neigh GN, Arnold HM, Rabenstein RL, Sarter M, Bruno JP (2004) Neuronal activity in the nucleus accumbens is necessary for performance-related increases in cortical acetylcholine release. *Neuroscience* 123: 635–645
66. Neigh GN, Arnold HM, Sarter M, Bruno JP (2001) Dissociations between the effects of intra-accumbens administration of amphetamine and exposure to a novel environment on accumbens dopamine and cortical acetylcholine release. *Brain Res* 894: 354–358
67. Zmarowski A, Sarter M, Bruno JP (2004) Modulation of cortical acetylcholine release *via* glutamatergic and D1 interactions in the nucleus accumbens. *Society for Neuroscience Annual Meeting*. Society for Neuroscience Abstracts, San Diego, CA, 950.914
68. Miner LA, Sarter M (1999) Intra-accumbens infusions of antisense oligodeoxynucleotides to one isoform of glutamic acid decarboxylase mRNA, GAD65, but not to GAD67 mRNA, impairs sustained attention performance in the rat. *Cogn Brain Res* 7: 269–283
69. Himmelheber AM, Bruno JP, Sarter M (2000) Effects of intra-accumbens infusions of amphetamine or cis-flupenthixol on sustained attention performance in rats. *Behav Brain Res* 116: 123–133

70. Gaykema RP, Zaborszky L (1996) Direct catecholaminergic-cholinergic interactions in the basal forebrain. II. Substantia nigra-ventral tegmental area projections to cholinergic neurons. *J Comp Neurol* 374: 555–577
71. Smiley JF, Subramanian M, Mesulam MM (1999) Monoaminergic-cholinergic interactions in the primate basal forebrain. *Neuroscience* 93: 817–829
72. Rodrigo J, Fernandez P, Bentura ML, de Velasco JM, Serrano J, Uttenthal O, Martinez-Murillo R (1998) Distribution of catecholaminergic afferent fibres in the rat globus pallidus and their relations with cholinergic neurons. *J Chem Neuroanat* 15: 1–20
73. Napier TC (1992) Contribution of the amygdala and nucleus accumbens to ventral pallidal responses to dopamine agonists. *Synapse* 10: 110–119
74. Momiyama T, Sim JA (1996) Modulation of inhibitory transmission by dopamine in rat basal forebrain nuclei: activation of presynaptic D1-like dopaminergic receptors. *J Neurosci* 16: 7505–7512
75. Momiyama T, Sim JA, Brown DA (1996) Dopamine D1-like receptor-mediated presynaptic inhibition of excitatory transmission onto rat magnocellular basal forebrain neurones. *J Physiol* 495 (Pt 1): 97–106
76. Johnson PI, Napier TC (1997) GABA- and glutamate-evoked responses in the rat ventral pallidum are modulated by dopamine. *Eur J Neurosci* 9: 1397–1406
77. Nelson CL, Sarter M, Bruno JP (2000) Repeated pretreatment with amphetamine sensitizes increases in cortical acetylcholine release. *Psychopharmacol* 151: 406–415
78. Day JC, Tham CS, Fibiger HC (1994) Dopamine depletion attenuates amphetamine-induced increases of cortical acetylcholine release. *Eur J Pharmacol* 263: 285–292
79. Arnold HM, Fadel J, Sarter M, Bruno JP (2001) Amphetamine-stimulated cortical acetylcholine release: role of the basal forebrain. *Brain Res* 894: 74–87
80. Roberts AC, De Salvia MA, Wilkinson LS, Collins P, Muir JL, Everitt BJ, Robbins TW (1994) 6-Hydroxydopamine lesions of the prefrontal cortex in monkeys enhance performance on an analog of the Wisconsin Card Sort Test: possible interactions with subcortical dopamine. *J Neurosci* 14: 2531–2544
81. Passetti F, Dalley JW, Robbins TW (2003) Double dissociation of serotonergic and dopaminergic mechanisms on attentional performance using a rodent five-choice reaction time task. *Psychopharmacol* 165: 136–145
82. Granon S, Passetti F, Thomas KL, Dalley JW, Everitt BJ, Robbins TW (2000) Enhanced and impaired attentional performance after infusion of D1 dopaminergic receptor agents into rat prefrontal cortex. *J Neurosci* 20: 1208–1215
83. Chudasama Y, Robbins TW (2004) Dopaminergic modulation of visual attention and working memory in the rodent prefrontal cortex. *Neuropsychopharmacol* 29: 1628–1636
84. Zahrt J, Taylor JR, Mathew RG, Arnsten AF (1997) Supranormal stimulation of D1 dopamine receptors in the rodent prefrontal cortex impairs spatial working memory performance. *J Neurosci* 17: 8528–8535
85. Goldman-Rakic PS, Muly III EC , Williams GV (2000) D(1) receptors in prefrontal cells and circuits. *Brain Res Rev* 31: 295–301
86. Ragozzino M (2000) The contribution of cholinergic and dopaminergic afferents in the rat prefrontal cortex to learning, memory, and attention. *Psychobiology* 28: 238–247.
87. Broersen LM, Heinsbroek RP, de Bruin JP, Olivier B (1996) Effects of local application of dopaminergic drugs into the medial prefrontal cortex of rats on latent inhibition. *Biol Psychiatry* 40: 1083–1090
88. Broersen LM, Heinsbroek RP, de Bruin JP, Uylings HB, Olivier B (1995) The role of the medial prefrontal cortex of rats in short-term memory functioning: further support for involvement of cholinergic, rather than dopaminergic mechanisms. *Brain Res* 674: 221–229

89. Izaki Y, Hori K, Nomura M (1998) Dopamine and acetylcholine elevation on lever-press acquisition in rat prefrontal cortex. *Neurosci Lett* 258: 33–36
90. Laplante F, Srivastava LK, Quirion R (2004) Alterations in dopaminergic modulation of prefrontal cortical acetylcholine release in post-pubertal rats with neonatal ventral hippocampal lesions. *J Neurochem* 89: 314–323
91. Sarter M, Parikh V (2005) Choline transporters, cholinergic transmission and cognition. *Nature Rev Neurosci* 6: 48–56
92. Apparsundaram S, Martinez V, Parikh V, Kozak R, Sarter M (2005) Increased capacity and density of choline transporters situated in synaptic membranes of the right medial prefrontal cortex of attentional task-performing rats. *J Neurosci* 15: 3851–3856
93. Ferguson SM, Blakely RD (2004) The choline transporter resurfaces: new roles for synaptic vesicles? *Mol Intervent* 4: 22–37
94. Ferguson SM, Savchenko V, Apparsundaram S, Zwick M, Wright J, Heilman CJ, Yi H, Levey AI, Blakely RD (2003) Vesicular localization and activity-dependent trafficking of presynaptic choline transporters. *J Neurosci* 23: 9697–9709
95. Gates J, Jr., Ferguson SM, Blakely RD, Apparsundaram S (2004) Regulation of choline transporter surface expression and phosphorylation by protein kinase C and protein phosphatase 1/2A. *J Pharmacol Exp Ther* 310: 536–545
96. Guermonprez L, O'Regan S, Meunier FM, Morot-Gaudry-Talarmain Y (2002) The neuronal choline transporter CHT1 is regulated by immunosuppressor-sensitive pathways. *J Neurochem* 82: 874–884
97. Xie J, Guo Q (2004) Par-4 inhibits choline uptake by interacting with CHT1 and reducing its incorporation on the plasma membrane. *J Biol Chem* 279: 28266–28275
98. Parikh V, Pomerleau F, Huettl P, Gerhardt GA, Sarter M, Bruno JP (2004) Rapid assessment of *in vivo* cholinergic transmission by amperometric detection of changes in extracellular choline levels. *Eur J Neurosci* 20: 1545–1554
99. Zhuang X, Oosting RS, Jones SR, Gainetdinov RR, Miller GW, Caron MG, Hen R (2001) Hyperactivity and impaired response habituation in hyperdopaminergic mice. *Proc Natl Acad Sci USA* 98: 1982–1987
100. Pecina S, Cagniard B, Berridge KC, Aldridge JW, Zhuang X (2003) Hyperdopaminergic mutant mice have higher "wanting" but not "liking" for sweet rewards. *J Neurosci* 23: 9395–9402
101. Berridge KC, Aldridge JW, Houchard KR, Zhuang X (2005) Sequential super-stereotypy of an instinctive fixed action pattern in hyper-dopaminergic mutant mice: a model of obsessive compulsive disorder and Tourette's. *BMC Biol* 3: 4
102. Westerink BH, de Boer P, Damsma G (1990) Dopamine-acetylcholine interaction in the striatum studied by microdialysis in the awake rat: some methodological aspects. *J Neurosci Methods* 34: 117–124
103. Steiner H, Kitai ST (2000) Regulation of rat cortex function by D1 dopamine receptors in the striatum. *J Neurosci* 20: 5449–5460
104. Calabresi P, Centonze D, Gubellini P, Pisani A, Bernardi G (2000) Acetylcholine-mediated modulation of striatal function. *Trends Neurosci* 23: 120–126
105. Ward NM, Brown VJ (1996) Covert orienting of attention in the rat and the role of striatal dopamine. *J Neurosci* 16: 3082–3088
106. Kitabatake Y, Hikida T, Watanabe D, Pastan I, Nakanishi S (2003) Impairment of reward-related learning by cholinergic cell ablation in the striatum. *Proc Natl Acad Sci USA* 100: 7965–7970
107. Burmeister JJ, Palmer M, Gerhardt GA (2003) Ceramic-based multisite electrode array for rapid choline measures in brain tissue. *Analytica Chimica Acta* 481: 65–74

108. Lockman PR, Allen DD (2002) The transport of choline. *Drug Dev Ind Pharm* 28: 749–771
109. Acquas E, Di Chiara G (2001) Role of dopamine D1 receptors in the control of striatal acetylcholine release by endogenous dopamine. *Neurol Sci* 22: 41–42
110. Acquas E, Di Chiara G (1999) Local application of SCH 39166 reversibly and dose-dependently decreases acetylcholine release in the rat striatum. *Eur J Pharmacol* 383: 275–279
111. Johnson BJ, Bruno JP (1995) Dopaminergic modulation of striatal acetylcholine release in rats depleted of dopamine as neonates. *Neuropharmacology* 34: 191–203
112. DeBoer P, Heeringa MJ, Abercrombie ED (1996) Spontaneous release of acetylcholine in striatum is preferentially regulated by inhibitory dopamine D2 receptors. *Eur J Pharmacol* 317: 257–262
113. Maurice N, Mercer J, Chan CS, Hernandez-Lopez S, Held J, Tkatch T, Surmeier DJ (2004) D2 dopamine receptor-mediated modulation of voltage-dependent Na+ channels reduces autonomous activity in striatal cholinergic interneurons. *J Neurosci* 24: 10289–10301
114. Bickerdike MJ, Abercrombie ED (1997) Striatal acetylcholine release correlates with behavioral sensitization in rats withdrawn from chronic amphetamine. *J Pharmacol Exp Ther* 282: 818–826
115. Steinberg R, Souilhac J, Rodier D, Alonso R, Emonds-Alt X, Le Fur G, Soubrie P (1998) Facilitation of striatal acetylcholine release by dopamine D1 receptor stimulation: involvement of enhanced nitric oxide production *via* neurokinin-2 receptor activation. *Neuroscience* 84: 511–518
116. Mandel RJ, Leanza G, Nilsson OG, Rosengren E (1994) Amphetamine induces excess release of striatal acetylcholine *in vivo* that is independent of nigrostriatal dopamine. *Brain Res* 653: 57–65
117. Sarter M, Bruno JP (1999) Abnormal regulation of corticopetal cholinergic neurons and impaired information processing in neuropsychiatric disorders. *Trends Neurosci* 22: 67–74
118. Goldman-Rakic PS, Castner SA, Svensson TH, Siever LJ, Williams GV (2004) Targeting the dopamine D1 receptor in schizophrenia: insights for cognitive dysfunction. *Psychopharmacol* 174: 3–16
119. Stip E, Chouinard S, Boulay LJ (2005) On the trail of a cognitive enhancer for the treatment of schizophrenia. *Prog Neuropsychopharmacol Biol Psychiat* 29: 219–232
120. Robinson TE, Berridge KC (2001) Incentive-sensitization and addiction. *Addiction* 96: 103–114
121. Robinson TE, Berridge KC (2003) Addiction. *Ann Rev Psychol* 54: 25–53
122. Robinson TE, Berridge KC (2000) The psychology and neurobiology of addiction: an incentive-sensitization view. *Addiction* 95 Suppl 2: S91–117
123. Everitt BJ, Dickinson A, Robbins TW (2001) The neuropsychological basis of addictive behaviour. *Brain Res Rev* 36: 129–138
124. Rogers RD, Everitt BJ, Baldacchino A, Blackshaw AJ, Swainson R, Wynne K, Baker NB, Hunter J, Carthy T, Booker E et al. (1999) Dissociable deficits in the decision-making cognition of chronic amphetamine abusers, opiate abusers, patients with focal damage to prefrontal cortex, and tryptophan-depleted normal volunteers: evidence for monoaminergic mechanisms. *Neuropsychopharmacol* 20: 322–339
125. Smith JE, Co C, Yin X, Sizemore GM, Liguori A, Johnson III WE, Martin TJ (2004) Involvement of cholinergic neuronal systems in intravenous cocaine self-administration. *Neurosci Biobehav Rev* 27: 841–850

126. Smith JE, Vaughn TC, Co C (2004) Acetylcholine turnover rates in rat brain regions during cocaine self-administration. *J Neurochem* 88: 502–512
127. Martinez V, Parikh V, Sarter M (2005) Sensitized attentional performance and Fos-immunoreactive cholinergic neurons in the basal forebrain of amphetamine-pretreated rats. *Biol Psychiat* 57: 1138–1146.
128. Bednar I, Friberg L, Nordberg A (2004) Modulation of dopamine release by the nicotinic agonist epibatidine in the frontal cortex and the nucleus accumbens of naive and chronic nicotine treated rats. *Neurochem Int* 45: 1049–1055
129. Cao YJ, Surowy CS, Puttfarcken PS (2005) Different nicotinic acetylcholine receptor subtypes mediating striatal and prefrontal cortical [3H]dopamine release. *Neuropharmacol* 48: 72–79
130. Zhang L, Zhou FM, Dani JA (2004) Cholinergic drugs for Alzheimer's disease enhance *in vitro* dopamine release. *Mol Pharmacol* 66: 538–544
131. Levin ED, M.W. D, Butcher LL (1992) Neurotransmitter interactions and cognitive function. In: Levin ED, Decker MW, Butcher LL (eds): *Neurotransmitter interactions and cognitive function.* Birkhäuser, Boston, 355–357

Neurotransmitter Interactions and Cognitive Function
Edited by Edward D. Levin
© 2006 Birkhäuser Verlag/Switzerland

Intraseptal cholinergic infusions alter memory in the rat: method and mechanism

James J. Chrobak, Helen R. Sabolek and Jamie G. Bunce

Department of Psychology, University of Connecticut, Storrs, CT 06269, USA

Introduction

The medial septum and the hippocampal formation are neural substrates for memory in mammals. Medial septal/vertical limb of the diagonal band (MS) neurons innervate the entire hippocampal formation and control hippocampal physiology and memory function. These neurons are an important relay for brainstem and hypothalamic regulation of the hippocampus [1, 2] and provide a feedback circuit for hippocampal self-regulation [3, 4]. Medial septal neurons receive brainstem and hypothalamic cholinergic, noradrenergic, serotonergic, dopaminergic, histaminergic and orexin input among others [1, 5–7]. Thus, MS neurons contain a number of potential drug targets for modulating the hippocampal formation. Dynamic changes among MS, notably cholinergic, neurons have been linked to the earliest stages of age-related memory dysfunction in humans [8–11]. Whether such changes are causative or a consequence of developing neuropathology is often debated. However, therapeutic strategies for ameliorating memory dysfunction directly target or inadvertently exert potent influences on MS neurons [12, 13]. The present review focuses on a series of studies [14–18] that ask under what conditions can direct cholinergic manipulation of MS neurons enhance or disrupt hippocampal dependent memory performance? While these studies relate to whether enhancing cholinergic tone with cholinesterase inhibitors or direct acting muscarinic agonists can ameliorate memory dysfunction, they ask more directly what happens if you artificially enhance cholinergic tone directly within the MS. MS cholinomimetics treatments are also a potent means to enhance hippocampal theta [19, 20] and ask questions about whether and how enhancing theta may enhance memory.

Theta: The physiologic target

One major function of MS neurons is to implement theta rhythmicity in hippocampal circuits and suppress the occurrence of hippocampal sharp waves [21, 22]. Neurons in the MS innervate all regions of the hippocampal formation, including the dentate gyrus, CA3, CA1, the subiculum and entorhinal cortex [23–26]. These MS efferents transform subcortical (brainstem and diencephalic) input into the well described theta

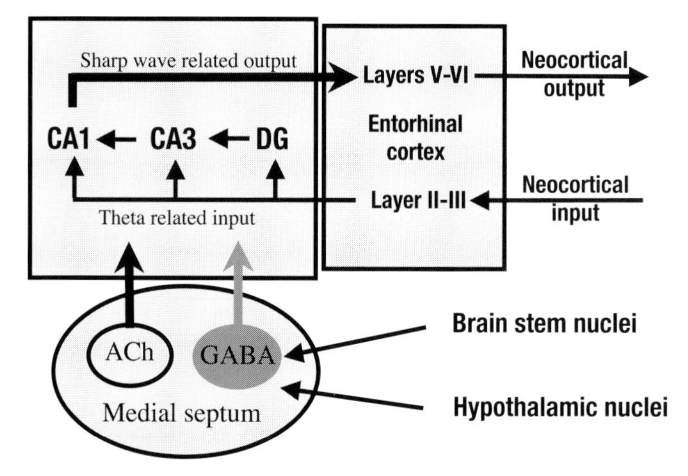

Figure 1. Hippocampal neurons (CA1, CA3, DG) receive information from the neocortex via neurons in layers II, III of the entorhinal cortex. CA1 neurons provide the main output of the hippocampus back to the neocortex via relays in the deep layer of the entorhinal cortex. These circuits participate in two distinct network patterns. During exploratory behavior, "attentive" immobility and REM sleep, hippocampal neurons participate in theta. The theta state promoted by both MS cholinergic (ACh) and GABA neurons involves activation of hippocampal targets by rhythmic input from neurons in layers II and III of the entorhinal cortex. In the absence of the MS input, which occurs during "quiet" immobility, consummatory behavior and slow-wave sleep, hippocampal neurons participate in sharp waves. The sharp wave state involves aperiodic bursts of the CA1 network that engages neuron in layers V and VI of the entorhinal cortex [22, 38]. In many respects the MS neurons serve as a switch or rheostat, regulating both the mode and intensity of each state. Infusion of cholinomimetics drugs into the MS can induce a dose-related increase in theta amplitude and the occurrence of theta irrespective of behavioral state.

modulation of hippocampal excitability and suppress the generation of hippocampal sharp waves [1, 2, 27–29].

Theta is at first appearance a rhythmic field potential, observed continuously during exploratory activity, "attentive" behavior and rapid-eye-movement (REM) sleep in the rat [1, 27, 29]. While most prominent in the rat because of the architecture of the rodent hippocampus, theta is evident in mammals including non-human primates and in the human hippocampus [30, 31]. During theta, entorhinal inputs into the hippocampus rhythmically excite dentate granule (DG), CA3 and CA1 neurons (see Fig. 1). The theta field potentials reflect the summation of current flow induced by this rhythmic excitation into the laminarly arranged dendritic fields of hippocampal neurons [29]. More importantly, all hippocampal neurons discharge in temporal relation to this rhythm.

Theta's primary function is foremost to synchronize the activity of individual neurons into population volleys and consequentially allow select subgroups to act on their targets in a coordinated manner. For example, CA1 neurons within the hippocampus receive two primary excitatory (glutamatergic) afferents. Their most dominant input is from CA3 neurons, but they also receive excitatory input from layer III entorhinal cortical neurons [32]. Theta coordination allows subsets of CA3 neurons to discharge as an ensemble at a slightly different phase of the theta cycle than subsets of entorhinal cortical neurons. As a consequence, CA1 neurons are influenced by CA3 and entorhinal afferents in a coordinated sequence. Slight changes in the frequency of the rhythm and thus the timing of excitatory inputs may bias the response of a CA1 neuron to either the CA3 or entorhinal input. At the computational level, theta may be considered a dynamic filter providing bias to different synaptic inputs at different time periods (phases) of each theta cycle [33–37]. Hasselmo and colleagues have suggested that such a biasing system could allow for the preferential encoding of new sensory representations (e.g. entorhinal input) within hippocampal circuitry, while preserving and integrating them with existing representations (e.g. recurrent input in CA3).

We have suggested that theta reflects an operational state whereby information is "written" into the circuitry of the hippocampus and the alternate hippocampal pattern, sharp waves, reflects a state whereby "written" information is progressively consolidated and transferred to neocortical stores [38]. This model suggests that different neurophysiological states are distinct computational states and that each may be important for the short and/or long-term retention of memories. Further, we have hypothesized that enhancing theta immediately after a single-trial event could, by suppressing the occurrence of sharp waves, be amnestic. Disrupting theta is invariably associated with memory dysfunction [39–41]. In contrast, enhancing theta under some conditions can enhance memory [41, 42]. Intraseptal treatment with cholinomimetic compounds increases the amplitude of theta field potentials [19, 41], most likely by increasing the number of entorhinal inputs activating hippocampal targets on each wave of theta [43]. The question is then, if and when this might enhance memory?

MS/DB: The pharmacologic targets

Medial septal/diagonal band (mS/DB) neurons are important targets for the development of cognitive enhancers [12, 13, 44]. Cholinergic, GABAergic, and glutamatergic neurons within the MS innervate the dentate gyrus, CA3, CA1, the subiculum as well as the entorhinal cortex [23–26]. Most experimental studies have focused on the cholinergic neurons, which are the dorsal most extension of the basal forebrain cholinergic column [45]. The cholinergic neurons exhibit pathologic plasticity and subsequently degenerate in Alzheimer's dementia [9–11] and a pharmacologic replacement strategy for restoring cholinergic tone has been well studied [44, 46, 47].

The MS cholinergic neurons promiscuously innervate all hippocampal targets providing muscarinic presynaptic modulation of afferent input and direct postsynaptic activation of hippocampal neurons [48, 49]. Equally important, however, half the MS projection is composed of GABAergic neurons that selectively innervate only GABAergic interneurons in the hippocampus [25]. Both MS cholinergic and MS GABAergic neurons play key roles in the generation of the hippocampal theta rhythm [21, 50, 51].

Both cholinergic and GABAergic MS/DB neurons are rich in cholinergic receptors [52] and intraseptal infusion of cholinergic agonists (e.g., physostigmine, carbachol, oxotremorine) can enhance or induce hippocampal theta [19–21, 53]. A number of laboratories have been examining the localization of specific receptors subtypes on MS/DB cholinergic and GABAergic neurons (e.g., [12, 54–56], thus future studies can explore the contribution and consequences of modulating specific receptor subtypes.

Memory: The cognitive target

Memory is a dynamic process that likely depends on a series of neurobiological processes occurring over unknown time periods. Minimally, it involves the "on-line" acquisition of information, the short-term retention of information and in some cases a protracted period of consolidation [57]. Different neurobiological processes within septohippocampal circuits, involving different levels of MS activation, may mediate the acquisition of information as compared with the short-term (minutes, hours or days) retention and/or consolidation of information. Other evidence also suggests that distinct neurobiological circuits operate in parallel to support memories with different time-courses [58]. Thus, pharmacologic strategies that may enhance certain neurobiologic processes may concurrently weaken alternate processes.

Pre-acquisition intraseptal cholinomimetics alter memory performance

Studies examining the effect of MS infusion of cholinomimetics on memory have produced somewhat inconsistent results with reports of either promnestic or amnestic effects [14–18, 20, 41, 42, 59–62]. Any number of variables (e.g., dose, age of animal, integrity of septohippocampal circuits, task difficulty) may contribute to these differences. In many respects these results mirror the inconsistencies observed following experimental insult to forebrain cholinergic nuclei and the effects of cholinomimetics treatment strategies to ameliorate memory dysfunction (eg., [63–70]. Our laboratory has been interested in defining the conditions under which direct MS cholinergic treatments enhance ([17, 18]; see Fig. 2) and/or disrupt ([14–16]; see Figs. 3 and 4) performance in hippocampal-dependent "episodic" memory tasks.

Pre-acquisition MS infusion of cholinomimetics, all of which induce/enhance theta, can in specific instances enhance spatial memory. The effectiveness of this treatment seems to work best when given prior to task performance (preacquisition) and particularly in aged and cognitively impaired rats [17, 42, 59]. Generally preacquisition, or "drug on board" intraseptal treatment with cholinomimetics have a limited effect on accuracy when given to normal young rats prior to testing [20, 41, 42], although see [61].

Medial septal infusion of the cholinesterase inhibitor tacrine can either enhance performance or impair performance (see Fig. 2) on the standard RAM task depending upon "cognitive status" of the rat [17, 18]. Thus, a dose of MS tacrine that enhanced performance of young rats performing very poorly on a radial maze task, impaired

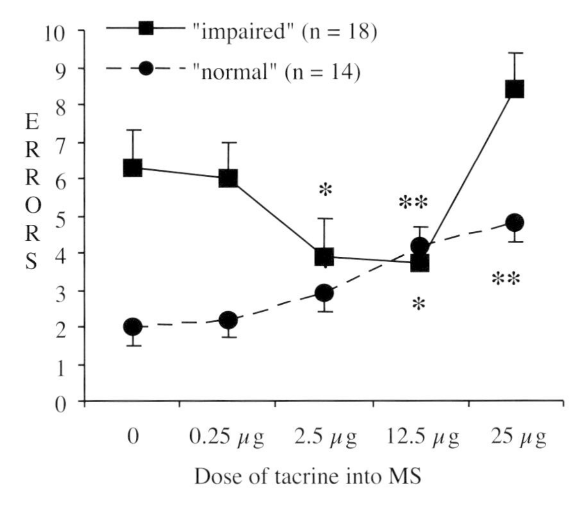

Figure 2. Medial septal (MS) cholinomimetic treatment (the acetylcholinesterase inhibitor tacrine) can enhance spatial memory performance when given prior to encoding (with drug "on-board" at testing) under select conditions. Effects of MS tacrine on young (< 1 year) "impaired" rats as compared to young normal rats on performance of a standard radial arm maze task. "Impaired" young rats performed > 2 standard deviations below the mean: note the substantial difference in baseline performance (0 saline dose). All data are within-subject with rats receiving doses once/week in random order. MS infusion of tacrine (2.5 μg and 12.5 μg) significantly decreased the number of errors (*, ** $p < 0.05, 0.01$) in "impaired rats". A higher (25 μg) had no significant effect, but exhibited a trend toward impairing performance. In contrast, normal young rats exhibited an impairment following 12.5 and 25 μg MS tacrine (*, ** $p < 0.05, 0.01$). Note that at 12.5 μg, "impaired" rats performance improved, while "normal" rats were impaired. All treatments were given 5 min prior to testing. Figure based on data presented in references [17] and [18].

the performance of normal young rats. Aged and impaired rats may have sub-optimal septohippocampal circuits for any number of reasons and the MS cholinomimetics treatment may boost the signaling capability of the MS input. In normal young rats it may be very difficult to increase the signaling capacity of the MS input and increasing doses of cholinomimetics not only induce theta and suppress sharp waves, but transform the theta rhythmicity of hippocampal networks into epileptiform activity ([71], Sabolek and Chrobak, *unpublished observations*).

Post-acquisition intraseptal carbachol is amnestic

A few studies have examined the effects of intraseptal infusion of cholinergic drugs on memory when treatments were administered post-acquisition or after the "to-be-remembered" event. Flood and colleagues [60, 62] reported an inverted "U" effect on retention in passive avoidance tasks following intraseptal cholinomimetics. In contrast, we have observed only amnestic effects in spatial memory following either carbachol or oxotremorine in a delayed-non-match-to-sample radial maze (RAM) task.

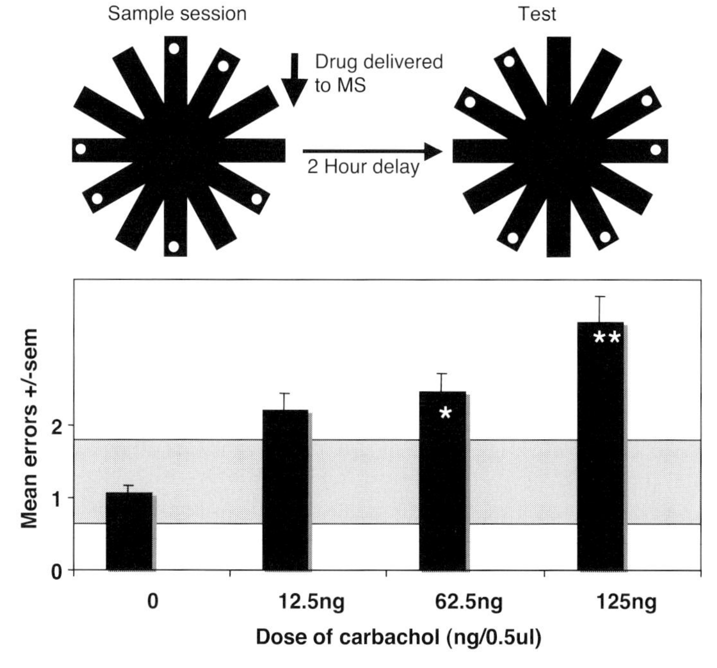

Figure 3. Top figure illustrates aerial view of the 12-arm radial maze. Six baited arms (white circles) are accessed during the sample session. Clear Plexiglas barriers prevent access to the alternate six arms. Following completion of the sample session (approximately 30–60 s) rats are returned to their home cage. During the test session, all arms are available, while only non-match arms are baited. Entry into sample session baited arms (retroactive errors) or repeat entry into any arm during the test (proactive errors) constitute errors. The delay interval on Mondays, Tuesdays, Thursdays and Fridays was 1.5 h (90 min). The delay was extended to 2 h (120 min) on the day of carbachol infusions (Wednesdays). Once trained at the 1–2 h delay, no differences are observed at 3–6 h delay intervals. The longer delay allows for a longer drug washout period, a longer time frame for delayed infusions and is logistically simpler given the extra time needed to infuse each animal. Cannulated rats received four carbachol infusions (0, 12.5, 62.5 and 125 ng/0.5 ul at rate 0.125 ul/min) and a sham treatment in pseudorandom order. Carbachol infusions were administered immediately following sample (within 2–5 min). Figure based on data presented in reference [15].

In this task, hungry rats had access to six sample arms during the sample session (see Fig. 3) which they sample for food reward. After a delay (typically 1 h), rats are allowed access to all arms with food reward located only in the "non-match" arms during a test session. The rats are given one sample and one test session each day with the sample session arms being different on each day. Entry into one of the sample session arms during the test session is considered a "retroactive error," while any re-entry into any arm during the test session is considered a "proactive error." After several weeks of training, most rats perform this task at a level of accuracy making 0–2 retroactive errors (mean = $\sim$ 1.0) and very few proactive

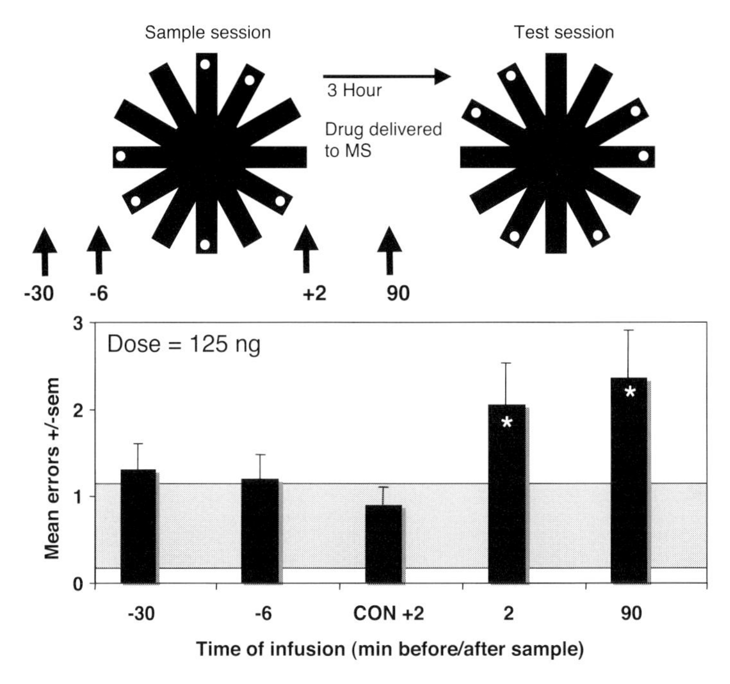

Figure 4. Top figure illustrates aerial view of the 12-arm radial maze and timing of drug infusions with respect to the sample session. For this experiment, the delay was extended to 3 h (180 min) on the day of carbachol infusions and a single does of MS carbachol (125 ng/0.5 ul) was infused once a week at different time points. MS infusion of 125 ng/0.5 ul carbachol had no effect when administered immediately prior to the sample session, but produced an increase in retroactive errors when administered immediately after the sample (+2) or midway between the sample and test session (+90). Horizontal gray field illustrates the range of mean retroactive errors for all Tuesdays prior to any infusions. $*$ indicates significantly different from saline control and following pre-sample session (−6) infusion. Data as adapted from reference [16].

errors (mean = ~0.2). Once the rats are trained, they are outfitted with a chronic cannulae implant that allows acute infusion of drug into the MS. Treatments are administered once a week, typically Wednesdays, allowing for an examination of any changes in performance on the day of treatment (acute treatment effect), or the days after treatment (subchronic/chronic insult).

Medial septal carbachol or oxetremorine, both muscarinic agonists, induce a dose-dependence amnestic effect when treatment is administered immediately after the sample session (Fig. 3A; see [14–16]). In a follow-up experiment, we examined the effects of a single dose of MS carbachol (125 ng/0.5 ul) administered once a week, but with the timing of drug administration varying. Infusion of carbachol prior to the sample session had no effect on performance 3 h later. The same dose administered anytime during the delay induced a deficit during the test session, indicating a retention or consolidation deficit. Importantly, for both amnestic doses

(3 h prior to the test or 90 min prior to the test) rats could remember locations visited within the test session (no increase in proactive errors). Thus there is no effect of this treatment on either sensorimotor processing or short-term retention at the time of test. In contrast intraseptal carbachol impaired rats' ability to avoid arms entered during the prior sample session. Such findings are consistent with an acute amnesia that is operationally a consolidation deficit (drug administered after acquisition).

When is intraseptal cholinomimetic treatment amnestic or promnestic?

Several studies have demonstrated that MS infusion of cholinomimetics can enhance memory if administered prior to task performance. This treatment seems to work best in aged and cognitively impaired rats [17, 42, 59]. Aged and "impaired" young rats may have sub-optimal septohippocampal circuits for any number of reasons and the cholinomimetics treatment may boost the signaling capability of the MS input. It may be difficult to increase the signaling capacity of the MS signal in "normal" young rats and increasing doses of MS cholinomimetics can transform the regulated excitability of hippocampal theta waves into epileptiform activity ([71], Sabolek and Chrobak, *unpublished observations*).

In contrast, MS cholinomimetics can disrupt memories for events occurring immediately prior to treatment. We have suggested that one mechanism underlying the amnestic effects of MS carbachol may involve the inappropriate induction of hippocampal theta. Inducing theta, under appropriate conditions, during information acquisition may enhance encoding, while inducing theta after acquisition may impair the retention or consolidation of information. The MS signal, which in some way is reflected or represented by the theta modulation of hippocampal neurons, reflects the importance/relevance of information and appears to be critical to the plasticity of select hippocampal synapses. Hasselmo and colleagues posit that theta reflects a state that optimizes synaptic modification of HPC circuits by allowing new information (patterns of EC input conveying current sensory input) to be integrated with existing information (patterns of intrinsic activity reflecting past input). In this context, an experimentally induced theta signal (MS carbachol) after encoding may induce inappropriate synaptic changes (e.g., [34, 36, 37, 72]). The heightened MS signal may modify HF circuits to an irrelevant pattern of EC input, or in cognitive terms induce retroactive interference. Alternately, MS treatments that induce theta also suppress hippocampal sharp waves that may be critical to the retention and consolidation of information [38]. Understanding the conditions when MS treatments affect memory in specific well-controlled behavioral paradigm sets the stage for future studies focusing on the mechanism(s) by which activation of cholinergic receptors are promnestic or amnestic. The fact that the cholinergic receptors are (e.g., [73]) a primary target for cognitive enhancing cholinomimetics treatments in both Alzheimer's demential and mild cognitive impairment (MCI) underscores the importance of understanding both conditions and the underlying neurobiological mechanisms that underlie these memory effects.

References

1. Petsch H, Stumpf G, Gogolak C (1962) The significance of the rabbit's septum as a relay station between the midbrain and the hippocampus. I. The control of hippocampus arousal activity by the septum cells. *Electroenceph Clin Neurophysiol* 14: 202–211
2. Vertes RP, Kocsis B (1997) Brainstem-diencephalo-septohippocampal systems controlling the theta rhythm of the hippocampus. *Neurosci* 81: 893–926
3. Toth K, Borhegyi Z, Freund TF (1993) Postsynaptic targets of gabaergic hippocampal neurons in the medial septum-diagonal band of Broca complex. *J Neurosci* 13: 3712–3724
4. Gulyas AI, Hajos N, Katona I, Freund TF (2003) Interneurons are the local targets of hippocampal inhibitory cells which project to the medial septum. *Eur J Neurosci* 17: 1861–1872
5. Vertes RP (1981) An analysis of ascending brain stem systems involved in hippocampal synchronization and desynchronization. *J Neurophysiol* 46: 1140–1159
6. Jakab R, Leranth C (1995) Septum. In: Paxinos G (ed): *The rat nervous system*. Academic Press, San Diego, 405–442
7. Gerashchenko D, Salin-Pascual R, Shiromani PJ (2001) Effects of hypocretin-saporin injections into the medial septum on sleep and hippocampal theta. *Brain Res* 913: 106–115
8. Coyle JT, Price DL, DeLong MR (1983) Alzheimer's disease: a disorder of cortical cholinergic innervation. *Science* 219: 1184–1190
9. Butcher LL, Woolf NJ (1989) Neurotrophic agents exacerbate the pathologic cascade of Alzheimer's disease. *Neurobiol Aging* 10: 557–570
10. DeKosky ST, Ikonomovic MD, Styren SD, Beckett L, Wisniewski S, Bennett DA, Cochran EJ, Kordower JH, Mufson EJ (2002) Upregulation of choline acetyltransferase activity in hippocampus and from cortex of elderly subjects with mild cognitive impairment. *Ann Neurol* 51: 145–155
11. Ikonomovic MD, Mufson EJ, Wuu J, Cochran EJ, Bennett DA, DeKosky ST (2003) Cholinergic plasticity in hippocampus of individuals with mild cognitive impairment: correlation with Alzheimer's neuropathology. *J Alzheimers Dis* 5: 39–48
12. Alreja M, Wu M, Liu, W, Atkins JB, Leranth C, Shanabrough M (2000) Muscarinic tone sustain impulse flow in the septohippocampal GABA but not cholinergic pathway: implications for learning and memory. *J Neurosci* 20: 8103–8110
13. Barnes CA, Meltzer J, Houston F, Orr G, McGann K, Wenk GL (2000) Chronic treatment of old rats with donepezil or galantamine: effects on memory, hippocampal plasticity and nicotinic receptors. *Neuroscience* 99: 17–23
14. Bunce JG, Sabolek HR, Chrobak JJ 2003. Intraseptal infusion of oxotremorine impairs memory in a delayed-non-match-to-sample radial maze task. *Neuroscience* 121: 259–267
15. Bunce JG, Sabolek HR, Chrobak JJ (2004) Intraseptal infusion of the cholinergic agonist carbachol impairs delayed-non-match-to-sample radial arm maze performance in the rat. *Hippocampus* 14: 450–459
16. Bunce JG, Sabolek HR, Chrobak JJ (2004) Timing of administration mediates the memory effects of intraseptal carbachol infusion. *Neuoscience* 127: 593–600
17. Sabolek HR, Bunce JG, Chrobak JJ (2004). Intraseptal tacrine can enhance memory in cognitively impaired young rats. *Neuroreport* 15: 181–183
18. Sabolek HR, Bunce JG, Giuliana D, Chrobak JJ (2005) Intraseptal tacrine-induced disruptions of spatial memory performance. *Behav Brain Research* 158: 1–7
19. Monmaur P, Breton P (1991) Elicitation of hippocampal theta by intraseptal carbachol injection in freely-moving rats. *Brain Research* 544: 150–155

20. Givens BS, Olton DS (1995) Bidirectional modulation of scopolamine-induced working memory impairments by muscarinic activation of the medial septal area. *Neurobiol Learning Memory* 63: 269–276
21. Lee MG, Chrobak JJ, Sik A, Wiley RG, Buzsaki G (1994) Hippocampal theta activity following selective lesion of the septal cholinergic system. *Neuroscience* 62: 1033–1047
22. Chrobak JJ, Buzsaki G (1996) High-frequency oscillations in the output networks of the hippocampal-entorhinal axis of the freely behaving rat. *J Neurosci* 16: 3056–3066
23. Rye DB, Wainer BH, Mesulam MM, Mufson EJ, Saper CB 1984. Cortical projections arising from the basal forebrain: A study of cholinergic and noncholinergic components employing combined retrograde tracing and immunohistochemical localization of choline acetyltransferase. *Neuroscience* 13: 627–643
24. Amaral DG, Kurz J, (1985) An analysis of the origins of the cholinergic and noncholinergic septal projections to the hippocampal formation of the rat. *J Comp Neurol* 240: 37–59
25. Freund T, Antal M (1988) GABA-containing neurons in the septum control inhibitory interneurons in the hippocampus. *Nature* 336: 170–173
26. Manns ID, Mainville L, Jones BE (2001) Evidence for glutamate, in addition to acetylcholine and GABA, neurotransmitter synthesis in basal forebrain neurons projecting to the entorhinal cortex. *Neuroscience* 107: 249–263
27. Bland BH (1986) Physiology and pharmacology of hippocampal formation theta rhythms. *Prog Neurobiol* 26: 1–54
28. Vinogradova OS (1995) Expression, control, and probable functional significance of the neuronal theta-rhythm. *Prog Neurobiol* 45: 523–83
29. Buzsaki G (2002) Theta oscillations in the hippocampus. *Neuron* 33: 325–340
30. Stewart M, Fox SE (1991) Hippocampal theta activity in monkeys. *Brain Res* 538: 59–63
31. Kahana MJ, Sekuler R, Caplan JB, Kirschen M, Madison JR (1999) Human theta oscillations exhibit task dependence during virtual maze navigation. *Nature* 399: 781–784
32. Amaral DG, Witter M (1995) The three dimension organization of the hippocampal formation: a review of anatomical data. *Neuroscience* 31: 571–591
33. Hasselmo ME, Bower JM (1993) Acetylcholine and memory. *Trends Neurosci* 16: 218–222
34. Hasselmo ME, Schnell E (1994) Laminar selectivity of the cholinergic suppression of synaptic transmission in rat hippocampal region CA1: computational modeling and brain slice physiology. *J Neurosci* 14: 3898–3914
35. Hasselmo ME, Wyble BP, Wallenstein GV (1996) Encoding and retrieval of episodic memories: role of cholinergic and GABAergic modulation in the hippocampus. *Hippocampus* 6: 693–708
36. Hasselmo ME (2000) What is the function of the theta rhythm? In: Numan R (ed): *The behavioral neuroscience of the septal region*, Springer-Verlag, New York, 92–114
37. Hasselmo ME, Hay J, Ilyn M, Gorchetchnikov A (2002) Neuromodulation, theta rhythm and rat spatial navigation. *Neural Networks* 15: 689–707
38. Chrobak JJ, Lorincz A, Buzsaki G (2000) Physiological patterns in the hippocampo-entorhinal cortex system. *Hippocampus* 10: 457–465
39. Winson J (1978) Loss of hippocampal theta rhythm results in spatial memory deficit in the rat. *Science* 201: 160–163
40. Mitchell SJ, Rawlins JN Steward O, Olton DS (1982) Medial septal area lesions disrupt theta rhythm and cholinergic staining in medial entorhinal cortex and produce impaired radial arm maze behavior in rats. *J Neurosci* 2: 292–302
41. Givens BS, Olton DS (1990) Cholinergic and GABAergic modulation of medial septal area: effect on working memory. *Behav Neurosci* 104: 849–855

42. Markowska A, Olton DS, Givens B (1995) Cholinergic manipulations in the medial septal area: age-related effects on working memory and hippocampal electrophysiology. *J Neurosci* 15: 2063–2073
43. Dickson CT, Magistretti J, Shalinsky M, Hamam B, Alonso A (2000) Oscillatory activity in entorhinal neurons and circuits. Mechansisms and function. *Ann NY Acad Sci* 911: 127–150
44. Bartus RT, Dean RL 3rd, Beer B, Lippa AS (1982) The cholinergic hypothesis of geriatric memory dysfunction. *Science* 217: 408–414
45. Mesulam MM, Mufson EJ, Wainer BH, Levey AL (1983) Central cholinergic pathways in the rat: an overview based on an alternative nomenclature (Ch1-Ch6). *Neuroscience* 10: 1185–1201
46. Dunnett SB, Fibiger HC (1993) Role of forebrain cholinergic systems in learning and memory: relevance to the cognitive deficits of aging and Alzheimer's dementia. *Prog Brain Res* 98: 413-420
47. Sarter M, Bruno JP (1997) Cognitive functions of cortical acetylcholine: toward a unifying hypothesis. *Brain Res Brain Res Rev* 23: 28–46
48. Madison DV, Lancaster B, Nicoll RA (1987) Voltage clamp analysis of cholinergic action in the hippocampus. *J Neurosci* 7: 733–741
49. Colgin LL, Kramar EA, Gall CM, Lynch G (2003) Septal modulation of excitatory transmission in hippocampus. *J Neurophys* 90: 2358–2366
50. Hajos M, Hoffmann WE, Orban G, Kiss T, Erdi P (2004) Modulation of septo-hippocampal theta activity by GABA-A receptors: an experimental and computational approach. *Neuroscience* 126: 599–610
51. Yoder R, Pang KCH (2005) Involvement of GABAergic and cholinergic medial septal neurons in hippocampal theta rhythm. *Hippocampus* 15: 381–392
52. Rouse ST, Levey AI (1996) Expression of m1-m4 muscarinic acetylcholine receptors immunoreactivity in septohippocampal neurons and other identified hippocampal afferents. *J Comp Neurol* 375: 406–416
53. Lawson VH, Bland BH (1993) The role of the septohippocampal pathway in the regulation of hippocampal field activity and behavior: analysis by the intraseptal microinfusion of carbachol, atropine and procaine. *Exp Neurol* 120: 132–144
54. Birthelmer A, Lazaris A, Riegert C, Marques Pereira P, Koenig J, Jeltsch H, Jackisch R, Cassel JC (2003) Does the release of acetylcholine in septal slices originate from intrinsic cholinergic neurons bearing p75(NTR) receptors? A study using 192 IgG-saporin lesions in rats. *Neuroscience* 122: 1059–1071
55. Luttgen M, Ogren SO, Meister B (2005) 5-HT1A receptor mRNA and immunoreactivity in the rat medial septum/diagonal band of broca-relationships to GABAergic and cholinergic neurons. *J Chem Neuroanat* 29: 93–111
56. Kosis B, Li S (2004) In vivo contribution of h-channels in the septal pacemaker to theta rhythm generation. *Eur J Neurosci* 20: 1249–2158
57. Nader K (2003) Memory traces unbound. *Trends Neurosci* 26: 65–72
58. Lee I, Kesner RP (2003) Time-dependent relationship between the dorsal hippocampus and the prefrontal cortex in spatial memory. *J Neurosci* 23: 1517–1523
59. Frick KM, Gorman LK, Markowska AL (1996) Oxotremorine infusions into the medial septum of middle-aged rats affect spatial reference memory and ChAT activity. *Behav Brain Res* 80: 99–109
60. Flood JF, Farr SA, Uezu K, Morley JE (1998) The pharmacology of post-trial memory processing in septum. *Eur J Pharmacol* 350: 31–38

61. Pang KC, Nocera R (1999). Interactions between 192-IgG saporine and intraseptal cholinergic and GABAergic drugs: role of cholinergic medial septal neurons in spatial working memory. *Behav Neurosci* 113: 265–275
62. Farr SA, Flood JF, Morley JE (2000) The effect of cholinergic, GABAergic, serotonergic, and glutamatergic receptor modulation on posttrial memory processing in the hippocampus. *Neurobiol Learn Mem* 73: 150–167
63. Hagan JJ, Salamone JD, Simpson J, Iversen SD, Morris RGM (1988) Place navigation in rats is impaired by lesions of medial septum and diagonal band but not nucleus basalis magnocellularis. *Behav Brain Res* 27: 9–20
64. Wenk GL Harrington CA, Tucker DA, Rance NE, Walker LC (1992) Basal forebrain lesions and memory: a biochemical, histological and behavioral study of differential vulnerability to ibotenate and quisqualate. *Behav Neurosci* 106: 909–923
65. Baxter MG, Bucci DJ, Gorman LK, Wiley RG, Gallagher M (1995) Selective immunotoxic lesions of basal forebrain cholinergic cells:effects on learning and memory in rats. *Behav Neurosci* 4: 714–722
66. Shen J, Barnes CA, Wenk GL, McNaughton B (1996) Differential effects of selective immunotoxic lesions of medial septal cholinergic cells on spatial working and reference memory. *Behav Neurosci* 110: 1181–1186
67. Walsh Tj Herzon CD, Gandhi C, Stackman RW, Wiley RG (1996) Injection IgG 192-saporin into the medial septum produces cholinergic hypofunction and dose-dependent working memory deficits. *Brain Res* 726: 69–78
68. McMahon RW, Sobel TJ, Baxter MG (1997) Selective immunolesions of hippocampal cholinergic inpu fail to impair spatial working memory. *Hippocampus* 7: 130–136
69. Chang O, Gold PE (2004) Impaired and spared cholinergic functions in the hippocampus after lesions of the medial septum/vertical limb of the diagonal band with 192 IgG saporin. *Hippocampus* 14: 170–179
70. Mulder J, Harkany T, Czollner K, Cremers TI, Keijser JN, Nyakas C, Luiten PG (2005) Galantamine-induced behavioral recovery after sublethal excitotoxic lesions of medial septum. *Behav Brain Res* 67: 117–125
71. Wallenstein GV, Hasselmo ME (1997) Functional transitions between epileptiform-like activity and associative memory in hippocampal region CA3. *Brain Res Bull* 43: 485–93
72. Wallenstein GV, Eichenbaum H, Hasselmo ME (1998) The hippocampus as an associator of discontiguous events. *Trends Neurosci* 21: 317–323
73. Gauthier SG (2005) Alzheimer's disease: the benefits of early treatment. *E J Neurol* 12: 11–16

Neurotransmitter Interactions and Cognitive Function
Edited by Edward D. Levin
© 2006 Birkhäuser Verlag/Switzerland

Modulation of visual perception and action by forebrain structures and their interactions in amphibians

Jörg-Peter Ewert and Wolfgang W. Schwippert

Department of Neurobiology, Faculty of Natural Sciences, University of Kassel, 34132 Kassel, Germany

Nervous systems of lower animals appraised simple prove to be relatively complex

Do amphibians "know" while humans "learn"?

Towards the middle of the last century, many ethologists suggested that behaviors displayed among so-called "lower" invertebrates or vertebrates – such as some crustaceans or amphibians, respectively – are predominantly innate and rigid in their performance (e.g., [1]). Moreover, responses to behaviorally significant sign stimuli should proceed reflex-like, for example, the lobster's *tailflip-reflex* in response to a threatening stimulus (e.g., [2]) and the toad's *snap-reflex* in response to a prey stimulus (e.g., [3]). In detail it was assumed, that a sign [key] stimulus activates an *innate releasing mechanism* (IRM), that – much like a safe is unlocked by a key – releases the corresponding action pattern (e.g., [4, 5]). According to the IRM concept, (re-)cognition and motor skills are innate. Neurophysiologically, the concept of IRM suggests an inborn sensorimotor interface that translates perception into action: recognizing ("detecting") the sign stimulus at its afferent input-side and activating ("commanding") the corresponding motor system at the efferent output-side. Hence, the concepts of "feature detector" and "command neuron" were born (e.g., [6, 7]). Theoretically, the simplest innate releasing mechanism would consist of a command neuron, CN, operating in a chain like:

sign-stimulus → **CN** → motor pattern generator → behavior

In this case, the commanding neuron should be a cell specifically tuned to the feature of the sign stimulus, and the axon of that neuron should have access to the motor pattern generating network.

Meanwhile, the concepts of command neuron and innate releasing mechanism were revisited and revised from many points of view (e.g., [8–13]). First of all, appetitive and consummative behavioral responses both in so-called lower and higher animals depend on state-dependent modulatory influences (referring to motivation and attention). For example, prey-orienting and snapping in a toad will not be elicited by prey either if the motivation (hunger), or if the attention of the animal is not appropriate. Thus, snapping is not readily elicited anytime, like a reflex, rather it presumes a process of decision-making. As to the analysis of the sign stimulus, regarding its features and its location in space, it was found in toads that different sensory filter systems structured in a parallel-distributed and partially converging fashion cooperate as *command systems*. Furthermore, releasing mechanisms – in so-called lower and higher animals – may take advantage of innate *and* acquired capabilities at different degrees, e.g., depending on the behavior, its intention and goal. Strictly speaking, a revised concept of releasing mechanism considers that environmental and genetic factors contribute to all behavioral responses (e.g., [14–17]). As to the provocative title of this paragraph, it should be rephrased: organisms are born with knowledge and can extend their knowledge by learning – regarding perception (cognition) and action (motor skills) – in adaptation to the different ecological constraints and requirements of the species. Prefixes in terms of "lower" and "higher" may be less appropriate in such context.

In vertebrates, the plasticity of sensorimotor processes significantly involves the functions of various forebrain structures.

Functions of the amphibian forebrain in terms of modulation, modification, specification

The significance of the functions of the amphibian forebrain for motivated behaviors were disputed vehemently in former times. Whereas Schrader [18] and Johannes [19] suggested from brain lesion studies that the forebrain has no influence on frog's behavior, Goltz [20], Blankenagel [21] and Diebschlag [22] suggested that the telencephalon is essential for the frog's spontaneity and for the release of prey catching. Actually, from a developmental point of view, the forebrain as the *prosencephalon* includes both the telencephalon and the diencephalon. The localized forebrain lesion studies by Ewert [23] thus showed that the opinions of both parties of researchers were correct in a certain sense. In the present review we provide ample evidence that various telencephalic and diencephalic structures – such as ventral striatum, ventral medial pallium, pretectal thalamus – interact with retinorecipient brainstem structures in different ways regarding the modulation of attention and the specification or modification of stimulus recognition.

Both amphibians and mammals share many homologous forebrain structures (e.g., [24–35]). More generally, studies investigating the connections of the basal forebrain provide strong evidence that tetrapod vertebrates share a common pattern of basal ganglia organization [28–32, 35–38]. This concerns dorsal and ventral striato-pallidal systems, reciprocal striatopallidal/diencephalic, and striatopallidal/mesencephalic connections, and descending striato-tectal, striato-pretecto-tectal, striato-nigro-tectal, and striato-reticular pathways. Whereas the anuran striatum is homologous to a portion of the amniote basal ganglia [33–35], the ventral medial pallium is homologous to the mammalian hippocampus [39], thus constituting a significant component of the limbic system (for recent data on the cytoarchitecture and the connectivity of the amphibian medial pallium see Roth and Westhoff [40]). For a discussion of pallial and basal ganglionic homologies in vertebrates we recommend Northcutt and Kass [27], Marín et al. [33, 34], and González et al. [35]. Studies concerning the chemoarchitecture of basal ganglionic limbic and mesolimbic structures are summarized by Marín et al. [28, 31–33, 41] and Marín and González [42].

In the present review we focus on forebrain interactions in visual *perception* (prey recognition) and *action* (prey catching) in anuran amphibians with special consideration of the neurotransmitters/modulators dopamine and neuropeptide Y.

Prey-recognition in the toad's brain involves parallel distributed and converging processing

Visual key feature in decision-making "prey vs nonprey"

If one watches the behavior of a toad in a terrarium, at a first glance the toad seems trying to prey on any moving object provided the object is not too big. For a long time it was commonly accepted that a toad interprets relatively "small" moving objects as prey and relatively "large" moving objects as predator (e.g., [43]). However, investigating the toad's prey catching activity quantitatively in an experimental paradigm (Fig. 1), in which different pieces of cardboard traverse the toad's visual field at

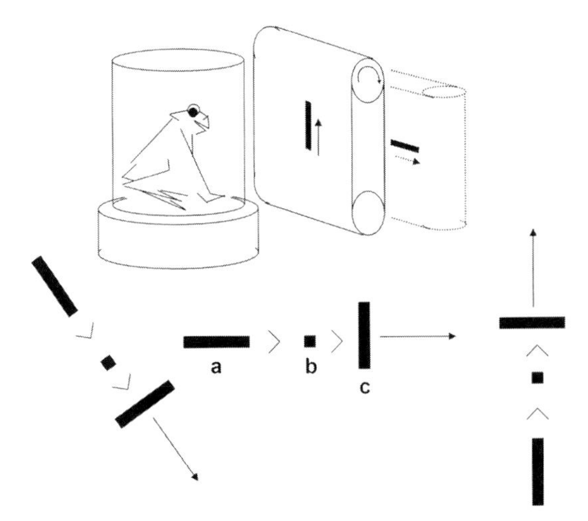

Figure 1. Prey selection. Principle of an experimental procedure suitable to investigate figurative prey selection in the common toad, *Bufo bufo*. The visual stimulus consists of a piece of rectangular cardboard attached to a moving belt in front of a toad sitting in a cylindrical glass vessel. The belt, carrying the stimulus, was moved to and fro at a velocity of $v = 25$ mm/s. The number of orienting and snapping responses towards the stimulus per time interval (30 s) determines the releasing value of the stimulus as prey. The direction of movement of the belt can be adjusted. Note that the configurative preferences **a** > **b** > **c** are independent of the direction (arrow) in which the stimulus traverses the toad's visual field. (Compiled from [46] with kind permission of Karger.)

constant speed, the toad reveals its capability of discriminating small objects figuratively [44, 45]. Actually, a moving square object of about 2.5 mm $\times$ 2.5 mm in size is treated as prey. A small bar – say 2.5 mm $\times$ 30 mm or 2.5 mm $\times$ 40 mm in size – elicits even stronger prey catching behavior, provided the bar moves in the direction of its longer axis. However, the bar is ignored as prey if its longer axis is oriented across the direction of movement, i.e. its configuration is altered. These configurative assignments – independent of (invariant under) the speed and the direction in which the bar traverses the toad's visual field (Fig. 1) – are decisive for the distinction between prey and nonprey [46].

More generally, among moving small objects toads determine prey by an analysis of two main figurative visual features: the object should have a reasonable length lp (parallel to the direction of its movement) and should have a relatively short width la (across the direction of its movement): $lp > la$. If objects are compact ($lp = la$), more circular or square-shaped, than the area $lp * la$ plays a decisive role, whereby the optimal diameter of an object suitable for prey catching corresponds to about 43% of the width of a toad's mouth.

The invariant principle of configurative prey selection is common (universal) among the investigated anuran genera; however, it displays species-specific variation [47]. Other visual cues, such as the direction of contrast (bright object against a dark background or *vice versa*) and other senses (olfactory, gustatory, somatosensory) may contribute to the recognition of prey.

A basic network hypothesis on prey recognition

How is prey recognized in terms of neurophysiological analysis? Single neuron recordings in response to behaviorally significant stimulus features suggest that the toad's neuronal network responsible for the recognition of prey and the neglect of nonprey involves parallel-distributed processing of visual input from the retina to its destination fields, optic tectum, and pretectal thalamus, and the interaction of these processing streams (e.g., [13, 44, 45]):

a) In the retina of common toad different types of ganglion cells (classes R2, R3, R4) provide a preprocessing of fundamental visual stimulus parameters: stimulus angular size, velocity of movement, stimulus-background-contrast, brisk change in diffuse illumination. The angular sizes of moving objects preferred by R-type neurons increase from R2 to R4 due to the different diameters of their excitatory receptive fields ($ERF_{R2} \sim 4°$diam, $ERF_{R3} \sim 8°$diam, $ERF_{R4} \sim 12°$diam). The movement-sensitivity increases from R4 to R2, the latter being movement-specific (for results in frogs see [48]).
b) Retinal signals are processed in the optic tectum by tectal T5-type neurons (subtypes T5.1; T5.2; T5.3; T5.4); the diameters of their ERFs are between 20 and 35°. These neurons display different sensitivities to the aforementioned figurative features, for example, T5.1 neurons are sensitive to the stimulus area and preferably to its length (lp) *parallel to* the direction of movement.
c) Retinal signals are processed in parallel in the pretectal thalamus, its lateral posterodorsal (Lpd) and lateral dorsal posterior (P) nuclei, by TH-type neurons.

Type TH3 neurons (ERF $\sim$ 40°–50°) and TH4 neurons (ERF $>$ 90°) display a sensitivity to the stimulus area and preferably to its length (la) *across* the direction of movement.

d) Discrimination of the stimulus features lp and la can be explained by an interaction of the described retinotectal and retinopretectal processing streams. It is hypothesized [13, 44] that excitatory ($\rightarrow$) retinotecto-tectal and inhibitory ($\vdash$) retinopretecto-tectal influences converge in the optic tectum to determine the response property of figurative selective tectal T5.2 neurons:

$$\boxed{\text{R2, R3} \rightarrow \text{T5.1} \rightarrow \textbf{T5.2} \vdash \text{TH3} \leftarrow \text{R3, R4}}$$

What does this mean? Towards a small bar in prey configuration ($lp > la$), the retinotectal activation will be stronger than the retinopretectal activation, so that T5.2 cells will be strongly excited and the toad responds with prey catching. Towards the same bar in nonprey configuration ($la > lp$), conversely, retinopretectal activation will be stronger than retinotectal, so that pretectal inhibitory influences on T5.2 will override tectal excitatory influences and the object will be ignored as prey. Towards a compact object ($la = lp$) of a large area $lp * la$ signalling predator, the retinopretectal activity and thus pretectal inhibitory influences on T5.2 will be strong, so that prey catching fails to occur while the escape system is excited.

We hasten to emphasize that this simple scheme is only acceptable with the restriction that it illustrates a basic idea. By no means should it imply (cf. [11, 16]) that a T5.2 cell operates as a *command neuron*. Actually, we have shown in detail that the toad's "releasing mechanism" of prey catching is relatively complex and involves various feature monitoring neuronal systems partly organized in retinotopic visual maps (cf. concept of *command releasing systems*, see [12, 49]). Applying the antidromic stimulation/recording method at the criterion of the collision test, Satou and Ewert [50] and Ewert et al. [51] demonstrated that T5.2 as well as other visual feature monitoring cells could be backfired by electrical stimuli applied to the medulla oblongata. This proves that such cells project their axons towards the bulbar/spinal motor systems. In certain combinations ("sensorimotor codes") they are suitable to shape complex sensorimotor interfaces.

Since the extension of an object across its direction of movement (la) provides the decisive "figurative key feature" in the decision-making prey *vs* nonprey, we shall focus on the putative pretectotectal inhibitory influences.

Test of the hypothesis: the importance of pretectotectal inhibitory influences

If figurative prey selection actually depends on inhibitory influences from pretectum to tectum, then tectal visual responses should be "disinhibited" and prey selection should be abolished after pretectal lesions. This can be tested [23, 45, 52]. Indeed, following pretectal lesions, the toad interprets any moving object – irrespective of its size and configuration – as prey, both neuronally (e.g., expressed by disinhibited responses of T5.2 cells) and behaviorally (shown by disinhibited prey catching). This was most convincingly demonstrated by single cell recordings with a chronically implanted electrode in free-moving common toads by checking for correlations between

- stimulus features prey *vs.* nonprey
- discharge activity of a tectal T5.2 neuron
- prey catching behavior

prior to an ipsilateral pretectal lesion and thereafter [53, 54]. Towards a small bar traversing the neuron's ERF in prey configuration, a strong burst of spikes of the recorded T5.2 neuron preceded, and so to speak, predicted, the onset of the toad's prey catching orienting or snapping, respectively. If the same bar moved in nonprey configuration, it elicited very weak spike activity and prey catching failed to occur. Shortly after an electrolytic lesion to the pretectal thalamic Lpd/P region – applied with a second implanted electrode – the same T5.2 neuron discharged strongly either to the prey or nonprey stimulus, and the toad responded hyperactively to either stimulus with prey catching. In addition the ERF of the T5.2 neuron increased post lesion. Thus, the pretectal lesion impaired the capability to distinguish between prey and nonprey, both neuronally and behaviorally (see also [55]).

Various studies suggest that the excitatory retinal input to the pretectal thalamic Lpd nucleus is glutaminergic. The fact that administration of glutamate to the Lpd nucleus leads to an attenuation of tectal field potentials evoked by electrical optic nerve stimulation is consistent with pretectotectal inhibitory influences. Administration of the glutamate agonist kainic acid or ibotenic acid to Lpd – each drug acting on cells as an excitotoxin, thus silencing them – leads to a visual disinhibition of tectal neurons and to disinhibited prey catching [54].

Pretectotectal connections are mediated by TH3 neurons

What does pretectum tell its tectum? In immobilized toads, *Bufo americanus*, different types of pretectal thalamic TH-type neurons were recorded extracellularly in response to visual stimuli ([45, 56]; for morphological features see [57, 58]). Most of these neurons were activated by moving stimuli known to elicit various types of avoidance behavior, such as ducking, turning away or jumping (in response to a moving predator) or sidestepping and detouring (in response to a stationary obstacle). Interestingly, such patterns of avoidance could be activated in free-moving toads by focal electro-stimulation of caudal dorsal thalamic areas [45].

Testing the descending character of TH-type cells with the aid of the antidromic stimulation/recording method, TH3 and TH4 cells could be backfired by electrical stimuli applied to the ipsilateral dorsal optic tectum [59]. The recording sites of these cells were identified in the Lpd and lateral P nuclei and the lateral posteroventral (Lpv) pretectal thalamic nucleus. Since both types of pretectotectal projection neurons, TH3 and TH4, responded preferably to visual stimuli resembling threat (e.g., a bar oriented across its direction of movement) or predator (e.g., a moving big square), their signals to the optic tectum may communicate "be cautious", i.e. "don't approach that stimulus". In addition stationary object detecting TH10 Neurons [45, 56] of a subdivision of the P nucleus – projecting to motor systems that elicit detour behavior [60, 61] – may inform the tectum on obstacles. A mode of addressing such messages to the tectum is inhibiting tectal structures responsible for prey catching. Which neurotransmitters/neuromodulators are involved in such "alarm channels"?

Pretectotectal influences mediated by neuropeptide Y

NPY attenuates retinotectal transfer
The investigations on neuropeptide-like immunoreactivity in the frog's brain substantiate pretectotectal projections that involve neuropeptide Y (NPY) as neurotransmitter/modulator [62–65]. Kozicz and Lázár [66] convincingly showed in *Rana esculenta* that NPY immunoreactive fibers in the superficial tectum (layer 9: lamina C dense, D-F sparse) originate from the ipsilateral Lpd and Lpv nuclei, i.e. areas of the recording sites of TH3 and TH4 pretectotectal projecting cells [59]. In addition, Tuinhof et al. [67] described in *Xenopus laevis* a NPY-mediated projection of the pretectal thalamic P nucleus to deeper tectal layers 2–6 and 8.

We are thus tempted to suggest that activation of TH3 and TH4 pretectotectal projecting neurons leads to the release of NPY in the superficial tectum. We tested in cane toads, *Bufo marinus*, the effects of the activation of pretectotectal efferents – or of the administration of NPY to the tectal surface – on tectal visual responses:

Experiment-1. Tectal field potentials, evoked by electrical stimulation of the contralateral optic nerve were recorded from an area of the tectal surface. The pretectal Lpd/P region – ipsilaterally to the tectal recording site – could be stimulated. If pretectal stimulation preceded the optic nerve stimulation by a delay of about 15 ms, the excitatory N1 wave of the tectal field potential was strongly attenuated (Fig. 2B) [68].

Experiment-2. Tectal field potentials were evoked by electrical stimulation of the contralateral optic nerve and 10^{-4} mol/l of porcine NPY was administered to the tectal surface at the area from where the field potentials were recorded. After NPY administration the excitatory N1 wave of the tectal field potential was strongly attenuated, reaching its minimum after an incubation time of 20–30 min (Fig. 2A). After removal of the NPY-pipette, the effect of NPY faded (for differences in retention times between frog NPY and porcine NPY see [63].

Experiment-3. Tectal field potentials were evoked by brisk changes in the diffuse illumination at light-*off* or light-*on* and synthetic porcine NPY was administered to the tectal surface at the area from where the field potentials were recorded. After NPY administration the excitatory *off*- and *on*-responses of the tectal field potentials were strongly attenuated. The effect of NPY faded in the course of time. Testing NPY fragments in this procedure, it was found that NPY13-36 (Y_2 receptor-agonist), but not NPY18-36 (Y_2 receptor-antagonist), attenuated the excitatory *off*- and *on*-responses of the tectal field potentials [69]. Y_2 receptor mediated NPY effects are known to presynaptically inhibit glutamate release [70]. In fact, spike amplitudes of single axon terminals from retinal R2 and R3 ganglion cells, in response to moving objects, were reduced under NPY.

Experiment-4. Studies examining the regional cerebral glucose utilization by means of the ^{14}C-2-deoxyglucose ^{14}C-2DG imaging method in red-bellied toads, *Bombina orientalis*, showed strong ^{14}C-2DG uptake in the superficial layers of both lobes of

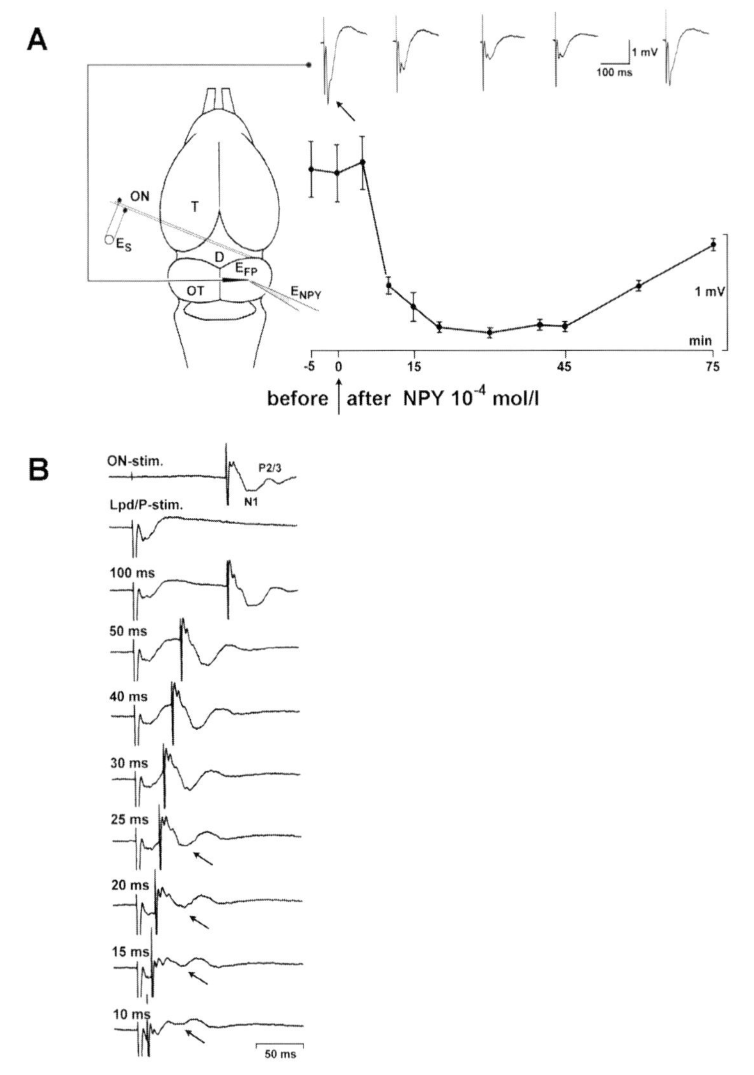

Figure 2. Pretectotectal control via neuropeptide Y (NPY). (A) Influence of (NPY) administered to the surface of the cane toad's, *Bufo marinus*, optic tectum on field potentials recorded from the tectal surface in response to electrical stimulation of the contraleteral optic nerve. The amplitude of the excitatory N1 wave of the field potential (see arrow, upper panel) is attenuated by NPY. The field potentials are assigned to the corresponding graph below (averages ±SDM, $n = 6$ immobilized toads). D, diencephalon; ON, optic nerve; OT, optic tectum; T, telencephalon; E_S, stimulation electrode; E_{FP}, field potential recording electrode; E_{NPY}, capillary with NPY. (B) Influence of electrical pretectal Lpd/P-stimulation on a tectal field potential evoked by electrical optic nerve ON-stimulation; the former stimulation preceded the latter by a variable delay. Note the strong attenuation of the N1 wave of the field potential at a delay of 10 to 25 ms. Representative example of an immobilized cane toad; *Bufo marinus*. (After [68]; figure reprinted with kind permission of Elsevier.)

the optic tectum when both eyes were stimulated with a prey dummy [71]. More specifically, Fig. 3Aa shows the distribution of ^{14}C-2DG uptake in the dorsal tectal laminae of a pharmacologically untreated fire-bellied toad towards a prey-dummy presented to each eye. The uptake was similar in both tectal lobes (representative example of $n = 10$ animals). In contrast Fig. 3Ab demonstrates that – in the same stimulation paradigm – unilateral administration of porcine NPY to the surface of the left tectal lobe (where a small agarose gel-pad soaked with NPY was applied to the tectal surface) diminished ^{14}C-2DG uptake in the left dorsal tectum. The left-to-right differences in ^{14}C-2DG uptake are statistical significant ($p < 0.01$; sign test; $n = 5$). Quantitative data are presented in Figs. 3Aa′ and Ab′, respectively [71]. Behavioral observations showed that animals ignored prey objects presented to the (right) eye corresponding to the NPY-treated contralateral (left) optic tectum. Control experiments in which animals were prepared with tecta exposed and where agarose gel-pads lacking NPY were applied to the tectum unilaterally, revealed no statistically significant differences in ^{14}C-2DG uptake if both tecta were compared. This shows as well that the reduction in ^{14}C-2DG uptake in the above experiments is caused by NPY rather than by the physical presence of the gel-pad. Comparative experiments in cane toads *Bufo marinus* yielded comparable data.

In an extension of the above experiment, fire-bellied toads received after the ^{14}C-2DG injection a subcutaneous injection of the dopamine agonist apomorphine (APO) at a dose of 50 mg/kg body weight. Systemically applied APO is known to strongly boost retinal input to the optic tectum (cf. Section *Neurophysiological recordings: APO boosts retinal visual responses*). As a result of APO treatment, both tecta showed very strong ^{14}C-2DG uptake (Fig. 3Ba, without NPY). In contrast Fig. 3Bb demonstrates that – in the same stimulation paradigm under APO treatment – unilateral administration of NPY to the surface of the left tectal lobe strongly attenuated ^{14}C-2DG uptake in the left dorsal tectum. With unilaterally applied NPY the left-to-right differences in ^{14}C-2DG uptake are statistically significant ($p < 0.01$; $n = 10$) [71].

Developmental aspects on pretectal thalamic tuning functions

In the evaluation of pretectal thalamic structure and function, the available developmental neuroanatomic studies in anuran amphibians are of particular interest. Clairambauld [72] provided evidence that the dorsal thalamus starts to pacellate/differentiate into Lpd and P before metamorphosis and is completed six months to one year thereafter. This brings up the question of the timed action of NPY. In fact, D'Aniello et al. [73] showed that the NPY immunoreactivity in frog caudal dorsal thalamic cells occurs in tadpoles, becomes very conspicuous in advanced larval stages and shows maximal values during and after metamorphosis.

Following the parcellation theory by Ebbesson [74], the ontogenetic parcellation of the caudal dorsal thalamus in Lpd and P nuclei should result in a finer tuning of the circuits involved in visual computation.

By comparison with the structural/pharmacological changes, the toad's preference for food actually alters with metamorphosis. Tadpoles are vegetarian (*Bufo*) or vegetarian with carrion (*Bombina*), whereas the juveniles and adults of both gen-

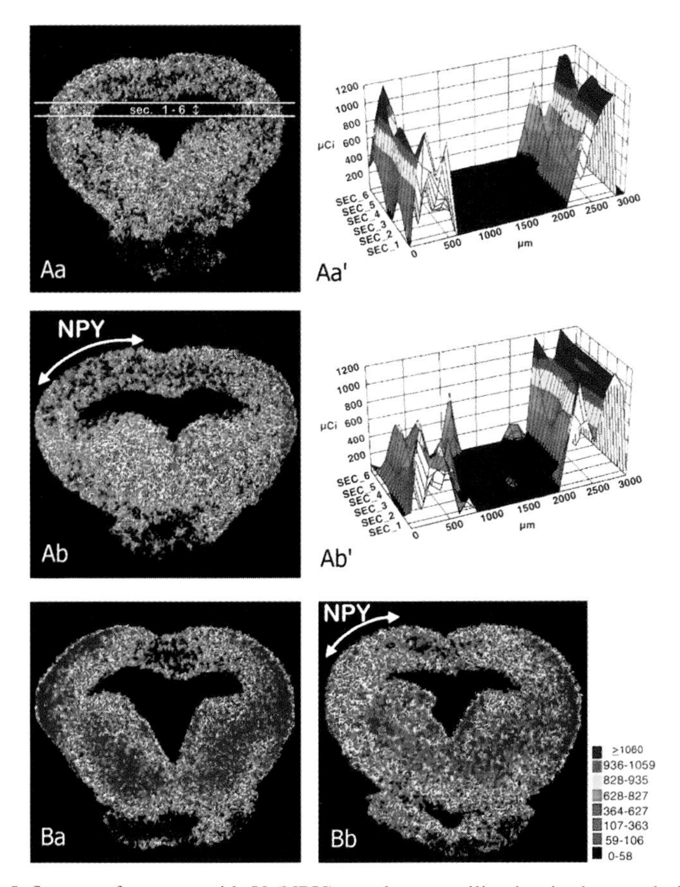

Figure 3. Influence of neuropeptide Y (NPY) on glucose utilization in the tectal visual map. (Aa) and (Ab) Color-coded autoradiographic images of transverse sections through the medial midbrain of *Bombina orientalis* illustrating the regional distribution of ^{14}C-2DG uptake (after its subcutaneuos injection at a dose of 0.23 μCi/kg body weight) in response to visual prey stimulation of the left and right eye, respectively. Prey dummy: 2×24 mm^2 horizontal bar of black card board – for a period of 45 min – mechanically moved in the horizontal plane at $v = 8°$/s in the left and right visual field of the animal sitting in a cylindrical glass vessel; the direction of movement was reversed every 2 min. In anaesthetized animals the brains were removed from the skull, cross-sectioned on a cryostat microtome, and exposed to x-ray film for 21 days in combination with eight calibrated [^{14}C]methyl-metacrylat standards. With the aid of a computer-assisted image analyzing system the average ^{14}C-concentration values of each brain region were measured, corrected (clearfield equalization) and converted into a false-color code image. (Aa′) and (Ab′) Applying the SigmaScan Pro5 software, histograms are demonstrating the local change of radioactivity (μCi) across six equidistant (15 μm) horizontal virtual lines through transverse sections of both tectal lobes. (a,a′) No NPY administration; (b,b′) after administration of NPY (applied in a cube agarose gel-pad of 0.5×0.5 mm^2) to the surface of the left optic tectum; NPY was dissolved in frog-Ringer at a concentration of 10^{-4} mol/l. (Ba) and (Bb) Comparable experiments in *B. orientalis* under systemic treatment of apomorphine at a dose of 50 mg/kg body weight. Representative examples. (From [71]; figure reprinted with kind permission of Elsevier.)

era are carnivorous and prefer living prey, i.e. objects that move. The configurative principle of prey selection – detecting prey by its shape relative to its locomotive direction – is an interesting adaptation to the transition to terrestrial life. The fine tuning in prey selection – acuity in configurative discrimination and estimation of absolute sizes – is subject to a maturation process that begins during the first postmetamorphic days, independent of experience with prey [75, 76], and is fully established about 6 months to 1 year after metamorphosis [77, 78].

Other inhibitory influences on tectal neurons

We emphasize that other neurotransmitter/modulator systems are involved in the inhibitory control of retinotectal and intratectal excitatory transmission (cf. also [79]). Kozicz and Lázár [80] provided evidence of co-localizations of NPY, enkephalin and GABA in the optic tectum. Furthermore, a pretectotectal inhibitory dopaminergic transmission may exist. The pretectal thalamic P nucleus – containing dopamine immunoreactive cells [81] obtains direct retinal input [82–84] and projects to the optic tectum [85–87].

Since focal administration of the acetylcholine antagonist curare to the surface of the optic tectum in frogs [88] or toads [52] leads to a visual disinhibition of prey catching behavior accompanied with an impairment of prey/nonprey discrimination – in a manner similar after pretectal lesions – a nicotinic cholinergic process may be involved. Interestingly, the nucleus isthmi obtains visual input and feeds back cholinergic output to the optic tectum [89–92]. It is hypothesized that cholinergic drive from nucleus isthmi activates presynaptic nicotinic receptors of retinal terminals in the superficial tectum, thereby initiating a process that facilitates the release of glutamate onto GABAergic inhibitory tectal interneurons.

Modulating pretectotectal inhibition by associative learning

The genus-universal figurative prey selection in anurans is based on an evaluation of the stimulus features ℓp and ℓa and the area $\ell p * \ell a$. If these features are within a certain range, the object signals prey. We suggest that during evolution, objects extended along ℓa (or with large $\ell p * \ell a$ areas) analyzed particularly in the pretectal thalamus, became linked with threat. If this linkage – ultimately stored in pretectotectal connections – were broken off, the discrimination between prey and threat should be impaired. In the following we show that this linkage can be broken off in a learning process where a large moving area is associated with prey, the so-called hand-feeding paradigm:

If an experimenter offers a mealworm (unconditioned stimulus, US) to a toad by hand, the mealworm will be snapped. If mealworm and hand are presented on consecutive days, the hand that is initially threatening comes to be associated with the prey. Then the hand, as conditioned stimulus CS, will alone come to elicit prey catching behavior [78, 93]. In other words, the hand that initially offered the prey, finally itself is treated as prey. This conditioning is generalized, so that other nonprey items such as a large black moving square or a bar moving in nonprey configuration will elicit prey catching, too.

A hand-conditioned toad thus behaves similar to a toad after a lesion in the pretectal thalamic Lpd region. Actually, hand-conditioned toads – whose cerebral glucose utilization was mapped with the ^{14}C-DG imaging method – showed very weak ^{14}C-2DG uptake in the pretectal thalamic Lpd nucleus when stimulated with a predator object. Comparably the Lpd of unconditioned toads displayed strong ^{14}C-2DG uptake toward the predator stimulus [94, 95].

What silenced structures of Lpd in the course of hand-conditioning? We hypothesize that during conditioning, the posterior ventral medial pallium (vMP) is involved in a process that reduces pretectotectal inhibitory influences. More specifically, during repetitive combined presentation of a prey stimulus (mealworm) and a threatening stimulus (experimenter's hand) the information related to each kind of visual stimulus coincide in the vMP:

$$\boxed{\text{mealworm} \rightarrow \text{R} \rightarrow \text{OT} \rightarrow \text{A} \rightarrow \textbf{vMP} \leftarrow \text{A} \leftarrow \text{OT} \leftarrow \text{R} \leftarrow \text{hand}}$$

(R= retina, OT= optic tectum, A= anterior thalamus). This may sensitize vMP neurons which – becoming responsive to threat features – would alter prey selective properties of tectal neurons by a disinhibitory pathway via A and Lpd:

$$\boxed{\textbf{vMP} \dashv \text{A} \rightarrow \text{Lpd} \dashv \text{OT}}$$

The putative inhibitory influence of vMP on A probably develops in the course of conditioning. There is experimental evidence that the posterior vMP is significantly involved in hand-conditioning:

(1) Mapping the glucose utilization in the brains of hand-conditioned toads, which were preying large objects, showed increases of ^{14}C-2DG uptake in the vMP, the lateral amygdalae, and the medial layers mOT, whereas the pretectal Lpd nucles displayed decreases, by comparison of unconditioned controls [94, 95]; cf. [96].
(2) After bilateral vMP lesions in hand-conditioned toads the original genus-universal configurative prey selection capabilities re-emerged, demonstrating that the forebrain loop responsible to mediate the modification processes acquired during hand-conditioning was not functioning anymore. The table 1 below shows the average prey catching rates (responses per 36 s ±SDM, $n = 10$ toads) of *Bufo marinus* towards three configurative different stimuli moving at $v = 10°/\text{s}$ – (a) prey, (b) nonprey, (c) threat – prior to hand-conditioning, after hand-conditioning and after hand-conditioning with bilateral vMP lesions [97]:

Table 1. Prey catching rates of *Bufo marinus* towards different visual stimuli (a)–(c): prior to hand-conditioning, after conditioning and after conditioning following vMP lesions

Stimulus	Prior to conditioning	Conditioned	Conditioned, vMP lesions
(a) $5 \times 50\,\text{mm}^2$	19 ± 2	21 ± 3	18 ± 2
(b) $50 \times 5\,\text{mm}^2$	0 ± 0	6 ± 1	0 ± 0
(c) $50 \times 50\,\text{mm}^2$	0 ± 0	17 ± 3	0 ± 0

Directed attention: gating prey catching by a forebrain loop involving the striatum

A hungry toad is not necessarily attentive and therefore not responsive to prey. The readiness of a toad to respond to a prey stimulus is subject to state-dependent determinants, such as attention and motivation (e.g., [98, 99]). The brain modulates its own responsiveness (see [100–102]). This involves decision-making. One such mechanism is responsible for *directed attention* and thus for gating the translation of perception (prey recognition) into action (prey catching) in an effort to localize the stimulus source.

Forebrain structures are essential for survival

After bilateral ablation of the telencephalic hemispheres *and* the diencephalic rostrocaudal dorsal thalamus, toads respond to moving stimuli with prey catching [23, 45]. Are these forebrain structures dispensable for prey catching? Although these structures are not serially connected to the retino-tectal/tegmental/bulbar/spinal processing stream, if toads are lacking those forebrain structures, their prey catching responses to moving stimuli display various peculiarities: they react *not selectively* (no stimulus discrimination), they react *readily* (no precautious hesitation), and they react *continually* (no stimulus habituation). Hence, these forebrain structures are involved in *recognition*, *attention*, and *learning* and are thus essential for survival.

The previous section showed that diencephalic pretectal thalamic areas of the forebrain are involved in prey recognition and that telencephalic ventral medial pallial areas are involved in the modification of prey recognition by learning. The present section will focus on forebrain structures participating in the toad's directed attention, i.e., its readiness to respond to a prey stimulus.

Focusing on influences of the telencephalon, it was shown that after total ablation of both hemispheres in toads or frogs, prey oriented turning behavior failed to occur [23, 45, 103]. Unilateral telencephalic ablation led to a neglect of prey moving in the visual field of the contralateral eye. Since small lesions to the toad's or frog's caudal ventral striatum (vSTR) – but not, for example, to the ventral medial pallium – also showed this prey neglect [94, 104], it was concluded that the visual release of prey oriented behavior depends on stimulating caudal striatal areas that gate the corresponding ipsilateral tectal output to the motor systems. This is in agreement with data from experiments in which caudal ventral striatal areas were stimulated electrically with implanted electrodes: a train of electrical impulses did not elicit a specific behavioral response in a toad, however, it facilitated prey-oriented turning towards a visual stimulus that – without striatal stimulation – was subliminal [45].

Disinhibition: properties of a striato-pretecto-tectal pathway

At a first glance, the lesion studies taken together are hard to understand, since bilateral striatal ablation leads to the neglect of prey, whereas bilateral striatal *and* pretectal thalamic ablations stimulate the toad's readiness to catch prey. This "dilemma" perfectly illustrates the historical controversy between Schrader [18] and Diebschlag [22] mentioned in the introduction.

How may striatum influence tectum?
Neuroanatomic studies by Marín et al. [28] showed that striatal efferents to the optic tectum may be acting *via* a direct (monosynaptic) and various indirect (di-/polysynaptic) pathways. Among these the characteristic of the disynaptic striato-pretecto-tectal pathway (cf. also [36, 37]) offers interesting perspectives with respect to the available neurobehavioral data. For example, studies mapping the local cerebral glucose utilization showed a positive correlation between the toad's visually elicited prey catching orienting activity and increases in ^{14}C-2DG uptake in the caudal ventral striatum and the optic tectum, whereas the pretectal Lpd nucleus revealed a decrease [96, 105]. Accordingly, in immobilized toads electrical stimulation of the ascending reticular formation evoked a comparable pattern of ^{14}C-2DG uptake in the three brain structures [106].

If the decrease in ^{14}C-2DG uptake in the Lpd nucles resulted from inhibitory striatal influences, this might suggest that prey catching can be gated by "double inhibition" involving two sequential inhibitory pathways (⊣), a striato(STR)-to-pretectal(Lpd) and a pretecto(Lpd)-to-tectal(OT) one:

STR ⊣ Lpd ⊣ OT → prey catching

Such putative disinhibitory pathway would explain the apparently controversial results of forebrain lesion studies. The following paragraph examines striatopretectal inhibitory connections.

Synaptology: striatal stimulation evokes pretectal inhibitory postsynaptic potentials
Lázár and Kozicz [107] showed that striatal efferents travelling in the lateral forebrain bundle, LFB, terminate in the ipsilateral pretectal Lpd nucleus. Immunocytochemical studies demonstrated that LFB fibers terminating in the pretectum contain the inhibitory neurotransmitter/neuromodulator enkephalin [62, 108]. Distinct sets of striatal GABAergic cells contain enkephalin, too [109].

Intracellular recording combined with Cobalt-lysine labeling showed that visually sensitive pretectal cells respond to electrical stimuli applied to the striatum (Fig. 4A) with inhibitory postsynaptic potentials (IPSPs) (Fig. 4Ba) or combined IPSPs and excitatory potentials (EPSPs), however, no pure EPSPs were observed. The fastest striatal input to pretectal cells was inhibitory [110]. These data indicate that pretectal activity can be regulated by inhibitory or inhibitory and excitatory striatal input. The ventral striatum itself obtains inputs from various structures, such as the posterior lateral pallium, the anterior entopeduncular nucleus, and the lateral anterior thalamic nucleus which itself receives tectal input [25, 28, 107].

Since striatopretectal fibers contribute to the LFB, pretectal postsynaptic potentials were recorded intracellularly also to electro-stimulation of the LFB at the level of the rostral diencephalon. These responses were IPSPs and/or EPSPs; the majority of pretectal cells displayed only IPSPs that mostly showed short latencies (Figs. 4Bb,

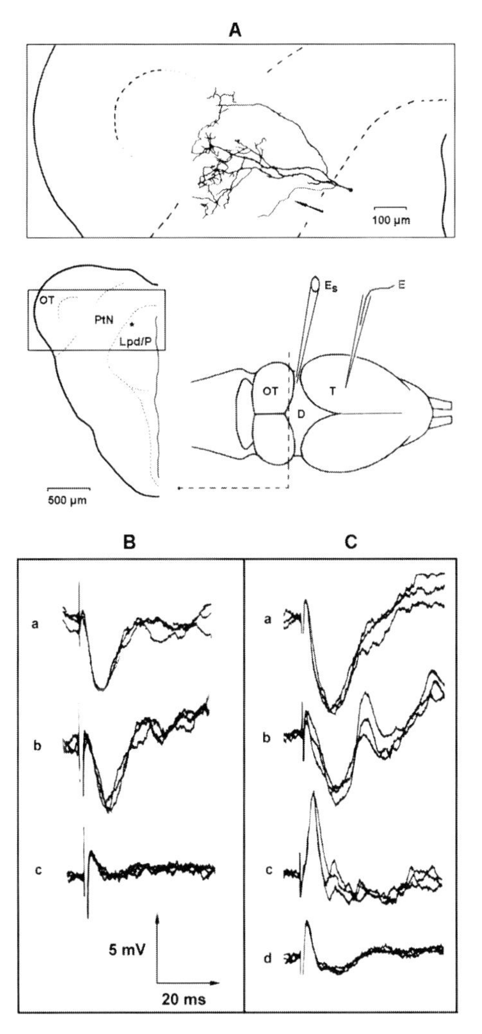

Figure 4. Striatopretectal inhibitory connections. Inhibitory postsynaptic potentials recorded from pretectal thalamic cells. Representative examples in immobilized cane toads, *Bufo marinus*. (A) Intracellular recording and Co^{3+}-lysine staining a pretectal thalamic neuron of the Lpd/P region; camera-lucida reconstruction (arrow points to the axon). E_S, electrical stimulation electrode; E, recording electrode; Lpd/P, lateral posterodorsal and lateral posterior thalamic nucleus; Ptn, pretectal neuropil; D, diencephalon; OT, optic tectum; T, telencephalon. (B) Records of inhibitory postsynaptic potentials in response to electrical stimulation of (a) the ipsilateral caudal ventral striatum or (b) the ipsilateral lateral forebrain bundle LFB at diencephalic level; (c) field potential in response to LFB-stimulation. (C) Example of testing the nature of hyperpolarizing activity by passing current through the recording electrode. Effect of intracellularly applied current on the inhibitory postsynaptic potential of a pretectal cell in response to LFB-stimulation: (a) depolarizing current (+2 nA) enhanced the inhibitory wave; (b) control, no current application; (c) hyperpolarizing current (−2 nA) reversed the polarity of the wave; (d) field potential. (After [110]; figure reprinted with kind permission of Elsevier.)

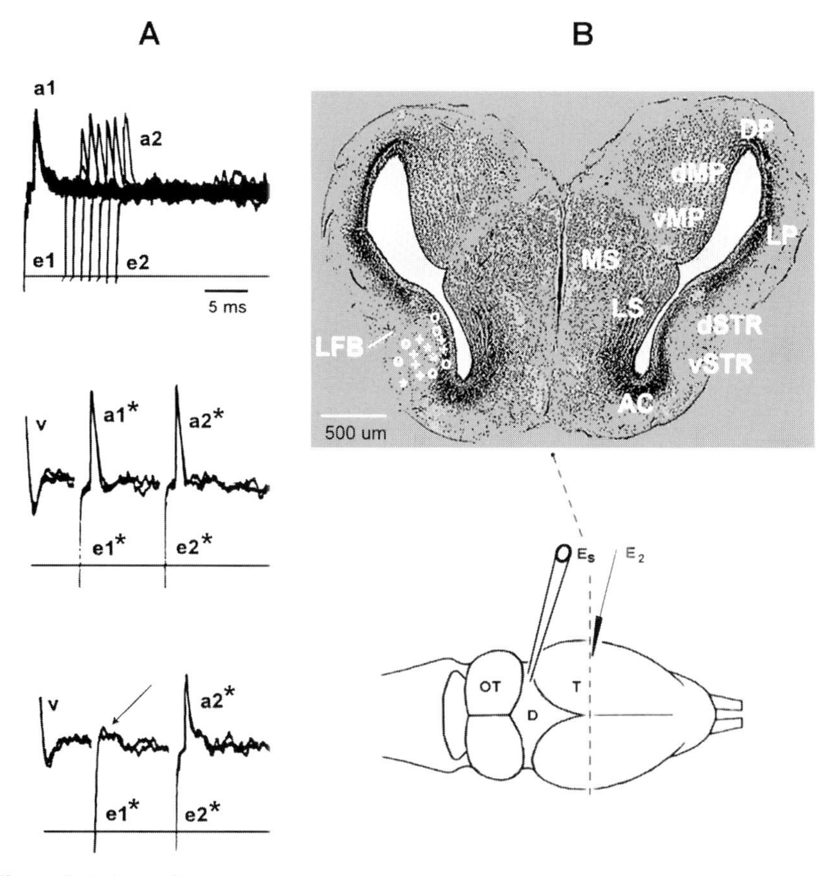

Figure 5. Striatal efferents. Neurons of the caudal ventral striatum projecting their axons in the lateral forebrain bundle LFB were examined by their antidromic activation in response to electrical stimuli, e_1–e_2, applied to the LFB at diencephalic level. Representative examples of extracellular recordings in immobilized cane toads, *Bufo marinus*. (A) Top: To determine the absolute refractory period R_A of a visual "motion detector" neuron recorded from the striatum, its antidromic spikes a_1 and a_2 were evoked in response to e_1 and e_2, whereby e_2 followed e_1 by a variable interval, cf. superimposed records ($R_A = 4.2$ ms; constant latency at $L_C = 2.4$ ms). Note that one a_2-spike is missing in this record because the corresponding e_2-stimulus fell into the absolute refractory period. Middle: The visually elicited spike v triggered two electrical stimuli e_{1*}–e_{2*}, both separated by a constant interval ($> 2R_A$). Each stimulus elicited an antidromic spike a_{1*} and a_{2*}, respectively (two superimposed records). Bottom: Collision test. At a critical delay $D_c = 6.2$ ms between v and e_{1*} the spike a_{1*} was extinguished (cf. arrow) but not a_{2*} serving as control (two superimposed records); however, spike a_{1*} was not extinguished at $D = 6.5$ ms. Bottom, right: Arrangement of recording (E) and stimulation (E_S) electrodes. D, diencephalon; OT, optic tectum; T, telencephalon. (B) Klüver-Barrera stained transverse section through the telencephalon at the level of the caudal ventral striatum. Symbols in the left hemisphere mark recording sites of striatal efferent neurons, + referring to "visual motion detectors". AC, nucleus accumbens; d,vSTR, dorsal/ventral striatum; d,vMP, dorsal/ventral medial pallium; DP, dorsal pallium; LP, lateral pallium; LS, lateral septum; MS, medial septum. (After [111]; figure reprinted with kind permission of Karger.)

Cb) [110]. This suggests a significant monosynaptic inhibition in the pretectum mediated by LFB fibers, to which striatal efferents contribute (Fig. 4Ba). The nature of the PSPs was determined by intracellular injection of depolarizing or hyperpolarizing current, respectively (Figs. 4Ca,c).

What kind of visual information leaves the toad's caudal ventral striatum? Buxbaum-Conradi and Ewert [111] recorded spikes of striatal cells in response to visual stimulation and in response to antidromic electro-stimulation of the ipsilateral LFB at diencephalic level (Fig. 5A). It was shown that most striatal visual cells descending in LFB are visual "motion detectors", i.e. responding readily and very sensitively to objects traversing the toad's field of vision. The excitatory receptive field of such a neuron encompassed the visual field of the contralateral eye or the entire field of vision. Other types of visual neurons – sensitive to "prey," "threat" or "compact" objects – contribute to striatal efferents running in the LFB. The recording sites of all these cells are consistent with neuroanatomic data showing that most efferent striatal neurons are found in the caudal ventral striatum (Fig. 5B) [26]. Lázár and Kozicz [107] characterized the majority of striatal neurons – projecting in the ipsilateral LFB – as piriform and pyramidal cells.

Hypothesis: how striatum may be involved in directed attention

Given that striatopretectal and pretectotectal inhibitory influences exist, the toad's readiness to orient towards prey would depend on striatal activity. The behavior of toads after forebrain lesions can be explained in this context: toads after striatal lesions "permanently hesitate to respond," while toads after pretectal (or pretectal and striatal) lesions are "permanently ready to respond." In intact toads the orienting response towards prey is mediated by the retino-pretectal/tectal/tegmental processing stream; we call this *the stimulus-response mediating pathway*. According to our hypothesis, this processing stream can be gated by striatopretectal activity due to telencephalic intrinsic input and to visual tectal input relayed by the lateral anterior thalamic nucleus [25]. Gating the translation of perception (prey recognition) into action (prey catching) may be executed by a disinhibitory striato-pretecto-tectal loop; we call this *the pathway that modulates stimulus-response mediation*. Striatal channels sensitive to "visual motion" and "prey" may be involved in this modulating loop. Striatal efferent neurons increase their steady tonic discharges if a stimulus is *novel*. Accordingly most striatal neurons show strong habituation towards repeated *familiar* stimuli [111, 112].

This substantiates the postulation by Blankennagel [21] and Diebschlag [22] (cf. also [113]) that orienting responses towards prey are elicited only if (certain) striatal neurons – as components of the basal ganglia – are excited.

Network modulations suitable to modify prey catching patterns

Common toads, *Bufo bufo*, and water frogs, *Rana esculenta*, display different prey catching strategies [1]. Common toads are active foragers; they hunt prey by orienting, stalking, fixating (directed locomotive responses), and snapping (consummative response). Whereas the same prey recognition process precedes each of these responses, the type of response is released depending on the location of prey in space: prey outside the frontal visual field → oriented turning movement; prey within the frontal field of vision, far away → stalking; prey within the frontal field of vision, close → bending forward and binocular fixating; prey in snapping position → snapping and swallowing. In common toads, the preying success takes advantage of the hunter's locomotive mobility but – leaving shelter – at the risk of being itself attacked by predators. In the hunting strategy, the discrimination between prey and nonprey is relatively selective.

Rana esculenta, a sit-and-wait predator, prefers motionless waiting for prey in a shelter at a pond's bank where prey density is relatively high. If prey crosses the visual field, the frog suddenly reacts with an aimed snapping or turning and snapping. In the waiting strategy, the preying success depends on a relatively low snapping threshold at the disadvantage of catching nonprey items occasionally also.

The "hunting" and "waiting" strategies differ mainly in the variety of *directed locomotive responses* (turning, stalking, approaching, bending forward) in relation to the *consummative responses* (snapping, gulping). Both in common toads and water frogs, transitions from waiting to hunting, and *vice versa*, will occur. For example, frogs may catch prey by a directed leap-snap response. In any case, once triggered by an appropriate prey stimulus, the behavioral response proceeds to completion, i.e. in a ballistic fashion without feedback from the stimulus.

Behavioral studies: systemic application of the dopamine-agonist APO confines locomotive patterns and facilitates consummative patterns of prey catching

The retino-tectal/tegmental/bulbar/spinal processing streams – e.g., responsible for prey catching – are integrated in a macro-network involving striatal, limbic, pretectal, preoptic/hypothalamic, and solitary/reticular structures that contain dopaminergic cell bodies or fibers [28–32, 81]. Hence, it is reasonable to anticipate significant dopaminergic influences on visual responses.

Glagow and Ewert [114] showed in common toads that systemic (intralymphatic) administration of the dopamine D_2/D_1-receptor agonist apomorphine, APO, affects both the locomotive and consummative components of prey catching in an opposite manner (Fig. 6): with increasing dose of APO, rates of prey-oriented turning and stalking progressively decreased, whereas snapping rates were progressively facilitated at the same time and reaching a maximum about 15 to 35 min after APO administration. Toads previously hunting, that is pursuing prey, after APO administration were sitting motionless just waiting for and snapping at prey. In other words, APO facilitated the consummative component and reduced the directed locomotive components of prey catching. The prey selective property – measured by the snapping rate – was maintained after APO treatment. About 70 to 90 min after APO administration, the prey-oriented turning behavior was restored and then displayed an intermediate rebound activity, while the snapping rate settled towards the level before the administration of APO. After pre-treatment with the dopamine antagonist haloperidol, administration of APO showed no measurable effect on prey catching.

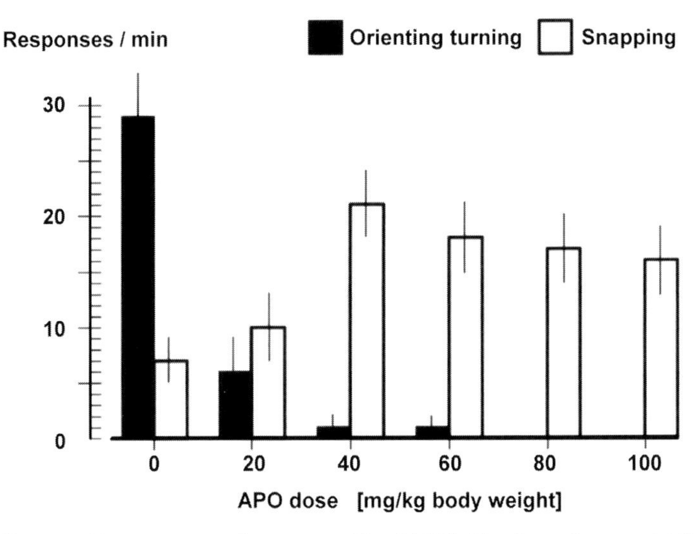

Figure 6. Prey catching patterns under apomorphine (APO). The dopamine agonist APO after its systemic (intralymphatic) administration in common toads, *Bufo bufo*, influences prey-oriented turning and snapping activities in an opposite manner. The dose/effect relationships – measured 20 min after APO treatment – show decreasing turning rates with rising dose of APO, while at the same time the snapping rates increase. Averages ±SDM, $n = 15$ toads. (After [114, 120]; figure reprinted with kind permission of Springer Science and Business Media.)

^{14}C-2DG mapping: systemically applied APO affects the pattern of local cerebral glucose utilization in prey catching toads

The ^{14}C-2DG imaging technique was employed to map the glucose utilization in the brains of APO-treated toads, in comparison with untreated controls, during visual release of prey catching (cf. Figs. 7 and 8) [115]. The retinal projection fields – e.g., dorsal optic tectum (dOT), pretectal thalamic nucleus (Lpd), and anterior thalamic nucleus (A) – showed increases in ^{14}C-2DG uptake. The medial tectal layers (mOT) and the ventral striatum (vSTR), both involved in visuomotor functions related to prey-oriented turning and approaching, displayed APO-induced decreases in ^{14}C-2DG uptake. These data suggest APO-induced increases in retinal output towards the central retinal projection fields (dOT), on the one hand, and APO-induced decreases in tectal structures (mOT) that mediate motor-related output, on the other hand. As a matter of fact, a reduced tectal output would explain the lack of directed locomotive prey catching responses. But how may this be achieved in the presence of an increased retinal input to the tectum?

Leaving this question open awhile, let us now focus on structures related to snapping. APO-induced increases in ^{14}C-2DG uptake were observed in the medial reticular formation (RET) and the hypoglossal nucleus (HGL) which are involved in the motor pattern generation of snapping (e.g., [51, 116]). APO-induced increases in ^{14}C-2DG uptake were also detected in the limbic ventromedial pallium (vMP)

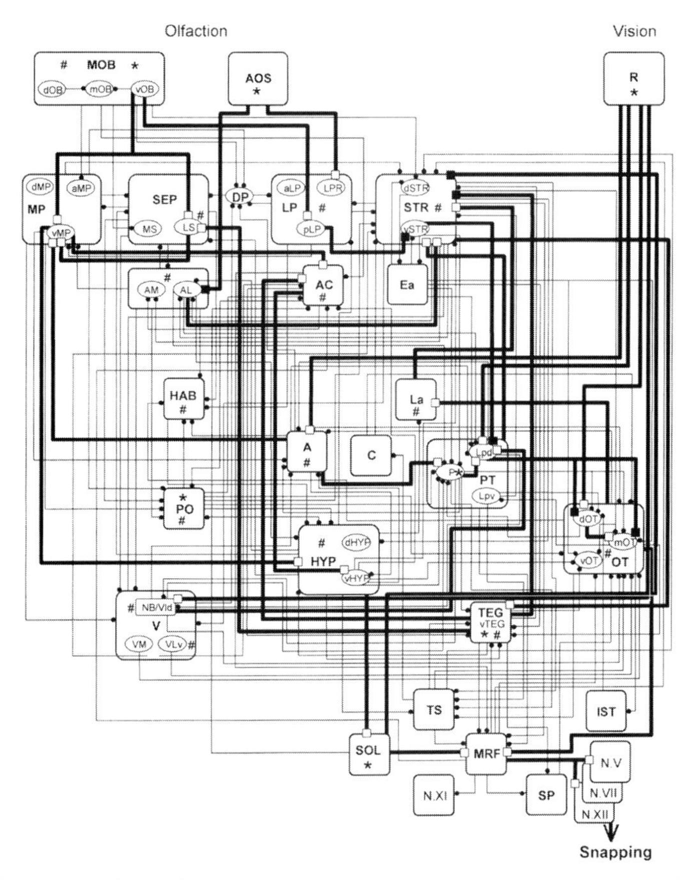

Figure 7. Multiple effects of apomorphine (APO) on the neural macronetwork. Influences of systemically administered APO – referring to the accentuated thick labeled lines – on the common toad's macronetwork in which structures of the forebrain interact with structures of the brainstem (compiled from [115]). Each morphological connection is verified by anatomic techniques, considering a sample of 73 references concerning anuran amphibains in the literature from 1969–1999; (∗) indicates structures harboring dopaminergic cell somata and (#) indicates structures harboring dopaminergic fibers. Abbreviations of structures so far mentioned in the text and in Fig. 8. Telencephalon: MOB, main olfactory bulb; vOB, ventral olfactory bulb; AOS, accessory olfactory system; AC, nucleus accumbens; AL, lateral amygdala; DP, dorsal pallium; dMP, dorsal medial pallium; vMP, ventral medial pallium; LP, lateral pallium; LS, lateral septum; MS, medial septum; vSTR, ventral striatum. Diencephalon: R, retina; A, anterior thalamus; La, lateral anterior thalamic nucleus; Lpd, lateral posterodorsal thalamic nucleus; P, posterior thalamic nucleus; Lpv, lateral posteroventral thalamic nucleus; PT, pretectal region; dHYP dorsal hypothalamus; vHYP, ventral hypothalamus. Mesencephalon: dOT, dorsal layers of the optic tectum; mOT, medial layers of the optic tectum; sOT, snapping evoking area of the optic tectum; IST, nucleus isthmi; TS, torus semicircularis; TEG, tegmentum; III, third ventricle. Medulla oblongata: HGL, nucleus hypoglossus; MRF, medullary medial reticular formation; SOL, nucleus of the solitary tract; SP, spinal cord; N.V, N.VII, N.XII, nuclei of the Vth, VIIth, and XIIth cranial nerve, respectively. (After [160]; figure reprinted with kind permission of Elsevier.)

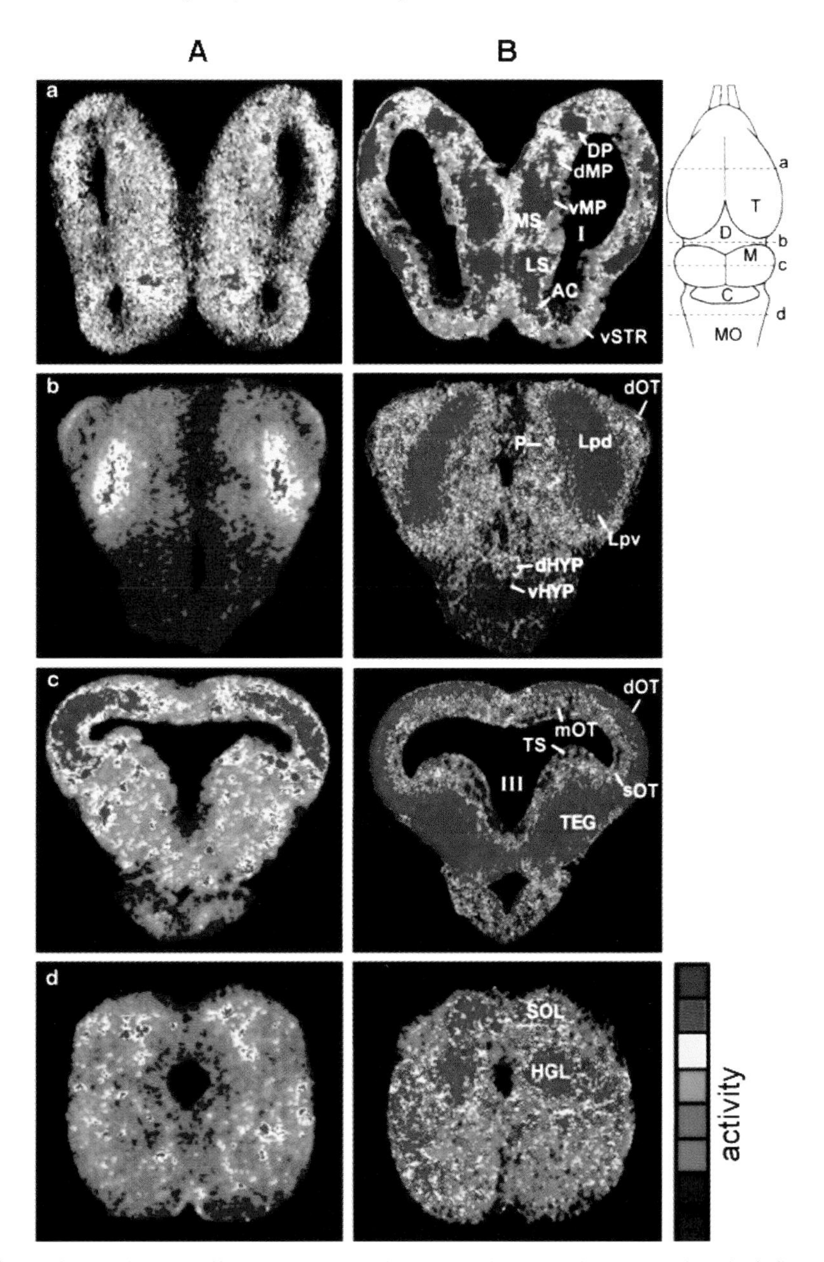

Figure 8. Imaging the effects of apomorphine (APO) in the brain. Monitoring the influence of systemically administered APO on the local cerebral glucose utilization by means of the ^{14}C-2DG method in common toads, *Bufo bufo*, during the release of prey catching behavior towards a prey dummy prior to (A) and after (B) intralymphatic administration of APO. The radioactivity, as a measure of ^{14}C-2DG uptake, increases from cold (blue) to warm (red) colors. For abbreviations see Fig. 7. (After [115]; figure reprinted with kind permission of Karger.)

and medial septum (MS) as well as the mesolimbic nucleus accumbens (AC) and ventral tegmentum (vTEG). Since limbic and mesolimbic structures, in connection with the hypothalamus (dHYP), contribute to the control of motivational state, the APO-induced increases in glucose utilization in these structures may explain APO-facilitated snapping, due to a reduction in the snapping threshold.

Various other brain structures showed APO-induced increases in ^{14}C-2DG uptake, such as the olfactory bulb (MOB), lateral pallium (LP), suprachiasmatic nucleus (SCN), nucleus of the periventricular organ (NPO), the paraventricular nucleus and the nucleus of the solitary tract (SOL). The lateral amygdala (AL) displayed APO-induced decreases in ^{14}C-2DG uptake.

Although the APO-induced alterations in glucose utilization display a correlation with dopaminergic structures and with structures connected to these [28–32, 81], in the interpretation we must consider that APO is a D_2/D_1-receptor agonist, whereby the effects of D_1- and D_2-receptor activation are opposite to each other. The distribution of both receptor types in the anuran brain [117] is not yet well understood. Therefore, we regard the results obtained from the ^{14}C-2DG imaging studies as interesting clues towards electrophysiological studies involving field potential and single cell recordings pre and post APO treatment. In the following we focus on APO-induced changes in ^{14}C-2DG uptake in the retinal projection fields dOT, mOT and Lpd, reported above.

Neurophysiological recordings: APO boosts retinal visual responses

Effect of intraocular administered APO on visual tectal field potentials
In the vertebrate retina dopamine plays an important role, since it uncouples the gap junctions of horizontal cells thus enhancing the efficacy of photoreceptor input (e.g., see [118]). To investigate in toads dopaminergic influences on retinal output, we recorded the summated field potentials from the retinal maps, optic tectum or pretectum, in response to diffuse light-*off* or -*on* stimulation, prior to and after contralateral intraocular administration of either dopamine or APO (10^{-2} mol/l). Upon drug administration, in both cases the initial excitatory N1 wave of the tectal field potential increased strongly, suggesting a dopaminergic enhancement in retinotectal output [119]. Intraocular administration of the dopamine antagonist haloperidol (2×10^{-3} mol/l) attenuated the amplitude of the N1 wave. Recordings from the retinorecipient pretectal neuropil region showed an APO-induced increase in the initial excitatory wave of the pretectal field potential (Schwippert, *unpublished observation.*).

Effect of systemically administered APO on visual tectal field potentials
Tectal and pretectal field potentials were recorded in response to diffuse light-*off* or -*on* stimuli prior to and after intralymphatic systemic administration of APO at a dose of 40 mg/kg body wt. In both types of field potentials the amplitudes of the initial excitatory waves displayed APO-induced increases. Although *systemically* applied APO has wide-spread effects on the brain, these data – in comparison with the ones

obtained from *intraocularly* applied APO – can be explained by APO-enhanced retinal output to the optic tectum and pretectal thalamus, respectively.

Effect of systemically applied APO on visual responses of single retinal ganglion cells
To tie in with the above suggestion single fibers of toad's R2 and R3 retinal ganglion cells were recorded from the optic tectum in response to a visual object traversing the centers of their ERFs. In the time segment of 15 to 35 min after systemic administration of APO (40 mg/kg body wt), the neuronal discharge rates increased strongly (Figs. 9Ab, Bb). Furthermore, the ERF diameters of R2 and R3 neurons approximately doubled their sizes [120]. These effects were independent of the recording site in the retinotectal map. The APO-induced increases in the firing rates of retinal ganglion cells are consistent with the strong APO-induced ^{14}C-2DG uptake in dorsal tectal layers (dOT), the termination field of retinal axons [121].

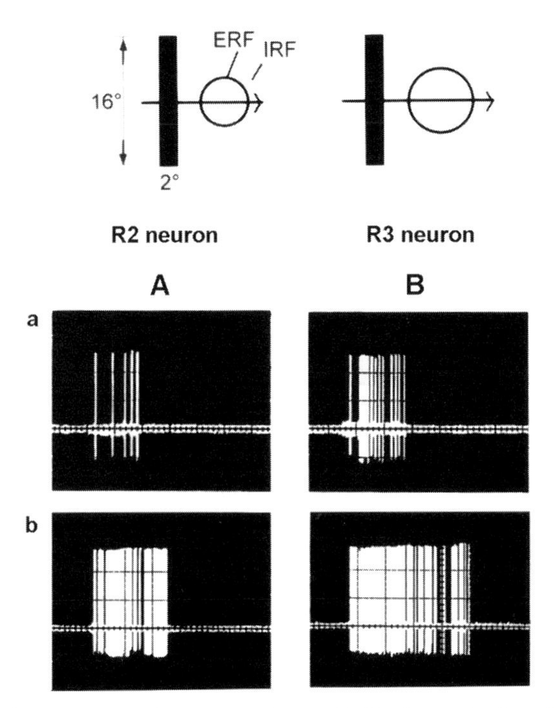

Figure 9. Retinal responses under apomorphine (APO). Systemically administered APO boosts the visual discharge rates of retinal ganglion cells, recorded extracellularly from their fiber endings in the superficial optic tectum in immobilized common toads, *Bufo bufo*. Representative examples. The visual stimulus, a $16° \times 2°$ vertical bar, traversed at $v = 7.6°/s$ the receptive field of (A) a R2 neuron, (B) a R3 neuron, (a) prior to and (b) 25 min after administration of APO. Note that the APO-induced effect reached a maximum between 20 and 35 min after its application. Scale: 0.5 s/div. ERF, excitatory receptive field; IRF, inhibitory receptive field. (After [114, 120]; figure reprinted with kind permission of Springer Science and Business Media.)

Systemically applied APO attenuates visual responses of tectal neurons

Immunohistochemical studies of the optic tectum revealed a complex pattern of catecholaminergic innervation both in anurans and urodeles [122]. Dopaminergic fibers basically terminate in deep layers of the tectum while noradrenergic fibers exist in the superficial layers. Interestingly, the tectum is lacking catecholaminergic somata. Using retrograde tracers it was shown that the origins of dopaminergic tectal innervation are in suprachiasmatic, juxtacommissural, and pretectal posterior thalamic (P) nuclei.

Are single neurons in the toad's medial tectal layers (mOT) – known to mediate tectal output – influenced under APO? In the time segment of 15 to 35 min after systemic administration of APO, in which the retinal R2 and R3 ganglion cells increased their discharge rates towards moving visual stimuli, tectal T5.1 and T5.2 neurons remarkably decreased their discharge rates (Fig. 10A-C, cf.b) [120]. These decreases are consistent with the reduction in ^{14}C-2DG uptake in mOT from where these tectal neurons were recorded. During this "tectal reduction phase," orienting towards prey failed to occur, whereas the snapping rate was increased. About 70 to 90 min after APO administration the discharge rates of T5.1 and T5.2 neurons displayed a short-term rebound-like increase to moving visual stimuli. Interestingly, during this "tectal rebound phase" the orienting rate, too, showed a rebound-like increase. Hence, these results suggest correlations between APO-induced alterations in the discharge rates of prey-sensitive/selective T5.1 and T5.2 neurons – known to project from the medial tectal layers to tegmental/bulbar/spinal motor pattern generating systems, on the one hand, and the prey-oriented turning activity, on the other hand.

A comparison of the T5.2-responses to different visual stimuli prior to (Fig. 10A-C; cf.a) and after administration of APO (Fig. 10A-C; cf.b) brings two aspects in focus: first, the typical configurative stimulus discrimination was maintained during the tectal reduction phase; second, the temporal discharge pattern in response to the prey-like moving horizontal bar started with a short burst of spikes and was immediately silent thereafter (Fig. 10Ab). This discharge pattern can be explained by a postexcitatory inhibition. We shall renew this phenomenon later.

Let us now return to the previous question of dispute: how can we explain the APO-induced attenuation of tectal responses in face of APO-induced enhancements of retinal outputs? Remembering *inhibitory interacting* retino-tectal and retinopretecto-tectal processing streams, the APO-enhanced retinotectal *and* retinopretectal activations would balance each other in the tectum.

Systemically applied APO boosts visual responses of pretectal neurons

In fact, recordings under APO from the pretectal Lpd nucleus demonstrate that the visual discharge rates of pretectal thalamic TH3 and TH4 neurons strongly increase exactly in that time segment in which the visual discharge rates of tectal T5.1 and T5.2 neurons display a minimum [121]. The APO-enhanced firing in TH3 and TH4 neurons is consistent with APO-enhanced increases of ^{14}C-2DG uptake in the pretectal thalamic Lpd nucleus.

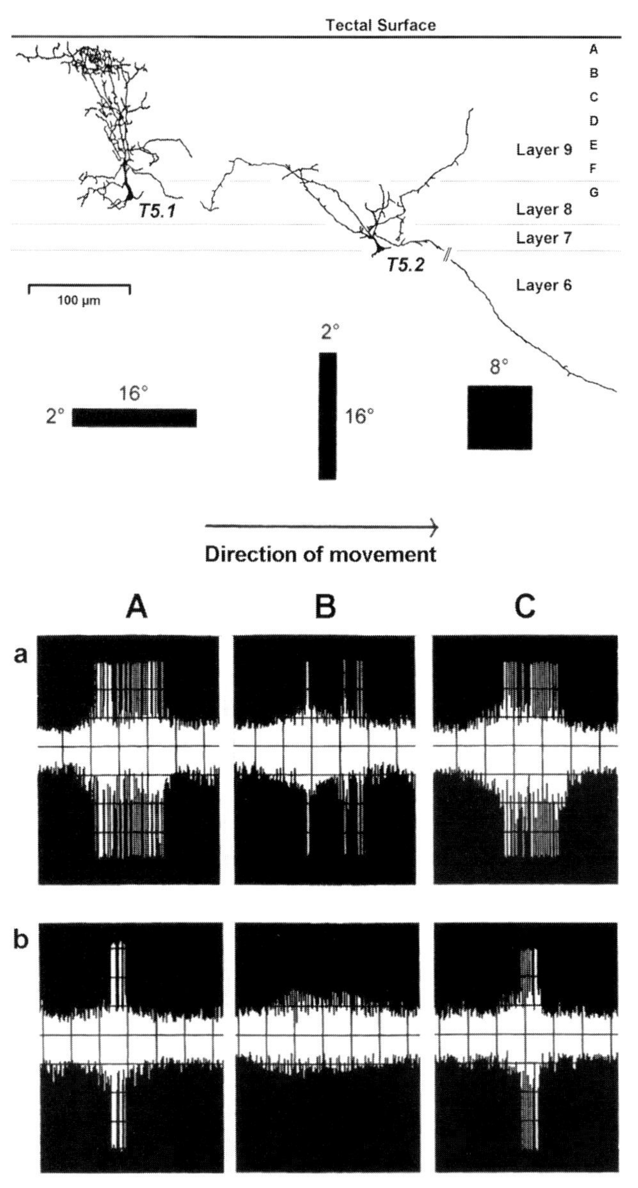

Figure 10. Tectal responses under apomorphine (APO). Systemically administered APO leads to an attenuation of the visual responses of tectal T5.2 neurons, extracellularly recorded from tectal layer 7 or top 6 in immobilized common toads, *Bufo bufo*. Representative examples. Figurative different stimuli traversed the receptive field at $v = 10°/s$: (A) $2° \times 16°$ horizontal bar, (B) $16° \times 2°$ vertical bar, (C) $8° \times 8°$ square, presented (a) prior to and (b) 35 min after APO administration. Scale: 1 s/div. (After [114, 120].) Top: Example of camera-lucida reconstructions of a T5.1 and a T5.2 neuron recorded intracellularly and stained with Co^{3+}-lysin; note that the axon of the T5.2 neuron is distorted in this representation. (After [159]; all figures reprinted with kind permission of Springer Science and Business Media.)

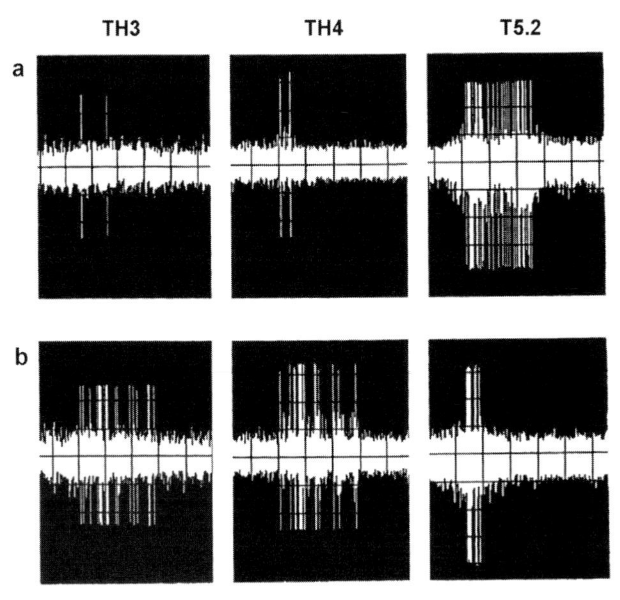

Figure 11. Pretectal and tectal responses under apomorphine (APO). Opposite effects of APO on the discharge patterns of common toad's pretectal thalamic TH3 and TH4 neurons, on the one hand, and a tectal T5.2 neuron, on the other hand. As stimulus served a $2° \times 16°$ horizontal bar traversing the receptive field at = 10°/s. (a) The prey-like moving bar is an ineffective stimulus for TH3 and TH4 neurons, however, optimal for the T5.2 neuron. (b) After administration of APO, the TH3 and TH4 neurons discharge strongly; the T5.2 neuron discharges a short burst of spikes and is then immediately silent, suggesting a postexcitatory inhibition (see text). Scale: 1 s/div. Representative examples of extracellular recordings in immobilized common toads. (After [121]; figure reprinted with kind permission of Elsevier.)

Of particular interest is a comparison of the discharge patterns of pretectal thalamic TH3, TH4, and tectal T5.2 neurons in response to a prey-like moving bar prior to and after systemic administration of APO. Prior to APO administration, both threat sensitive TH3 and TH4 neurons displayed a short weak burst of spikes (Fig. 11, TH3, TH4; cf.a), whereas the prey selective T5.2 neuron started with a strong burst followed by a relatively long train of spike activity (Fig. 11, T5.2; cf.a). In the tectal reduction phase 34 min after APO administration, TH3 and TH4 neurons increased their firing rates and discharged a relatively long spike train (Fig. 11, TH3, TH4; cf.b), whereas the T5.2 neuron began to fire a short burst of spikes and was immediately silent thereafter (Fig. 11, T5.2; cf.b). To tie in with the aforementioned aspect of postexcitatory inhibition, this silence can be explained by the APO-induced prolonged activity pattern of pretectal thalamic TH3 and TH4 neurons in combination with pretecto-tectal inhibitory influences.

Hypothesis: interactions in a macronetwork

The multiple effects of systemically administered APO – accompanied with a shifting from hunting prey to waiting for prey – suggest complex interactions in a macronetwork (Fig. 7, thick labelled lines).

APO-induced suppression of orienting and locomotion

The APO-induced increase in retinal output leads to increased retinotectal and retino-pretectal excitation and – in line with the previous arguments – to enhanced pretecto-tectal inhibitory influences and thus to a suppression of prey-oriented locomotive responses.

Straightforward as this explanation looks, questions remain open. If pretecto-tectal inhibitory influences are mediated by pretectal presynaptic inhibition of retinal axonal endings *via* NPY in superficial tectal layers, then the strong discharges in retinal fibers recorded from these tectal layers and the strong ^{14}C-2DG uptake in these tectal layers under systemically applied APO are hard to understand. However, as a matter of fact it could be shown that NPY experimentally administered to the tectal surface strongly reduced ^{14}C-2DG uptake in superficial tectal layers after systemic administration of APO in prey-catching red-bellied toads *Bombina orientalis* [71]. This again provides evidence of an inhibitory effect of NPY on retinotectal transfer. But the effect of NPY may be weaker under physiological conditions of its release. Therefore, further inhibitory influences on tectal cells should be considered as emphasized in Section *Other inhibitory influences on tectal neurons*.

We assume that the APO-induced attenuation of tecto-motor output is subthreshold to trigger the motor pattern generating networks responsible for prey-oriented turning and stalking, however, sufficient for the release of augmented snapping, provided that APO modulates the activity of structures that lower the trigger threshold in the snap-generating network.

APO-induced facilitation of snapping

APO-induced increases in ^{14}C-2DG uptake were observed in the bulbar medial reticular formation (RET) and in the hypoglossal nucleus (HGN), both of which are involved in motor pattern generation of snapping [51, 116]. The increases in glucose utilization in limbic structures (ventromedial pallium vMP, lateral septum LS, nucleus accumbens AC) and in structures that are connected to the limbic system (dorsal hypothalamus dHYP, ventral tegmentum vTEG) may be associated with setting the level of the snapping threshold.

The vMP integrates converging APO-enhanced visual input from the retina (R), relayed by the anterior dorsal thalamic nucleus (A), and APO-enhanced sustained input from the ventral olfactory bulb (vOB), relayed by the lateral septum (LS) (Fig. 7) (for anatomic pathways see, e.g., [24, 86, 123–125]). We speculate that the resulting vMP output

vOB → LS → **vMP** ← A ← R

in concert with the APO-enhanced strong influences from dopaminergic structures of the ventral tegmental area, stimulates AC and vHYP which, with the nucleus of the solitary tract (SOL), influence the reticular/branchiomotor/hypoglossal system to lower the threshold of snapping.

Concluding remarks

Three decades ago, Ploog and Gottwald [126] already pointed out that the phylogenetic origin of systems responsible for attention, approach and avoidance, reward and punishment, positive and negative reinforcement and for motor skills observed in mammals can partly be traced back to amphibian brain structures that are homologous to the limbic and mesolimbic systems, the diencephalic central grey, and the basal ganglia. This prediction becomes more and more substantiated. It suggests that information processing for a set of essential basic functional principles emerged early during evolution of tetrapod vertebrates, was stored in terms of neuronal connections and synaptic processes and was conserved during phylogeny in order to be modified or specified in adaptation to the different ecological constraints and requirements of a species.

Visuomotor functions involving NPY

Studies in phylogenetic basal tetrapods like toads and frogs show that visuomotor functions involving stimulus discrimination depend on retinotectal processing streams in connection with forebrain structures like the pretectal thalamus. In the optic tectum glutamate seems to be a dominant transmitter in the retinotectal information transfer. The monosynaptic retinotectal connections act through voltage-dependent NMDA-receptors, whereas polysynaptic transmissions are mediated by NMDA- and non-NMDA-receptors [127, 128]. In anurans it is suggested that pretectal NPYergic fibers contact presynaptically NPY-receptive retinal fiber terminals in the superficial tectum in order to modulate/specify tectal visual responses [59, 68, 111]. More explicitly, we suggest that NPY via presynaptic inhibition reduces glutaminergic retinotectal transmission. A comparable pretectal NPY-mediated inhibitory mechanism is discussed in pigeons by Gamlin et al. [129].

The pretectum of anamniotes and amniotes is composed of different – partially homologous – nuclei which in amphibians may be involved in different visual functions, involving sensitivities to moving threat or stationary obstacles, optokinetic nystagmus, pupillary light reflex, modulation of retinofugal transfer. Regarding the latter in mammals it is suggested that certain pretectal GABAergic cells project to inhibit thalamic cells which in turn would disinhibit geniculate relay cells, thereby facilitating the retinal information to the cortex [130].

A comparative treatment of the neurochemical structure of the different pretectal nuclei based on immunoreativity to monamines, neuropeptides (e.g., NPY) and GABA in reptiles, birds, and mammals is provided by Kenigfest et al. [131]. In fact, pretectal cells containing NPY are common in tetrapods from frogs to humans [132]. Kenigfest et al. [131] suggest that species of divergent lines of amniotes have evolutionary conservatism of the neurochemical organization of pretectal structures and their efferents, on the one hand, and have certain plasticity in terms of rearrangements during phylogenetic development, on the other hand.

NPY is widely distributed in the CNS and may, *via* G-protein coupled different receptors, participate in processes dealing with vasoconstriction, anxiolytic syndromes, sedation (Y_1 receptor), gastrointestinal functions (Y_4), appetite regulation (Y_5). The Y_2 receptor mediated NPY effects of presynaptic inhibition of glutamate release are the most abundant and may have an evolutionary conserved role in modulating visuomotor processing (e.g., see [65, 70, 133]).

In mammals functions concerning head/neck movements [134] involve retino-superiorcollicular processing streams in connection with geniculate and cortical forebrain structures. The "philosophy" of combining homologous mesencephalic structures – optic tectum or superior colliculus, respectively – with forebrain structures is comparable. Furthermore, connections with the ventromedial pallium (amphibians) or the homologous hippocampus (mammals) make the systems adjustable and adaptable to individual experience. The discussion by Foreman and Stevens [135] and Gonzalez-Lima [136] on the relationships between superior colliculus and hippocampus relayed by preoptic/hypothalamic structures in mammals reveals interesting homologies with corresponding relationships between optic tectum and ventral medial pallium in anuran amphibians [95, 137].

Gating directed attention

In anurans, gating the translation of visual perception into action – in a manner of directed attention – may be carried out by a loop involving a disinhibitory striato-pretecto-tectal pathway [12, 23, 104, 111]. Striatal influences are discussed also for mechanosensory and acoustic behaviors in anuran amphibians by Birkhofer et al. [138] Walkowiak et al. [139], and Endepols and Walkowiak [140].

Both in anurans and mammals the activity in (certain) striatal structures increases towards *novel* stimuli and decreases towards *familiar* stimuli, while (certain) structures of the ventral medial pallium and hippocampus, respectively, display an opposite pattern of activity [94, 95, 136]. Striatal efferent neurons in toads may be considered also in a striato-nigro-tectal pathway [26, 28, 107]. In mammals, it was shown that a disinhibitory striato-nigro-collicular pathway does play an attentional gating function [141, 142]. The important point we put forward here, however, is that directed visual attention in prey-oriented behavior in anurans may take advantage (also) of a striato-pretecto-tectal pathway, a channel whose homologue in mammals does not seem to exist (cf. [33, 37, 38]). The "philosophy" of disinhibition by combining striatal and tectal/collicular structures *via* a relay (pretectum or substantia nigra, respectively) is comparable in anurans and mammals.

Dopaminergic modulation of movement patterns

In anurans motor patterns in terms of hunting prey and waiting for prey involve basal ganglionic, pallial, limbic, mesolimbic, tectal, tegmental, and bulbar reticular systems. Most of these harbour dopaminergic cell bodies and/or dopaminergic fibers [81]. After systemic administration of the dopamine agonist apomorphine these nuclei show significant changes in ^{14}C-2DG uptake and this in turn may influence non-dopaminergic structures [115].

Monitoring the local cerebral glucose utilization in various structures of the toad's brain prior to and after systemic administration of APO, it appears that *waiting and augmented prey-snapping* (locomotive akinesia) in the present context requires much more distributed brain activity than *prey-oriented hunting* (locomotive agility). However, an interpretation that the APO-induced patterns of brain activity corresponds to the sit-and-wait strategy of prey catching would be inadequate, since there are a variety of APO-induced "side effects" in local cerebral metabolism.

The APO-induced facilitation of snapping in common toads is comparable to oral behavioral facilitation reported in other vertebrates after systemic administration of APO, such as biting in the tortoise [143], pecking in the pigeon (e.g. [144, 145]), sniffing, licking, and gnawing in mice and rats (e.g. [146–148]), and yawning in humans [149]. In regard to the multiple effects of APO in humans, the drug is used clinically as an emetic but also as an anti-Parkinson drug [150].

Regarding the dosage, in toads maximal snapping rates were obtained under APO at a dose of 40 mg/kg body wt which is relatively high by comparison, e.g., with a dose (1 mg/kg) eliciting compulsive pecking in birds after intramuscular administration of APO [151] or a dose (2–4 mg sublingual) eliciting effects in humans. The high dose of APO required in toads may be due to a poor lymphatic circulation in the ectothermic animals and a correspondingly weak access of APO to the brain. Regarding the time-course of APO-induced effects, they begin in toads and humans about 10 min after administration and last for at least 30 min.

Since it is known that the amphibian "hippocampal" pallium (vMP) is involved in conditioning, the increase in activity in the toad's vMP after administration of APO may also result from its action as a positive reinforcer. Systemically applied APO induces both enhanced olfactory resting activities and enhanced retinal visual responses that converge in vMP. Furthermore, it was shown in toads that APO-induced facilitation of snapping became sensitized after repeated administrations of APO (Glagow, unpublished). The mechanism of sensitization to APO is unknown in toads and is a disputed issue in birds and rodents (e.g., see [145, 152–158]). The role of APO in drug-state-dependent conditioning and findings related to the sensitization of psychostimulant drugs are discussed by Godoy and Delius [151].

All these data provide evidence of the enormous multiple – direct and indirect – dopaminergic effects on brain structures and their interactions after systemic application of the dopamine agonist apomorphine. The data also point to the various possible "side effects" of drugs that release dopaminergic actions involving motor co-ordinations, limbic processes, and sensory sensations.

References

1. Eibl-Eibesfeldt I (1951) Nahrungserwerb und Beuteschema der Erdkröte (*Bufo bufo* L). *Behaviour* 4: 1–35
2. Wiersma CAG, Ikeda K (1964) Interneurons commanding swimmeret movements in the crayfish, *Procambarus clarkii* (Girard). *Comp Biochem Physiol* 12: 509–525
3. Hinsche G (1935) Ein Schnappreflex nach "Nichts" bei Anuren. *Zool Anz* 111: 113–122
4. Tinbergen N (1951) *The study of instinct*. Clarendon Press, Oxford
5. Lorenz K (1954) Das angeborene Erkennen. *Natur und Volk* 84: 285–295
6. Barlow HB (1953) Summation and inhibition in the frog's retina. *J Physiol (Lond)* 173: 377–407
7. Lettvin JY, Maturana HR, McCulloch WS, Pitts WH (1959) What the frog's eye tells the frog's brain. *Proc Inst Radio Engin* 47: 1940–1951
8. Kupfermann I, Weiss KR (1978) The command neuron concept. *Behav Brain Sci* 1: 3–39
9. Eaton RC (1983) Is the Mauthner cell a vertebrate command neuron? A neuroethological perspective on an evolving concept. In: Ewert J-P, Capranica RR, Ingle DJ (eds): *Advances in vertebrate neuroethology*. Plenum, New York, 629–636
10. Eaton RC (2001) The Mauthner cell and other identified neurons of the brainstem escape network of fish. *Prog Neurobiol* 63: 467–485
11. Ewert J-P (1980) *Neuroethology: an introduction to the neurophysiological fundamentals of behavior*. Springer, Berlin
12. Ewert J-P (1997) Neural correlates of key stimulus and releasing mechanism: a case study and two concepts. *Trends Neurosci* 20: 332–339
13. Ewert J-P (2004) Motion perception shapes the visual world of amphibians. In: Prete FR (ed): *Complex worlds from simpler nervous systems*. MIT Press, Cambridge MA, 117–160
14. Hailman JP (1969) How an instinct is learnt. *Sci Amer* 221: 98–106
15. Bolhuis JE, Giraldeau L-A (2005) *The behavior of animals. Mechanisms, function, and evolution*. Blackwell, Malden MA
16. Ewert J-P (1985) The Niko Tinbergen Lecture: concepts in vertebrate neuroethology. *Animal Behav* 33: 1–29
17. Ewert J-P (2005) Stimulus perception. Chapter 2. In: Bolhuis JJ, Giraldeau L-A (eds): *The behavior of animals*. Blackwell, Malden MA, 13–40
18. Schrader MEG (1887) Zur Physiologie des Froschgehirns. *Pflügers Arch* 51: 11–21
19. Johannes T (1930) Zur Funktion des sensiblen Thalamus. *Pflüg Arch* 224
20. Goltz P (1869) Beiträge zur Lehre von den Funktionen der Nervenzentren des Frosches. In: Buddenbrock W v (1937) (ed): *Grundriß der vergleichenden Physiologie* Bd 1, Berlin
21. Blankenagel S (1931) Untersuchungen über die Großhirnfunktionen von *Rana temporaria* L. *Zool Jb Abteilung allgem Zool* 49: 272–322
22. Diebschlag E (1935) Zur Kenntnis der Großhirnfunktion einiger Urodelen und Anuren. *Z vergl Physiol* 21: 343–394
23. Ewert J-P (1967) Untersuchungen über die Anteile zentralnervöser Aktionen an der taxisspezifischen Ermüdung der Erdkröte (*Bufo bufo* L). *Z Vergl Physiol* 57: 263–298
24. Northcutt RG, Kicliter E (1980) Organization of the amphibian telencephalon. In: Ebbesson SOE (ed): *Comparative neurology of the telencenphalon*. Plenum Press, New York London, 203–255
25. Wilczynski W, Northcutt RG (1983) Connections of the bullfrog striatum: afferent organization. *J Comp Neurol* 214: 321–332

26. Wilczynski W, Northcutt RG (1983) Connections of the bullfrog striatum: efferent projections. *J Comp Neurol* 214: 333–343
27. Northcutt RG, Kaas H (1995) The emergence and evolution of mammalian neocortex. *Trends Neurosci* 18: 373–379
28. Marín O, González A, Smeets WJAJ (1997) Anatomical substrate of amphibian basal ganglia involvement in visuomotor behaviour. *Eur J Neurosci* 9: 2100–2109
29. Marín O, González A, Smeets WJAJ (1997) Basal ganglia organization in amphibians: afferent connections to the striatum and the nucleus accumbens. *J Comp Neurol* 378: 16–49
30. Marín O, González A, Smeets WJAJ (1997) Basal ganglia organization in amphibians: efferent connections of the striatum and the nucleus accumbens. *J Comp Neurol* 380: 23–50
31. Marín O, Smeets WJAJ, González A (1997) Basal ganglia organization in amphibians: catecholaminergic innervation of the striatum and the nucleus accumbens. *J Comp Neurol* 378: 50–69
32. Marín O, Smeets WJAJ, González A (1997) Basal ganglia organization in amphibians: development of striatal and nucleus accumbens connections with emphasis on the catecholaminergic inputs. *J Comp Neurol* 383: 349–369
33. Marín O, González A, Smeets WJAJ (1998) Basal ganglia organization in amphibians: chemoarchitecture. *J Comp Neurol* 392: 285–312
34. Marín O, Smeets WJAJ, González A (1998) Evolution of the basal ganglia in tetrapods: a new perspective based on recent studies in amphibians. *Trends Neurosci* 21: 487–494
35. González A, Smeets WJ, Marín O (1999) Evidences for shared features in the organization of the basal ganglia in tetrapods: studies in amphibians. *Eur J Morphol* 37(2-3): 151–154
36. Reiner A, Brecha NC, Karten HJ (1982) Basal ganglia pathways to the tectum: the afferent and efferent connections of the lateral spiriform nucleus of pigeon. *J Comp Neurol* 208: 16–36
37. Reiner A, Brauth SE, Karten HJ (1984) Evolution of the amniote basal ganglia. *Trends Neurosci* 7: 320–325
38. Reiner A, Medina L, Veenman CL (1998) Structural and functional evolution of the basal ganglia in vertebrates. *Brain Res Rev* 28: 235–285
39. Herrick CJ (1933) The amphibian forebrain. VIII: Cerebral hemispheres and pallial primordia. *J Comp Neurol* 58: 737–759
40. Roth G, Westhoff G (1999) Cytoarchitecture and connectivity of the amphibian medial pallium. *Eur J Morphol* 37: 166–171
41. Marín O, Smeets WJ, Munoz M, Sanchez-Camacho C, Pena JJ, Lopez JM, González A (1999) Cholinergic and catecholaminergic neurons relay striatal information to the optic tectum in amphibians. *Eur J Morphol* 37: 155–159
42. Marín O, González A (1999) Origin of tectal cholinergic projections in amphibians: a combined study of choline acetyltransferase immunohistochemistry and retrograde transport of dextran amines. *Vis Neurosci* 16: 271–283
43. Schneider D (1954) Beitrag zu einer Analyse des Beute- und Fluchtverhaltens einheimischer Anuren. *Biol Zbl* 73: 225–282
44. Ewert J-P (1974) The neural basis of visually guided behavior. *Sci Amer* 230: 34–42
45. Ewert J-P (1984) Tectal mechanisms that underlie prey-catching and avoidance behaviors in toads. In: Vanegas H (ed): *Comparative neurology of the optic tectum*. Plenum, New York, 247–416
46. Ewert J-P, Arend B, Becker V, Borchers H-W (1979) Invariants in configurational prey selection by *Bufo bufo* (L.). *Brain Behav Evol* 16: 38–51

47. Ewert J-P, Burghagen H (1979) Configurational prey selection by *Bufo, Alytes, Bombina*, and *Hyla*. *Brain Behav Evol* 16(3): 157–175
48. Grüsser O-J, Grüsser-Cornehls U, Finkelstein D, Henn V, Patutschnik M, Butenandt E (1967) A quantitative analysis of movement detecting neurons in the frog retina. *Pflügers Arch* 293: 100–106
49. Ewert J-P (1987) Neuroethology of releasing mechanisms: prey-catching in toads. *Behav Brain Sci* 10: 337–405
50. Satou M, Ewert J-P (1985) The antidromic activation of tectal neurons by electrical stimuli applied to the caudal medulla oblongata in the toad *Bufo bufo* (L.). *J Comp Physiol* 157: 739–748
51. Ewert J-P, Framing EM, Schürg-Pfeiffer E, Weerasuriya A (1990) Responses of medullary neurons to moving visual stimuli in the common toad: I) Characterization of medial reticular neurons by extracellular recording. *J Comp Physiol* A 167: 495–508
52. Ewert J-P, Hock FJ, Wietersheim A v (1974) Thalamus/Praetectum/Tectum: retinale Topographie und physiologische Interaktionen bei der Kröte (*Bufo bufo* L). *J Comp Physiol* 92: 343–356
53. Schürg-Pfeiffer E, Spreckelsen C, Ewert J-P (1993) Temporal discharge patterns of tectal and medullary neurons chronically recorded during snapping toward prey in toads *Bufo bufo spinosus*. *J Comp Physiol* A 173: 363–376
54. Ewert J-P, Schürg-Pfeiffer E, Schwippert WW (1996) Influence of pretectal lesions on tectal responses to visual stimulation in anurans: field potential, single neuron and behavior analyses. *Acta Biologica Acad Sci Hungaria* 47(2-4): 223–245
55. Ewert J-P, Wietersheim A v (1974) Der Einfluß von Thalamus/Praetectum-Defekten auf die Antwort von Tectum-Neuronen gegenüber bewegten visuellen Mustern bei der Kröte (*Bufo bufo* L). *J Comp Physiol* 92: 149–160
56. Ewert J-P (1971) Single unit response of the toad (*Bufo americanus*) caudal thalamus to visual objects. *Vergl Physiol* 74: 81–102
57. Lázár G (1989) Cellular architecture and connectivity of the frog's optic tectum and pretectum. In: Ewert J-P, Arbib MA (eds): *Visuomotor coordination*. Plenum, New York, 175–199
58. Matsumoto N (1989) Morphological and physiological studies of tectal and pretectal neurons in the frog. In: Ewert J-P, Arbib MA (eds): *Visuomotor coordination*. Plenum, New York, 201–222
59. Buxbaum-Conradi H, Ewert J-P (1995) Pretecto-tectal influences I. What the toad's pretectum tells its tectum: an antidromic stimulation/recording study. *J Comp Physiol* A 176: 169–180
60. Ingle DJ (1977) Detection of stationary objects by frogs (*Rana pipiens*) after ablation of optic tectum. *J Comp Physiol Psychol* 91: 1359–1364
61. Ingle DJ (1980) Some effects of pretectum lesions on the frog's detection of stationary objects. *Behav Brain Res* 1: 139–163.
62. Lázár G, Maderdrut JL, Trasti SL, Liposits Z, Tóth P, Kozicz T, Merchenthaler I (1993) Distribution of proneuropeptide Y-derived peptides in the brain of *Rana esculenta* and *Xenopus laevis*. *J Comp Neurol* 327: 551–571
63. Danger JM, Guy J, Benyamina M, Jegou S, Leboulenger F, Cote J, Tonon MC, Pelletier G, Vaudry H (1985) Localization and identification of neuropeptide Y (NPY)-like immunoreactivity in the frog brain. *Peptides* 6: 1225–1236
64. Chapman AM, Debski EA (1995) Neuropeptide Y immunoreactivity of a projection from the lateral thalamic nucleus to the optic tectum of the leopard frog. *Vis Neurosci* 12: 1–9

65. Lázár G (2001) Peptides in frog brain areas processing visual information. *Microsc Res Tech* 54(4): 201–219
66. Kozicz T, Lázár G (1994) The origin of tectal NPY immunopositive fibers in the frog. *Brain Res* 635: 345–348
67. Tuinhof R, Gonzalez A, Smeets WJAJ, Roubos EW (1994) Neuropeptide Y in the developing and adult brain of the South African clawed toad, *Xenopus laevis*. *J Chem Neuroanatom* 7: 271–283
68. Schwippert WW, Ewert J-P (1995) Effect of neuropeptide-Y on tectal field potentials in the toad. *Brain Res* 669: 150–152
69. Schwippert WW, Röttgen A, Ewert J-P (1998) Neuropeptide Y (NPY) or fragment $NPY_{13\text{-}36}$, but not $NPY_{18\text{-}36}$, inhibit retinotectal transfer in cane toads *Bufo marinus*. *Neurosci Lett* 253: 33–36
70. Carr JA, Brown CL, Mansouri R, Venkatesan S (2002) Neuropeptides and amphibian prey-catching behavior. *Comp Biochem Physiol* Part B 132: 151–162
71. Funke S, Ewert J-P (2006) Neuropeptide Y suppresses glucose utilization in the dorsal optic tectum towards visual stimulation in the toad *Bombina orientalis*: A $[^{14}C]$2DG study. *Neuroscience Lett* 392: 43–46
72. Clairambault P (1976) Development of the prosencephalon. In: Llinás R, Precht W (eds): *Frog neurobiology*. Springer, Berlin, 924–945
73. D'Aniello B, Imperatore C, Fiorentiono M, Vallarino M, Rastogi RK (1994) Immunocytochemical localization of POMC-derived peptides (adrenocorticotropic hormone, α-melanocyte-stimulating hormone and β-endorphin) in the pituitary, brain and olfactory epithelium of the frog, *Rana esculenta*, during development. *Cell Tissue Res* 278: 509–516
74. Ebbesson SOE (1987) Prey-catching in toads: an exceptional neuroethological model. *Behav Brain Sci* 10: 375–376
75. Traud R (1983) Einfluß von visuellen Reizmustern auf die juvenile Erdkröte (*Bufo bufo* L). Dr.rer.nat. Dissertation. Abt. Neurobiologie. Fachbereich Biologie/Chemie, Univ Kassel
76. Kuhn P (2003) Quantitative Untersuchungen über die visuelle Steuerung des Beutefangs der Chinesischen Rotbauchunke *Bombina orientalis* während der Ontogenese. Dr.rer.nat. Dissertation, Abt. Neurobiologie, Fachbereich Biologie/Chemie, Univ Kassel
77. Ewert J-P, Burghagen H (1979) Ontogenetic aspects of visual size constancy phenomenon in the midwife toad *Alytes obstetricans* (Laur.). *Brain Behav Evol* 16(2): 99–112
78. Ewert J-P, Burghagen H, Schürg-Pfeiffer E (1983) Neuroethological analysis of the innate releasing mechanism for prey-catching behavior in toads. In: Ewert J-P, Capranica RR, Ingle DJ (eds): *Advances in vertebrate neuroethology*. Plenum, New York, 413–475
79. Székely G, Lázár G (1976) Cellular and synaptic architecture of the optic tectum. In: Llinás R, Precht W (eds): *Frog neurobiology*. Springer, Berlin, 407–434
80. Kozicz T, Lázár G (2001) Colocalization of GABA, enkephalin and neuropeptide Y in the tectum of the green frog *Rana esculenta*. *Peptides* 22: 1071–1077
81. González A, Smeets WJAJ (1991) Comparative analysis of dopamine and tyrosine hydroxylase immunoreactivities in the brain of two amphibians, the anuran *Rana ridibunda* and the urodele *Pleurodeles waltlii*. *J Comp Neurol* 303: 457–477
82. Lázár G (1971) The projection of the retinal quadrants on the optic centers in the frog: a terminal degeneration study. *Acta Morph Acad Sci Hung* 19: 325–334
83. Lázár G (1979) Organization of the frog visual system. In: Lissák K (ed): *Recent developments of neurobiology in Hungary*, Vol 8. Akadémiai Kiadò, Budapest, 9–50

84. Fite KV, Scalia F (1976) Central visual pathways in the frog. In: Fite KV (ed): *The amphibian visual system: a multidisciplinary approach*. Academic Press, New York, 87–118
85. Wilczynski W, Northcutt RG (1977) Afferents to the optic tectum of the leopard frog: an HRP study. *J Comp Neurol* 173: 219–229
86. Neary TJ, Wilczynski W (1980) Descending inputs to the optic tectum in ranid frogs. *Soc Neurosci Abstr* 6: 629
87. Neary T, Northcutt RG (1983) Nuclear organization of the bullfrog diencephalon. *J Comp Neurol* 213: 262–278
88. Stevens RJ (1973) A cholinergic inhibitory system in the frog optic tectum: its role in visual electrical responses and feeding behavior. *Brain Res* 49: 309–321
89. Gruberg ER (1989) Nucleus isthmi and optic tectum in frogs. In: Ewert J-P, Arbib MA (eds): *Visuomotor coordination*. Plenum, New York, 341–356
90. Gruberg ER, Wallace M, Caine H, Mote M (1991) Behavioral and physiological consequences of unilateral ablation of the nucleus isthmi in the leopard frog. *Brain Behav Evol* 37: 92–103
91. Gruberg ER, Hughes TE, Karten HJ (1994) Synaptic interregulationships between the optic tectum and the ipsilateral nucleus isthmi in *Rana pipiens*. *J Com Neurol* 339(3): 353–364
92. Xiao J, Wang Y, Wang SR (1999) Effects of glutamatergic, cholinergic and GABAergic antagonists on tectal cells in toads. *Neuroscience* 90(3): 1061–1067
93. Brzoska J, Schneider H (1978) Modification of prey-catching behavior by learning in the common toad (*Bufo bufo* L, Anura, Amphibia): changes in response to visual objects and effects of auditory stimuli. *Behav Processes* 3: 125–136
94. Finkenstädt T (1989) Visual associative learning: searching for behaviorally relevant brain structures in toads. In: Ewert J-P, Arbib, MA (eds): *Visuomotor coordination*. Plenum, New York, 799–832
95. Finkenstädt T, Ewert J-P (1992) Localization of learning-related metabolical changes in brain structures of common toads: a 2-DG-study. In: Gonzalez-Lima F, Finkenstädt T, Scheich H (eds): *Advances in metabolic mapping techniques for brain imaging of behavioral and learning functions*. Kluwer Academic Publishers, Dordrecht, 409–445
96. Finkenstädt T, Adler NT, Allen TO, Ewert J-P (1986) Regional distribution of glucose utilization in the telencephalon of toads in response to configurational visual stimuli: a ^{14}C-2DG study. *J Comp Physiol* A 158: 457–467
97. Dinges AW, Ewert J-P (1994) Species-universal stimulus responses, modified through conditioning, re-appear after telencephalic lesions in toads. *Naturwissenschaften* 81: 317–320
98. Guha K, Jorgensen CB, Larsen LO (1980) Relationship between nutritional state and testes function, together with the observations on patterns of feeding, in the toad. *J Zool (London)* 192: 147–155
99. Laming PR, Cairns C (1998) Effects of food, glucose, and water ingestion on feeding activity in the toad (*Bufo bufo*). *Behav Neurosci* 112(5): 1266–1272
100. Laming PR (1989) Central representation of arousal. In: Ewert J-P, Arbib MA (eds): *Visuomotor coordination*. Plenum, New York, 693–727
101. Laming PR (1993) Slow potential shifts as indicants of glial activation and possible neuromodulation. In: McCallum WC, Curry SH (eds): *Slow potential changes in the human brain*. Plenum, New York, 35–46
102. Laming PR, Nicol AU, Roughan JV, Ocherashvili IV, Laming BA (1995) Sustained potential shifts in the toad tectum reflect prey-catching and avoidance behavior. *Behav Neurosci* 109(1): 150–160

103. Patton P, Grobstein P (1998). The effects of telencephalic lesions on the visually mediated prey orienting behavior in the leopard frog (*Rana pipiens*). I. The effects of complete removal of one telencephalic lobe, with a comparison to the effect of unilateral tectal lobe lesions. *Brain Behav Evol* 51: 123–143
104. Patton P, Grobstein P (1998). The effects of telencephalic lesions on the visually mediated prey orienting behavior in the leopard frog (*Rana pipiens*). II. The effects of limited lesions to the telencephalon. *Brain Behav Evol* 51: 144–161
105. Finkenstädt T, Adler NT, Allen TO, Ebbesson SOE, Ewert J-P (1985) Mapping of brain activity in mesencephalic and diencephalic structures of toads during presentation of visual key stimuli: a computer assisted analysis of ^{14}C-2DG autoradiographs. *J Comp Physiol* A 156: 433–445
106. Finkenstädt T, Ewert J-P (1985) Glucose utilization in the toad's brain during anesthesia and stimulation of the ascending reticular arousal system: a ^{14}C-2- deoxyglucose study. *Naturwissenschaften* 72: 161–162
107. Lázár G, Kozicz, T (1990) Morphology of neurons and axon terminals associated with descending and ascending pathways of the lateral forebrain bundle in *Rana exculenta*. *Cell Tissue Res* 260: 535–548
108. Merchenthaler I, Lázár G, Maderdrut, JL (1989) Distribution of proenkephalin-derived peptides in the brain of *Rana esculenta*. *J Comp Neurol* 281: 23–39
109. Schwerdtfeger WK, Germroth P (1990) *The forebrain in nonmammals*. Springer, Berlin, 57–65
110. Matsumoto N, Schwippert WW, Beneke TW, Ewert J-P (1991) Forebrain-mediated control of visually guided prey-catching in toads: investigation of striato-pretectal connections with intracellular recording/labeling methods. *Behav Processes* 25: 27–40
111. Buxbaum-Conradi H, Ewert J-P (1999) Responses of single neurons in the toad's caudal ventral striatum to moving visual stimuli and test of their efferent projection by extracellular antidromic stimulation/recording techniques. *Brain Behav Evol* 54: 338–354
112. Gruberg ER, Ambros VR (1974) A forebrain visual projection in the frog (*Rana pipiens*). *Exp Brain Res* 44: 187–197
113. Buddenbrock W v (1937) *Grundriß der vergleichenden Physiologie*. Borntraeger, Berlin
114. Glagow M, Ewert J-P (1997) Dopaminergic modulation of visual responses in toads. I. Apomorphine-induced effects on visually directed appetitive and consummatory prey-catching behavior. *J Comp Physiol* A 180: 1–9
115. Glagow M, Ewert J-P (1999) Apomorphine alters prey-catching patterns in the common toad: behavioural experiments and ^{14}C-2-deoxyglucose brain mapping studies. *Brain Behav Evol* 54: 223–242
116. Ewert J-P, Beneke TW, Schürg-Pfeiffer E, Schwippert WW, Weerasuriya A (1994) Sensorimotor processes that underlie feeding behavior in tetrapods. In: Bels VL, Chardon M, Vandevalle P (eds): *Advances in comparative and environmental physiology*, Vol. 18: Biomechanics of feeding in vertebrates. Springer, Berlin, 119–161
117. Chu J, Wilcox RE, Wilczynski W (1994) Pharmacological characterization of D1 and D2 dopamine receptors in *Rana pipiens*. *Soc Neurosci Abstr* 20: 167
118. Djamgoz MBA, Wagner, H-J (1992) Localization and function of dopamine in the adult vertebrate retina. *Neurochem Int* 20: 139–191
119. Röttgen A (1999) Über den Einfluß von Neuropharmaka auf die visuelle Ansprechbarkeit in der retino-tectalen Projektion der Agakröte. Dr.rer.nat. Dissertation, Abt. Neurobiologie, Fachbereich Biologie/Chemie, Univ Kassel.
120. Glagow M, Ewert J-P (1997) Dopaminergic modulation of visual responses in toads. II. Influences of apomorphine on retinal ganglion cells and tectal cells. *J Comp Physiol* A 180: 11–18

121. Glagow M, Ewert J-P (1996) Apomorphine-induced suppression of prey oriented turning in toads is correlated with activity changes in pretectum and tectum: ^{14}C-2DG studies and single cell recordings. *Neurosci Lett* 220: 215–218
122. Sanchez-Camacho C, Marín O, Lopez JM, Moreno N, Smeets WJ, Ten Donkelaar HJ, González A (2002) Origin and development of descending catecholaminergic pathways to the spinal cord in amphibians. *Brain Res Bull* 57(3-4): 325–330
123. Hoffmann A (1973) Stereotaxis atlas of the toad's brain. *Acta Anat* 84: 416–451
124. Kicliter E, Northcutt G (1975) Ascending afferents to the telencephalon of ranid frogs: an anterograde degeneration study. *J Comp Neur* 161: 239–254
125. Northcutt RG, Royce GJ (1975) Olfactory bulb projections in the bullfrog *Rana catesbeiana*. *J Morphol* 145: 51–268
126. Ploog D, Gottwald P (1974) Verhaltensforschung: Instinkt, Lernen, Hirnfunktion. Urban & Schwarzenberg, München
127. Nistri A, Sivilotti L, Welsh DM (1990) An electrophysiological study of the action of N-methyl-D-aspartate on excitatory synaptic transmission in the optic tectum of the frog in vitro. *Neuropharmacol* 29: 681–687
128. Hickmott PW, Constantine-Paton M (1993) The contributions of NMDA, non-NMDA, and GABA receptors to postsynaptic responses in neurons of the optic tectum. *J Neurosci* 13(10): 4339–4353
129. Gamlin PD, Reiner A, Keyser T, Brecha N, Karten HJ (1996) Projection of the nucleus pretectalis to a retinorecipient tectal layer in the pigeon (*Columba livia*). *J Comp Neurol* 368(3): 424–438
130. Cucchiaro JB, Bickford ME, Sherman SM (1991) A GABAergic projection from the pretectum to the dorsal lateral geniculate nucleus in the cat. *Neurosci* 41(1) 213–226
131. Kenigfest NB, Belekhova MG, Karamyan OA, Minakova MN, Rio J-P, Reperant J (2002) Neurochemical organization of the turtle pretectum: an immunohistochemical study. Comparative analysis. *J Evol Biochem Physiol* 38(6): 673–688
132. Borostyankoi-Baldauf Z, Herczeg L (2002) Parcellation of the human pretectal complex: a chemoarchitectonic reappraisal. *Neurosci* 110(3): 527–540
133. Ebersole TJ, Coulon JM, Goetz FW, Boy SK (2001) Characterization and distribution of neuropeptide Y in the brain of a caecilian amphibian. *Peptides* 22: 325–334
134. Bertoz A, Vidal PP, Graf W (1992) *The head-neck sensory motor system*. Oxford Univ Press, New York
135. Foreman N, Stevens R (1987) Relationships between the superior colliculus and hippocampus: Neural and behavioural considerations. *Behav Brain Sci* 10: 101–152
136. Gonzalez-Lima F (1989) Functional brain circuitry related to arousal and learning in rats. In: Ewert J-P, Arbib MA (eds): *Visuomotor coordination*. Plenum, New York, 729–765
137. Ewert J-P, Finkenstädt T (1987) Modulation of tectal functions by prosencephalic loops in amphibians. *Behav Brain Sci* 10(1): 122–123
138. Birkhofer M, Bleckmann H, Görner P (1994) Sensory activity in the telencephalon of the clawed toad, *Xenopus laevis*. *Eur J Morphol* 2-4: 262–266
139. Walkowiak W, Berlinger M, Schul J, Gerhardt HC (1999) Significance of forebrain structures in acoustically guided behavior in anurans. *Eur J Morphol* 37(2-3): 177–181
140. Endepols H, Walkowiak W (1999) Influence of descending forebrain projections on processing of acoustic signals and audiomotor integration in the anuran midbrain. *Eur J Morphol* 37(2-3): 182–184
141. Chevalier G, Vacher S, Deniau JM (1984) Inhibitory nigral influence on tectospinal neurons, a possible implication of basal ganglia in orienting behavior. *Exp Brain Res* 53: 320–326

142. Chevalier G, Deniau JM (1990) Disinhibition as a basic process in the expression of striatal functions. *Trends Neurosci* 13: 277–280
143. Andersen H, Bræstrup C, Randrup A (1975) Apomorphine-induced stereotyped biting in the tortoise in relation to dopaminergic mechanisms. *Brain Behav Evol* 11: 365–373
144. Dhawan B, Saxena PN, Gupta GP (1961) Apomorphin-induced pecking in pigeons. *Brit J Pharmacol* 15: 285–295
145. Burg B, Haase C, Lindenblatt U, Delius JD (1989) Sensitization to and conditioning with apomorphine in pigeons. *Pharmacol Biochem Behav* 34: 59–64
146. Fekete M, Kurti AM, Priubusz J (1970) On the dopaminergic nature of the gnawing compulsion induced by apomorphine in mice. *J Pharmacol* 22: 377–379
147. McCulloch J, Savaki HE, McCulloch MC, Jehle J, Sokoloff L (1982) The distribution of alterations in energy metabolism in the rat brain produced by apomorphine. *Brain Res* 243: 67–80
148. Blackburn JB, Pfaust JG, Phillips AG (1992) Dopamine functions in appetitive and defensive behaviours. *Prog Neurobiol* 39: 247–279
149. Szechman H, Cleghorn JM, Brown GM, Kaplan RD, Franco SW, Rosenthal K (1987) Sensitization and tolerance to apomorphine in men: yawning, growth hormone, nausea, and hypothermia. *Psychiatr Res* 23: 245–255
150. Ugwoke MI, Sam E, Van den Mooter G, Verbeke N, Kinget R (1999) Assessment of apomorphine nasal spray in Parkinson treatment. *Int J Pharmac* 181: 125–193
151. Godoy AM, Delius JD (1999) Sensitization to apomorphine in pigeons is due to conditioning, subject to generalization but resistant to extinction. *Behav Pharmacol* 10: 367–378
152. Baxter BL, Gluckman MJ, Stein L, Scerni RA (1974) Self-injection of apomorphine in the rat: positive reinforcement by a dopamine receptor stimulant. *Pharmacol Biochem Behav* 2: 387–392
153. Cools AR, Broekkamp CLE, van Rossum JM (1977) Subcutanous injections of apomorphine, stimulus generalization and conditioning: serious pitfalls for the examiner using apomorphine as a tool. *Pharmacol Biochem Behav* 6: 705–708
154. Woolverton WL, Goldberg LI, Ginos JZ (1984) Intravenous self-administration of dopamine receptor agonists by rhesus monkeys. *J Pharmacol Exp Ther* 230: 678–683
155. Möller, H-G, K. Nowak K, Kuschinsky K (1987) Studies on interactions between conditioned and unconditioned behavioural responses to apomorphine in rats. *Naudyn-Schmiedeberg's Arch Pharm* 335: 673–679
156. Lindenblatt U, Delius JD (1988) Nucleus basalis prosencephali, a substrate of apomorphine-induced pecking in pigeons. *Brain Res* 453: 1–8
157. Wynne B, Delius JD (1995) Sensitization to apomorphine in pigeons: unaffected by latent inhibition but still due to classical conditioning. *Psychopharmacology* 119: 414–420
158. Godoy AM, Delius JD, Siemann M (2000) Dose shift effects on an apomorphine-elicited response. *Med Sci Res* 28: 39–42
159. Ewert J-P, Matsumoto N, Schwippert WW (1985) Morphological identification of prey-selective neurons in the grass frog's optic tectum. *Naturwissenschaften* 72: 661–662
160. Ewert J-P, Buxbaum-Conradi H, Dreisvogt F, Glagow M, Merkel-Harff C, Röttgen A, Schürg-Pfeiffer E, Schiwppert WW (2001) Neural modulation of visuomotor functions underlying prey-catching behaviour in anurans: perception, attention, motor performance, Learning. *Comp Biochem Physiol* A 128: 417–461

Neurotransmitter Interactions and Cognitive Function
Edited by Edward D. Levin
© 2006 Birkhäuser Verlag/Switzerland

Neuromodulators of LTP and NCAMs in the amygdala and hippocampus in response to stress

Adi Guterman and Gal Richter-Levin

Department of Psychology, Faculty of Social Sciences and the Brain and Behavior Research Center, University of Haifa, Israel

Introduction

Emotional arousal, namely stress, induces structural changes in neurons of the adult central nervous system (CNS) involving a biphasic secretion of stress-related hormones, in which norepinephrin (NE) represents the first phase and glucocorticoids (i. e. corticosterone, [CORT]) represent the second phase [1–3]. These stress hormones are potent modulators of both learning and brain plasticity, mediating their effects presumably by involvement of the limbic system, namely the amygdala and the hippocampus [2, 3]. The basolateral nucleus of the amygdala (BLA) is specifically activated by an emotional experience and is a critical site of the converging modulating influences of adrenal stress hormones on memory consolidation [38, 91]. In turn the amygdala modulates hippocampal-dependent memory in a complex manner *via* the stress hormones NE and CORT [7]. Therefore, the BLA may be a critical locus of interaction between glucocorticoids and the noradrenergic system in modulating memory consolidation [8, 9]. The effects of both NE and CORT upon the amygdala and the hippocampus areas, which affect synaptic plasticity alterations, are exemplified by modulation of long-term potentiation (LTP) formation. In addition, the stress hormones effects on the neurochemical circuitry, leading to changes in intracellular and cell-matrix interactions, also affect neural cell adhesion molecules, exhibiting a change in their post-translational modification molecular form in following stress exposure, further affecting synaptic plasticity.

Although the adrenomedullary hormone NE (adrenaline) and the adrenocortical hormone CORT affect brain function through different specific mechanisms and pathways, they converge in regulating memory consolidation by influencing central noradrenergic mechanisms [1, 7], of which the prominent ones will be reviewed in this chapter.

The effects of norepinephrin in the amygdala and hippocampus

Noradrenergic projections originating in the nucleus of the solitary tract (NTS) innervate forebrain structures involved in learning and memory, including the amygdala.

Since BLA activation is required for modulation of memory consolidation in the hippocampus, activation of both α- and β-adrenoceptors in the BLA, followed by hippocampal activation, is critical in mediating noradrenergic influences on synaptic plasticity processes [14, 37, 40]. NE release plays an important, possibly critical, role in the amygdala, mediating emotional arousal effects on memory consolidation [36, 46]. *In vivo* microdialysis and high-performance liquid chromatography (HPLC) studies indicate that epinephrine released by emotionally arousing training experiences induces the release of NE within the amygdala. For example, footshock stimulation, such as that used in inhibitory avoidance training, also induces the release of NE in the amygdala. The amount of release varies directly with stimulus intensity [36, 46, 87]. The elevated NE levels observed in the amygdala following training, as well as the NE levels assessed within the individual animals correlate highly with later retention performance [41]. Conversely, systemic injections of epinephrine in the amygdala enhance NE release [94]. In addition, NE or β-adrenoceptors infusions in the amygdala block epinephrine effects on memory consolidation, while these same hormones agonists' infusions to the amygdala will enhance memory consolidation following training [14, 36, 39].

The effects of corticosterone in the amygdala and hippocampus

At the onset of an emotional event glucocortiocoid levels permissively mediate the cognitive stress response, whereas the subsequent stress-induced rise in the glucocorticoid concentrations suppresses this response [69, 88, 90]. Glucocorticoids enter the brain freely and bind to two intracellular types of adrenal steroid receptors [41], allowing their direct influence on hippocampal glucocorticoid receptors (GRs) in order to modulate LTP [12]. The low-affinity GRs are involved in mediating glucocorticoid effects on memory consolidation [27, 32, 38, 86]. Although the BLA contains a moderate density of GRs [88], the hippocampus exhibits a high-density level of these receptors [26, 53]. Emotional arousal also activates the hypothalamic-pituitary-adrenocortical (HPA) axis, resulting in elevated plasma levels of CORT. Indeed, glucocorticoid-induced impairment of declarative memory retrieval has also been observed in human subjects [51]. In addition, either post-training infusions of CORT, or administration of specific agonists or antagonists of GRs into the hippocampus, affect memory consolidation for both aversive and appetitive tasks [27, 32]. Specifically, acute post-training administration of low doses of glucocorticoids enhances memory consolidation [86]. Blockade of the CORT stress response using the CORT synthesis inhibitor metyrapone prevents inhibitory avoidance retention enhancement induced by post-training epinephrine injections or exposure to psychological stress [92, 95]. Similarly, metyrapone treatment prevented the stress-induced enhancement of spatial performance in the water maze [1]. Glucocorticoid effects on memory consolidation require activation of the BLA, signaling it as one of the glucocorticoid loci of action in modulation of memory consolidation [27, 32, 38, 86]. Infusions of the specific GR agonist RU 28362 into the BLA, immediately following inhibitory avoidance training, enhance retention performance. Intra-BLA infusions of the GR antagonist RU 38486 impair retention performance in a water-maze spatial task. Furthermore, selective lesions of the BLA block inhibitory avoidance retention enhancement induced by post-training systemic injections of the synthetic glucocorticoid dexamethasone [27, 36, 47]. Training on a water-maze spatial task also increases phosphrylation of the extracellular regulated kinase (ERK2), a subtype of the mitogen-activated protein kinases (MAPK cascade) [24, 35]. Phosphorylation of ERK2 in the amygdala was found only in rats trained under high stress conditions. The training conditions were accompanied by high plasma levels of training-induced CORT. ERK2 is considered critical for memory consolidation and long-term neuronal plasticity in both the amygdala and the hippocampus [35, 66]. Furthermore, it can be activated by noradrenergic stimulation and cAMP formation [24, 38].

The mediation of both NE and CORT stress hormones formulate diverse memory processes in the BLA, which in turn activate hippocampal memory [40, 42]. Noradrenergic activation within the BLA is essential for the memory modulating influences of systematically administered epinephrine and glucocorticoids as well as for the effects of glucocorticoids infused directly into the hippocampus. Thus, NE and CORT effects on the consolidation of memory for emotional experiences are intimately linked to noradrenergic activation in the BLA followed by hippocampal activation [53].

Stress hormones effects on long-term potentiation in the amygdala and hippocampus

Although emotional experiences can either enhance or impair hippocampal memory and plasticity [65], BLA activation was reported to enhance hippocampal LTP [45, 54]. Both NE and CORT are required for BLA modulation (enhancement or suppression) of DG-LTP. Ipsilateral BLA spaced activation (2 h prior to Perforant Path tetanization) suppressed DG-LTP. This suppressive effect was also mediated by NE and CORT. Thus, both NE and CORT seem to be involved in the enhancing as well as the inhibitory effects of the BLA. The involvement of both hormones could be attributed to time dependence, i. e., the effects of a brief exposure to these hormones are excitatory, whereas their prolonged presence in the spaced phase may lead to inhibitory effects [1].

Another possibility is the involvement of a third scaffold mediator, upstream of these stress hormones that will define the net effect of the cascade, be it excitatory or inhibitory. Such a mediator could be acetylcholine (ACh), which has been suggested to mediate the transition of early into late phase LTP by BLA activation [89], and there are indications that NE effects on memory involve subsequent cholinergic activation in the amygdala [74]. Furthermore, it has been suggested that ACh is involved in stress effects on hippocampal processing [83]. Another mediator could be corticotrophin-releasing factor (CRF) released from the hypothalamus in response to stress, thus leading to the secretion of the stress hormones [47]. CRF injected into the DG produced a dose-dependent and long-lasting enhancement in synaptic efficacy of these neurons [91], though sustained administration of CRF prevented the occurrence of LTP [69, 81]. Excitation or inhibition could also be accounted for by the exact ratio between the effects of the two hormones, i. e., both are required for the modulation, but the specific concentration of each will define the final outcome [1, 86].

NE has been shown repeatedly to be involved in memory reinforcement of various behavioral tasks [63, 76] and in the reinforcement of hippocampal LTP [23, 50]. The locus ceruleus may also be activated to induce NE release in the hippocampus and contribute to the facilitation of LTP [67, 79]. Specifically, it has been suggested that noradrenergic activation of the BLA may serve to modulate memory storage, enhancement and plasticity in the hippocampus [17, 23, 27, 36, 50]. Accordingly, NE-depleted rats showed no priming effect in the BLA. NE depletion also prevented inhibition of LTP by BLA spaced activation [1, 87].

CORT release and, by this, hippocampal LTP is modulated by the amygdala, which in turn affects the hypothalamus [54]. CORT has dose-dependent inverted U-shaped effects on hippocampal LTP and primed burst potentiation (PBP) [29, 54, 56–59]. In addition, amygdala electrical stimulation has been shown to increase plasma levels of CORT [56], suggesting that a functioning BLA is required for adrenal steroids to exert their influence on hippocampal memory storage [61, 80, 88]. Inhibitory effects of the spaced activation of the amygdala on DG-LTP are mediated by CORT, therefore inhibition was significantly suppressed in CORT-depleted rats and BLA priming was absent [1, 40], just as with NE. Moreover, it has been shown that administration of exogenous CORT in the appropriate temporal context, i. e., in close relation to training, potentiated memory for hippocampal-dependent tasks [19, 34, 52, 70]. Additionally, because the blocking of priming by metyrapone was evident only 30 min. post-HFS, it is possible that amygdala-induced increase in CORT levels is required for post post-tetanic potentiation mechanisms of LTP enhancement [1].

Noradrenergic activation of the BLA is required for the adrenal steroids to influence hippocampal memory storage [61]. Glucocorticoids seem to exert a permissive action on the efficacy of the noradrenergic system and vice versa [63, 90]. It is currently unclear whether an interaction between these two modulatory systems or their parallel action is required. It may be that lack of either system could affect BLA modulation of hippocampal LTP to the same degree [1, 26]. Results also suggest the existence of two distinctive pathways: an ipsilateral neural pathway that requires the involvement of NE and CORT and a contralateral pathway that presumably acts *via* mediation of another brain structure. The effects of this pathway are NE and CORT independent [1]. NE or CORT- depleted animals receiving priming stimulation of the contralateral BLA exhibited a significantly enhanced DG-LTP compared to the control LTP group. The contralateral effect however proved non dependent on neither noradrenergic nor corticosteroid activation. Thus, differential neural mechanisms probably underlie the ipsilateral and contralateral BLA priming effects on DG-LTP, i. e. hippocampal plasticity [1, 34, 67, 71]. However, the majority of findings, including pharmacological studies that do not differentiate and include effects on both ipsi- and contralateral pathways, indicate that the NE and CORT dependent ipsilateral pathway dominates with respect to effects on hippocampal dependent memory and plasticity. It can thus be concluded that these major hormonal systems – adrenergic and glucocorticoid – appear to interact to influence memory consolidation.

Stress hormones effects on neural cell adhesion molecules expression in the amygdala and hippocampus

Neurons in the hippocampus, the amygdala and in other brain areas, such as the prefrontal cortex, undergo neurite remodeling following chronic stress. In the hippocampus some of these effects can be mimicked with chronic administration of adrenal steroids, such as glucocorticoids, which – apart from LTP modulation – may also affect memory consolidation through trans-activation or protein-protein interactions with other transcription factors or effector systems. These changes in neuronal structure may be mediated by certain molecules related to plastic events such as the polysialylated form of the neural cell adhesion molecules (PSA-NCAM). NCAM is a membrane bound glycoprotein member of the immunoglobulin superfamily of adhesion molecules which, through homophilic and heterophilic binding, mediates cell to cell and cell to extracellular matrix interactions, thus is of critical prominence in morphogenesis and synaptic plasticity processes [13, 15, 33, 55, 60, 73, 84]. NCAM can be polysialylated by the attachment of long $\alpha 2$, 8-linked polysialic acid (PSA) homopolymer chains. Such a posttranslational modification confers the NCAM anti-adhesive properties [21, 73] and is believed to inhibit NCAM-mediated cell-cell and cell-matrix interactions [21, 25, 33, 44, 48]. It is also known to play key roles in activity-dependent synaptic remodeling [25, 33, 44] and memory storage [5] as well as developmental events, such as synaptogenesis [72] and axonal outgrowth and fasciculation [73]. PSA-NCAM may participate in the modification of mossy fibers ultrastructure [15] and in the reduction of synapse density in the hippocampal CA3 region [82]. In accordance, stress affects the expression of these adhesion molecules. Accumulating evidence show that chronic stress induces dendritic atrophy in hippocampal neurons [17, 50], alters mossy fiber synaptic terminal structure [15] and promotes a transient upregulation of PSA-NCAM expression [55, 75]. Specifically, the morphological changes induced by chronic restraint stress are accompanied by an upregulation of PSA-NCAM hippocampal expression [15, 49], increasing the number of PSA-NCAM immunoreactive neurons [75] in both the hippocampus [77] and the amygdala [25], already presenting elevated levels of these plasticity related molecules during adult life [75]. In particular, the frequency of PSA-NCAM neurons at the intragranular border of the hippocampal DG has been shown to present transient increases 10–12 h following training of rats in a variety of learning tasks, including the Morris water maze [4, 18, 77, 78]. Moreover, chronic restraint stress appears to downregulate the NCAM140 isoform but not NCAM180, which is believed to be the carrier of PSA. Since PSA-NCAM appears to be an important player in morphological plasticity in the nervous system [33], these increases in PSA-NCAM expression following chronic restraint stress may be related to the reported structural plasticity of the hippocampal DG and CA3 region. The timing of this transient increase in PSA-NCAM expression may play a role in hippocampal synapse selection, an ongoing process during this time period [28]. Specific removal of PSA from NCAM using endoneuraminidase (EndoN) impairs activity-induced synaptic potentiation [30, 60, 72] and spatial memory [4]. Evidence also indicates that PSA activation is involved in learning-associated synaptic remodeling [78]. Accordingly, a positive correlation was found between mean latency to learn the platform location at training in the Morris water maze task and the activated frequency of dentate polysialylated neurons, such that the higher polysialylation response was observed in rats showing the slower acquisition rate. Neural circuits subserving learning in fast and slow learners show a differential training-induced regulation of synaptic remodeling mechanisms. Accumulating evidence strongly suggest that a greater structural reorganization of neural circuits occurs in the hippocampus of animals that require a greater effort to learn the task [49], therefore PSA modulation occurring several hours following training appears to be related to the consolidation of long-term memory [8, 71]. Interestingly, *a priori* differences in hippocampal morphometry [83], neurochemistry [85], and expression of the cell adhesion molecule L1 have been related to differential performance in spatial learning tasks in adult rodents [20]. The correlation found between water maze performance and PSA regulation is related, not only to performance in a single session, but to a pattern of spatial learning and memory abilities, and moreover, learning-related glucocorticoid responsiveness [49, 62]. The learning-related neural circuits of fast learners are better suited to solving the water maze task than those of slow learners, the latter requiring structural reorganization to form memory as opposed to the relatively economic mechanism of altering synaptic efficacy used by the former. Indeed, synaptic density was shown to be higher in animals that showed a poorer acquisition curve [49, 64, 68].

There are several types of evidence suggesting a role of glucocorticoids in the regulation of PSA-NCAM expression. PSA is specifically attached to NCAM by sialyltransferases and glucocorticoids may be involved in facilitating this structural plasticity process [25, 30, 33, 43, 75]; in contrast, chronic CORT treatment also induces dendritic atrophy and structural changes in the hippocampal mossy fibers [6, 48, 74]. Thus, aversive training and post-training injections of glucocorticoids affect expression of NCAMs in the hippocampus [5, 15, 62]. The difference between chronic restraint stress and chronic CORT treatment effects on PSA-NCAM expression may be related to differential effects of the CORT: The exposure to elevated levels of CORT is more prolonged during chronic CORT treatment than in chronic restraint stress, in which the CORT response habituates over time. However, to explain the decrease of PSA-NCAM following CORT administration, other mediators such as excitatory amino acids should be considered [75]. Moreover, chronic stress is a complex scenario, in which not only the CORT response becomes activated but also other neurotransmitter systems [3, 9, 23, 58, 64, 92, 95].

The changes induced by chronic restraint stress on neuronal structure, and possibly on the expression of molecules related to structural plasticity, involve the participation of adrenal hormone CORT. This hormone seems necessary for the stress effects on dendritic morphology, because stress-induced dendritic atrophy is prevented by treatment with cyanoketone, a blocker of adrenal steroid synthesis, and chronic glucocorticoid treatment causes dendritic atrophy in the hippocampus [69, 74]. These effects of adrenal hormones in the hippocampus are mediated by glucocorticoid receptors, which are abundantly expressed in neurons of this limbic region [88, 90].

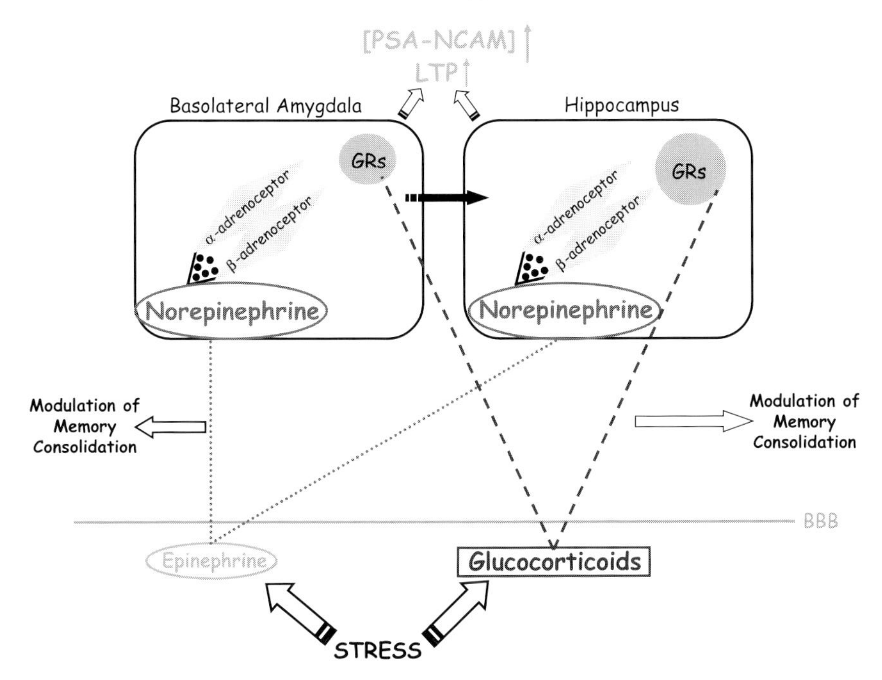

Figure 1. Interactions of adrenal stress hormones with the noradrenergic system in the basolateral amygdala (BLA) in modulating memory consolidation *via* long-term potentiation (LTP) mechanisms and the expression of polysialylated neural cell adhesion molecules (PSA-NCAM). Adrenal stress hormones are released following exposure to stress experiences and are known to enhance memory consolidation. Epinephrine, which does not cross the blood brain barrier (BBB), induces norepinephrine release in the BLA and the hippocampus limbic areas involved in the response to stress. Norepinephrine binds to both α- and β-adrenoceptors at postsynaptic sites. The noradrenergic activation of the BLA is required for the adrenal steroids to influence hippocampal memory storage. Glucocorticoids freely enter the brain and bind to glucocorticoid receptors (GRs), more abundant in the hippocampus than in the BLA, to potentiate norepinephrine release in the BLA, as well as postsynaptically in the BLA neurons to facilitate the norepinephrine signal cascade. These stress hormones effects on noradrenergic activation in the BLA are required for modulation of memory consolidation in other brain areas, causing enhancement of LTP and upregulation in the expression of the polysialylated synaptic plasticity neural adhesion molecules, mainly in the hippocampus.

Summary

Possibly, at the onset of an emotional event the stress hormones permissively mediate plasticity [1]. Specifically, CORT and NE stress hormones participate in modulation of memory consolidation processes in both the amygdala and the hippocampus. In addition, glucocorticoids and norepinephrin bound to adrenoceptors are also involved in modulating the regulation of NCAM polysialylation both in the amygdala and in the hippocampus [85]. PSA-related synaptic remodeling is mobilized for memory formation in particularly challenging circumstances [49].

References

1. Akirav I, Sandi C., Richter-Levin G. (2001) Differential activation of hippocampus and amygdala following spatial learning under stress *Eur J Neurosci* 14: 719–725
2. Conrad C D, Lupien S J, McEwen B S (1984) Support for a bimodal role of type II adrenal steroid receptors in spatial memory. *Neurobiol Learn Mem* 72: 39–46
3. McEwen B S, Sapolsky R M (1995) Stress and cognitive function. *Curr Opin Neurobiol* 5: 205–216
4. Murphy KJ, O'Connell AW, Regan CM (1996) Repetitive and transient increases in hippocampal neural cell adhesion molecule polysialylation state following multi-trial spatial training. *J Neurochem* 67: 1268–1274
5. Sadoul R, Hirn M, Deagostini H, Rougon G, Goridis C (1983) Adult and embryonic mouse neural cell adhesion molecules have different binding properties. *Nature* 304: 347–349
6. Sandi C, Loscertales M (1999) Opposite effects on NCAM expression in the rat frontal cortex induced by acute vs. chronic corticosterone treatments. *Brain Res* 828: 127–134
7. Kaufer D, Friedman A, Seidman S, Soreq H (1998) acute stress facilitates long-lasting changes in cholinergic gene expression. *Nature* 28: 373–377
8. Murphy KJ, Regan CM (1998) Contributions of cell adhesion molecules to altered synaptic weights during memory consolidation. *Neurobiol Learn Mem* 70: 73–81
9. Nacher J, Pham K, Gil-Fernandez V, McEwen BS (2004) Chronic restraint stress and chronic corticosterone treatment modulate differentially the expression of molecules related to structural plasticity in the adult rat piriform cortex. *Neuroscience* 126: 503–509
10. Gold PE, van Buskirk R (1975) Facilitation of time-dependent memory processes with posttrial epinephrine injections. *Behav Biol* 13: 145–155
11. Impey S, Obrietan K, Storm DR (1999) Making new connections: role of ERKqMAP kinase signaling in neuronal plasticity. *Neuron* 23: 11–14
12. Kerr DS, Huggett AM, Abraham WC (1994) Modulation of hippocampal long-term potentiation and long-term depression by corticosteroid receptor activation. *Psychobiology* 22: 123–133
13. Rougon G (1993) Structure, metabolism and cell biology of polysialic acid. *Eur J Cell Biol* 61: 197–207
14. Inroinin-Collison IB, Dalmaz C, McGaugh JL (1996) Amygdala beta-noradrenergic influences on memory storage involve cholinergic activation. *Neurobiol Learn Mem* 65: 57–64
15. Seki T, Rutishauser U (1998) Removal of polysialic acid-neural cell adhesion molecule induces aberrant mossy fiber innervation and ectopic synaptogenesis in the hippocampus. *J Neurosci* 18: 3757–3766
16. Gamaro GD, Denerdin JD Jr., Michalowski MB, Catelli D, Correa JB, Xavier MH, Dalmaz C (1997) Epinephrine effects on memory are not dependent on hepatic glucose release. *Neurobiol Learn Mem* 68: 221–229
17. McGaugh JL (1989) Involvement of hormonal and neuromodulatory systems in the regulation of memory storage. *Annu Rev Neurosci* 12: 255–287
18. Becker CG, Artola A, Gerardy-Schahn R, Becker T, Welzl H, Schachner M (1996) The polysialic acid modification of the neural cell adhesion molecule is involved in spatial learning and hippocampal long-term potentiation. *J Neurosci Res* 45: 143–152
19. Seidenbecher T, Reymann KG, Balschun D (1997) A post-tetanic time window for the reinforcement of long-term potentiation by appetitive and aversive stimuli. *PNAS USA* 18: 1494–1499

20. Merino JJ, Cordero MI, Sandi C (2001) Regulation of hippocampal cell adhesion molecules NCAM and L1 by contextual fear conditioning is dependent upon time and stressor intensity. *Eur J Neurosci* 14: 675–689
21. Cunningham BA, Hemperley JJ, Murray BA, Prediger EA, Brackenbury R, Edelman GM (1987) Neural cell adhesion molecule: structure immunoglobulin-like domains cell surface modulation and alternative RNA splicing. *Science* 236: 799–806
22. Diamond DM, Bennett MC, Fleshner M, Rose GM (1992) Inverted-U relationship between the level of peripheral corticosterone and the magnitude of hippocampal primed burst potentiation. *Hippocampus* 2: 421–430
23. McEwen BS, Albeck D, Cameron HA, Chao HM, Gould E, Hastings N, Kuroda Y, Luine V, Magarinos AM, McPatrick CR et al. (1995) Stress and the brain: a paradoxical role for adrenal steroids. In: *Vitamins and hormones* 51: 371–402. Academic Press
24. Roberson ED, English JD, Adams JP, Selcher JC, Kondratick C, Sweatt JD (1999) The mitogen-activated protein kinase cascade couples PKA and PKC to cAMP response element binding protein phosphorylation in area CA1 of hippocampus. *J Neurosci* 19: 4337–4348
25. Rønn LC, Berezin V, Bock E (2000) The neural cell adhesion molecule in synaptic plasticity and aging. *Int J Dev Neurosci* 18: 193–199
26. Pavlides C, Watanabe Y, Magarinos AM, McEwen BS (1995) Opposing roles of type I and type II adrenal steroid receptors in hippocampal long-term potentiation. *Neuroscience* 68: 387–394
27. McIntyre CK, Hatfield T, McGaugh JL (2000) Norepinephrine release in the rat amygdala during inhibitory avoidance training. *Soc Neurobisci Abstr* 26: 193
28. Doyle E, Bell R, Regan CM (1992) Hippocampal NCAM180 transiently increases sialylation during the acquisition and consolidation of a passive avoidance response in the adult rat. *J Neurosci Res* 31: 513–523
29. Reul JMHM, de Kloet ER (1985) Two receptor systems for corticosterone in the rat brain: microdistribution and differential occupation. *Endocrinology* 117: 2505–2512
30. Rutishauser U, Landmesser L (1996) Polysialic acid in the vertebrate nervous system: a promoter plasticity in cell-cell interactions. *Trends Neurosci* 19: 422–427
31. De Kloet ER (1991) Brain corticosteroid receptor balance and homeostatic control. *Front Neuroendocrinol* 12: 95–164
32. Roozendaal B (2000) Glococorticoids and the regulation of memory consolidation. *Psychoneuroendocrinology* 25: 213–238
33. Schachner M (1997) Neural recognition molecules and synaptic plasticity. *Curr Opin Cell Biol* 9: 627–634
34. Akirav I, Richter-Levin G (1999) Priming stimulation in the basolateral amygdale modulates synaptic plasticity in the rat dentate gyrus. *Neurosci Lett* 30: 83–86
35. Rouppe van der Voort C, Kavelaars A, van de Pol M, Heijnen CJ (2000) Noradrenaline induces phosphorylation of ERK-2 in human peripheral blood mononuclear cells after induction of alpha1-adrenergic receptors. *J Neuroimmunol* 108: 82–91
36. Liang KC, McGaugh JL (1986) Modulating effects of posttraining epinephrine on memory: involvement of the amygdala noradrenergic system. *Brain Res* 368: 125–133
37. Roozendaal B, Portillo-Marquez G, McGaugh JL (1996) Basolateral amygdala lesions block glucocorticoid-induced modulation of memory for spatial learning. *Behav Neurosci* 110: 1074–1083
38. Roozendaal B, Quirarte GL, McGaugh JL (2002) Glucocorticoids interact with the basolateral amygdale beta-adrenoceptor-cAMP/cAMP/PKA system in influencing memory consolidation. *Eur J Neurosci* 15: 553–560

39. Roozandaal B, McGaugh JL (1996) Amygdaloid nuclei lesions differentially affect glucocorticoid-induced memory enhacement in an inhibitory avoidance task. *Neurobiol Learn Mem* 65: 1–8
40. Roozandaal B, McGaugh JL (1997) Basolateral amygdale lesions block the memory-enhancing effect of glucocorticoids administration in the dorsal hippocampus of rats. *Eur J Neurosci* 9: 76–83
41. Ferry B, Roozendaal B, McGaugh JL (1999) Basolaterla amygdale noradrenergic influences on memory storage are mediated by an interaction between β- and α1-adrenoceptors. *J Neurosci* 19: 5119–5123
42. Akirav I, Richter-Levin G (2002) Mechanisms of amygdala modulation of hippocampal plasticity. *J Neurosci* 22(22): 9912–9921
43. Kiss JZ (1998) A role of adhesion molecules in neuroglial plasticity. *Mol Cell Endocrinol* 140: 89–94
44. Kiss JZ, Muller D (2001) Contribution of the neural cell adhesion molecule to neuronal and synaptic plasticity. *Rev Neurosci* 12: 297–310
45. Bliss TVP, Goddard GV, Riives M (1983) Reduction of long-term potentiation in the dentate gyrus of the rat following selective depletion of monoamines. *J Physiol (Lond)* 334: 475–491
46. Liang KC, McGaugh JL, Yao H (1990) Involvement of the amygdale pathways in the influence of posttraining amygdala norepinephrine and peripheral epinephrine on memory storage. *Brain Res* 508: 225–233
47. McEwen BS, Weiss JM, Schwartz LS (1970) Retention of corticosterone by cell nuclei from brain regions of adrenalectomized rats. *Brain Res* 17(3): 471–482
48. Muller D,Wang C, Skibo G, Toni N, Cremer H, Calaora V, Rougon G, Kiss JZ (1996) PSA-NCAM is required for activity-induced synaptic plasticity. *Neuron* 17: 413–422
49. Seki T, Arai Y (1991) The persistent expression of a highly polysialylated NCAM in the dentate gyrus of the adult rat. *Neurosci Res* 12: 503–513
50. van Bockstaele E, Colago E, Aicher S (1998) Light and electron microscopic evidence for topographic and monosynaptic projections from neurons in the ventral medulla to noradrenergic dendrites in the rat locus coeruleus. *Brain Res* 784: 123–138
51. Cahill L, McGaugh JL (1998) Mechanisms of emotional arousal and lasting declarative memory. *Trends Neurosci* 21: 294–299
52. Pugh CR, Tremblay D, Fleshner M, Rudy JW (1997) A selective role for corticosterone in contextual-fear conditioning. *Behav Neurosci* 111: 503–511
53. Lathe R (2001) Hormones and the hippocampus. *J Endocrinol* 169: 205–231
54. Rey M, Carlier E, Talmi M, Soumireu-Mourat B (1994) Corticosterone effects on long-term potentiation in mouse hippocampal slices. *Neuroendocrinology* 60: 36–41
55. Durbec P, Cremer H (2001) Revisiting the function of PSA-NCAM in the nervous system. *Mol Neurobiol* 24: 53–64
56. Feldman S, Conforti N, Siegel RA (1982) Adrenocortical responses following limbic stimulation in rats with hypothalamic deafferentations. *Neuroendocrinology* 35: 205–211
57. Ferry B, Roozendaal B, McGaugh JL (1999) Involvement of α1-adrenoceptors in the basolateral amygdala in modulation of memory storage. *Eur J Pharmacol* 372: 9–16
58. Liu L, Tsuji M, Takeda H, Takada K, Matsumiya T (1999) Adrenocortical suppression blocks the enhancement of memory storage produced by exposure to psychological stress in rats. *Brain Res* 821: 134–140
59. Roozendaal B, Carmi O, McGaugh JL (1996) Adrenocortical suppression blocks the memory-enhancing effects of amphetamine and epinephrine. *PNAS USA* 93: 1429–1433

60. Rutishauser U (1996) Polysialic acid and the regulation of cell interactions. *Curr Opin Cell Biol* 8: 679–684
61. Sandi C, Loscertales M, Guaza C (1997) Experience-dependent facilitating effect for corticosterone on spatial memory formation in the water maze. *Eur J Neurosci* 9: 637–642
62. Rodriguez JJ, Montaron MF, Petry KG, Aurousseau C, Marinelli M, Premier S, Rougon G, Le Moal M, Abrous DN (1998) Complex regulation of the expression of the polysialylated form of the neuronal cell adhesion molecule by glucocorticoids in the rat hippocampus. *Eur J Neurosci* 10: 2994–3006
63. Diamond DM, Bennett MC, Engstrom DA, Fleshner M, Rose GM (1989) Adrenalectomy reduces the threshold to hippocampal primed burst potentiation in the anesthesized rat. *Brain Res* 17: 356–360
64. Sandi C, Merino JJ, Cordero MI, Touyarot K, Venero C (2001) Effects of chronic stress on contextual fear conditioning and the hippocampal expression of the neural cell adhesion molecule its polysialiylation and L1. *Neuroscience* 102: 329–339
65. de Quervain DJ-F, Roozendaal B, Nitsch RM, McGaugh JL, Hock C (2000) Acute cortisone administration impairs retrieval of long-term declarative memory in humans. *Nat Neurosci* 3: 313–314
66. Schafe GE, Atkins CM, Swank MW, Bauer EP, Sweatt JD, LeDoux JE (2000) Activation of ERK/MAP kinase in the amygdala is required for memory consolidation of Pavlovian fear conditioning. *J Neurosci* 20: 8177–8187
67. Frey S, Bergado-Rosado J, Seidenbecher T, Pape HC, Frey JU (2001) Reinforcement of early long-term potentiation (early-LTP) in dentate gyrus by stimulation of the basolateral amygdala: heterosynaptic induction mechanisms of late-LTP. *J Neurosci* 15: 3697–3703
68. Sandi C, Davies HA, Cordero MI, Rodriguez JJ, Popov VI, Stewart MG (2003) Rapid reversal of stress induced loss of synapses in CA3 of rat hippocampus following water maze training. *Eur J Neurosci* 17: 2447–2456
69. Magarinos AM, McEwen BS (1995) Stress-induced atrophy of apical dendrites of hippocampal CA3c neurons: involvement of glucocorticoid secretion and excitatory amino acid receptors. *Neuroscience* 69: 89–98
70. Magarinos AM, Verdugo JM, McEwen BS (1997) Chronic stress alters synaptic terminal structure in hippocampus. *PNAS USA* 94: 14002–14008
71. O'Malley A, O'Connell C, Murphy KJ, Regan CM (2000) Transient spine density increases the mid-molecular layer of hippocampal dentate gyrus accompany consolidation of a spatial learning task in the rodent. *Neuroscience* 99: 229–232
72. Nothias F, Vernier P, von Boxberg Y, Mirman S, Vincent JD (1997) Modulation of NCAM polysialylation is associated with morphofunctional modifications in the hypothalamo-neurohypophysial system during lactation. *Eur J Neurosci* 9: 1553–1565
73. Finne J, Finne U, Deagostini-Bazin H, Goridis C (1983) Occurrence of α2-8 linked polysialosyl units in a neural cell adhesion molecule. *Biochem Biophys Res Commun* 112: 482–487
74. McGaugh JL, Roozendaal B (2002) Role of adrenal stress hormones in forming lasting memories in the brain. *Curr Opi in Neurobiol* 12: 205–210
75. Pham K, Nacher J, Hof PR, McEwen BS (2003) Repeated restraint stress suppresses neurogenesis and induces biphasic PSA-NCAM expression in the adult rat dentate gyrus. *Eur J Neurosci* 17: 879–886
76. Richter-Levin G (2004) The amygdala, the hippocampus, and emotional modulation of memory. *Neuroscientist* 10(1): 31–39
77. Venero C, Tilling T, Hermans-Borgmeyer I, Herrero AI, Schachner M, Sandi C (2004) Water maze learning and forebrain mRNA expression of the neural cell adhesion molecule L1. *J Neurosci Res* 75: 172–181

78. Foley AG, Hedigan K, Roullet P, Sara SJ, Murphy KJ, Regan CM (2003) Consolidation of odor-reward associative memory involves neural cell adhesion molecule polysialylation-mediated synaptic plasticity within the rodent hippocampus. *J Neurosci Res* 74: 570–576
79. Huang YY, Kandel ER (1996) Postsynaptic induction and PKA-dependent expression of LTP in the lateral amygdale. *Neuron* 21: 169–178
80. Huber G, Bailly Y, Marin JR, Mariani J, Brugg B (1997) Synaptic β-amyloid precursor proteins increase with learning capacity in rats. *Neuroscience* 80: 313–320
81. Rebaudo R, Melani R, Balestrino M, Izvarina N (2001) Electrophysiological effects of sustained delivery of CRF and its receptor agonists in hippocampal slices. *Brain Res* 13: 112–117
82. Bernasconi-Guastalla S, Wolfer DP, Lipp HP (1994) Hippocampal mossy fibers and swimming navigation in mice: Correlations with size and left-right asymmetries. *Hippocampus* 4: 53–63
83. Bhatnagar S, Costall B, Smythe JW (1997) Hippocampal cholinergic blockade enhances hypothalamic-pituitary-adrenal responses to stress. *Brain Res* 22: 244–248
84. Theodosis DT, Bonhomme R, Vitiello S, Rougon G, Poulain DA (1999) Cell surface expression of polysialic acid on NCAM is a prerequisite for activity-dependent morphological neuronal and glial plasticity. *J Neurosci* 19: 10228–10236
85. Nacher J, Lanuza E, McEwen BS (2002) Distribution of PSA-NCAM expression in the amygdala of the adult rat. *Neuroscience* 113: 479–484
86. McGaugh JL (2000) Memory – a century of consolidation. *Science* 14: 248–251
87. Williams CL, Men D, Clayton EC, Gold PE (1998) Norepinephrine release in the amygdala following systemic injection of epinephrine or escapable footshock: contribution of the nucleus of the solitary tract. *Behav Neurosci* 112: 1414–1422
88. Quirarte GL, Roozendaal B, McGaugh JL (1997) Glucocorticoids enhancement of memory storage involves noradrenergic activation in the basolateral amygdala. *PNAS USA* 9: 14048–14053
89. McIntyre CK, Marriot LK, Gold PE (2003) Patterns of brain acetylcholine release predict individual deifferences in preferred learning strategies in rats. *Neurobiol Learn Mem* 79: 177–183
90. Roozendaal B, Williams CL, McGaugh JL (1999) Glucocorticoid receptor activation in the rat nucleus of solitary tract facilitates memory consolidation: involvement of the basolateral amygdala. *Eur J Neurosci* 11: 1317–1323
91. Wang HL, Wayner MJ, Chai CY, Lee EH (1998) Corticotrophin-releasing factor produces a long-lasting enhancement of synaptic efficacy in the hippocampus. *Eur J Neurosci* 10: 3428–3437
92. Akirav I, Kozenicki M, Tal D, Sandi C, Venero C, Richter-Levin G (2004) A facilitative role for corticosterone in the acquisition of a spatial task under moderate stress. *Learn Mem* Mar-Apr 11(2): 188–95
93. Cordero MI, Sandi C (1998) A role for brain glucocorticoid receptors in contextual fear conditioning: dependence upon training intensity. *Brain Res* 786: 11–17
94. Quirarte GL, Galvez R, Roozendaal B, McGaugh JL (1998) Norepinephrine release in the amygdala in response to footshock and opioid peptidergic drugs. *Brain Res* 808: 134–140
95. de Kloet ER, Oitzl OS, Joëls M (1999) Stress and cognition: are corticosteroids good or bad guys? *Trends Neurosci* 22: 422–426
96. Richter-Levin G, Akirav I (2003) Emotional tagging of memory formation–in the search for neural mechanisms. *Brain Res Brain Res Rev* 43(3): 247–56, Review

Neurotransmitter Interactions and Cognitive Function
Edited by Edward D. Levin
© 2006 Birkhäuser Verlag/Switzerland

Central histaminergic system interactions and cognition

Patrizio Blandina and Maria Beatrice Passani

Dipartimento di Farmacologia Preclinica e Clinica, Università di Firenze, Italy

Introduction

In spite of early reports that histamine is present [1] and has a physiological role in the mammalian brain [2], only recently has attention has been paid to its role as neurotransmitter [3]. Curiously, despite the sparse studies of histamine in the brain, this amine led adventitiously to the development of psychotropic drugs [4]. Indeed, the phenotiazines were initially developed as antihistamines and the observation that one of them, chlorpromazine, affects mood and produces an "euphoric quietude" led to its use for treating schizophrenia [4]. Chemical modifications produced imipramine, which is effective in treating depression [4]. Yet, no attention was given to histamine receptors as sites of action for these drugs, or to the common side effects (sedation, drowsiness, slowed reaction time) shown by first-generation H_1-receptor antagonists. Therefore, the therapeutic potentials of histamine receptor ligands remain to be learned. The aim of this chapter is to evaluate their role for treatment of cognitive deficiencies.

Histamine neurons

The morphological features of the central histaminergic system, with a compact cell group and a widespread distribution of fibers, resembles that of other biogenic amines, such as norepinephrine or serotonin, thus suggesting that the histaminergic neurons may also act as a regulatory center for whole-brain activity [5]. All histamine cell bodies are localized in the tuberomammillary nucleui of the hypothalamus [6, 7] that is also the sole location of histidine decarboxylase immunoreactivity [8], an essential determinant of brain histamine levels [9, 10], and project mostly unmyelinated varicose fibers to most areas of the central nervous system [11]. Numerous synaptic contacts have been observed only in the mesencephalic trigeminal nucleus [12], while in the rest of the brain histaminergic fibers make relatively few synaptic contacts [13]. The peculiarity of the histaminergic axons that apparently do not form synaptic contacts, but rather varicosities containing synaptic vesicles [14, 15], suggests that histamine may act as a local hormone affecting not only neuronal, but also glial activity and blood vessel tone [5]. Indeed, cultured astrocytes from rat cerebral cortex display histamine receptors identical to those present on neuronal cells [16, 17].

Histamine receptors and constitutive activity

Three metabotropic, histaminergic receptor subtypes, H_1, H_2 and H_3, have been described in the mammalian central nervous system [18], whereas the presence of a fourth histaminergic receptor, demonstrated in the peripheral tissue [19], is still controversial [20–22]. All histaminergic receptors display a high degree of constitutive (agonist–independent) activity that occurs in human, rat and mouse recombinant receptors expressed at physiological concentrations [23–26]. Noteworthy is that constitutive activity of native H_3 receptors seems to be one of the highest among G-protein–coupled receptors in the brain [27]. Constitutively active H_3 receptors presumably regulate the release of neuronal histamine [25], therefore the classical H_3 receptor antagonists (e.g., clobenpropit, thioperamide and ciproxifan) that block constitutive activity are being reclassified as inverse agonists, a concept that may have clinical relevance. Indeed, either inverse agonists or neutral antagonists may be favorable for different therapeutic applications.

The role of histamine in arousal may affect cognition

Cognition is a complex phenomenon involving the integration of multiple neurological and behavioral activities among which arousal is crucial, being a prerequisite condition for responding to behavioral and cognitive challenges [28, 29]. Histamine seems to be required to mantain arousal, as histidine decarboxylase knock-out mice that lack histamine are unable to remain awake when high vigilance is required [30], and narcoleptic dogs show histamine deficiency [31]. Indeed, histaminergic neurons fire tonically and specifically during wakefulness [32] and are responsible for the maintenance of cortical activation (EEG desynchronization), a salient sign of wakefulness [33, 34]. The histaminergic system achieves cortical activation through excitatory interactions with cholinergic corticopetal neurons originating from the nucleus basalis magnocellularis [35] and the substantia innominata [36]. Moreover, histaminergic afferents elicit cortical activation also indirectly, through thalamo- and hypothalamo-cortical circuitries, as they excite cholinergic neurons in the mesopontine tegmentum projecting to the thalamus and the hypothalamus [37]. Moreover, the nature of the interactions between histamine and orexin neurons, which have a crucial role in sleep regulation [38], further supports the importance of histamine in arousal [39, 40]. Consequently, if arousal is the prerequisite for other brain functions like learning and memory, histamine, by increasing arousal, may affect cognitive processes. However, there is also much evidence suggesting that the histaminergic system may also influence biological processes underlying learning and memory directly [41, 42].

Histamine in cognition: good or bad?

Early observations obtained with several learning paradigms indicate that the histaminergic central system has a positive role in cognitive function, as histaminergic compounds enhance memory (recall) in both a passive [43] and an active avoidance tasks [44]. However, Huston and colleagues have suggested a negative influence of the histaminergic system on learning and memory, since bilateral lesions of the tuberomammillary nuclei improve performance in several learning paradigms [45–48]. Possible confounding factors may be the systemic administration of histaminergic compounds, or the extensive lesions of the histaminergic nuclei, which do not exclude effects on arousal, anxiety, perception or other homeostatic mechanisms in which histamine is involved [49], thus affecting learning and memory indirectly. Moreover, since the memory modulating action of histamine affects several brain regions differently (see review [50]), and same histaminergic compounds affect cognition in opposite ways depending on the behavioral task (see reviews [34, 42, 51, 52]), the procognitive or amnesic effects of histamine should be evaluated with experimental protocols that interfere with the exact timing of histamine release from discrete brain regions during the appropriate behavioral task.

H_3 receptors are potential targets for cognitive enhancers

Administration of H_3 antagonists/inverse agonists improves cognitive performance in the five-trial inhibitory avoidance task [53, 54], social memory in the rat [55], and enhances attention as evaluated in the five-choice, serial reaction time test [56]. H_3 receptor antagonists/inverse agonists exert procognitive effects also in cognitively impaired animals: as observed in senescence-accelerated mice or scopolamine-impaired rats challenged in a passive-avoidance response [57, 58], scopolamine-treated rats tested in the object recognition [58] or the elevated plus-maze paradigm [59], and MK-801-treated rats evaluated in the radial maze [60]. It is also worth pointing out the procognitive effect observed in spontaneously hypertensive rat (SHR) pups challenged in a five-trial avoidance test following administration of non-imidazole H_3 antagonists/inverse agonists [54]. Juvenile SHR rats are normotensive, but exhibit many cognitive impairments [53, 54]. The genetic origin of these deficits renders this model more clinically relevant than those requiring pharmacological or surgical intervention. Consequently, it is not surprising that so much effort is being directed at understanding the H_3-receptor physiology and at synthesizing ever more selective and potent ligands with therapeutic potentials [61]. A recent report, however, provides some contrasting data, as H_3-receptor antagonists impaired object recognition in wild-type and *Apoe*$^{-/-}$ mice [62].

H_3 receptors are members of the seven transmembrane receptor superfamily [63] and couple to $G_{i/o}$ proteins [64]. Their stimulation restricts the influx of calcium ions [65], inhibits adenylate cyclase [63], and increases extracellular signal-related kinase (ERK) phosphorylation in receptor-transfected cells [66]. Originally, H_3 receptors were detected as autoreceptors mediating inhibition of histamine release both *in vitro* [67], and *in vivo* [68–72]. In addition, stimulation of presynaptic H_3 receptors inhibits histamine synthesis [73–75]. Consequently, drugs selective for the H_3 autoreceptor may influence the functions of histamine in the brain through the modulation of endogenous histamine release and synthesis [3]. However, the presence of H_3 receptors is not restricted to histaminergic neurons [76–78]. Accordingly, H_3-receptors act also as heteroreceptors that modulate the release of other neurotransmitters, including ACh [79, 80], dopamine [81], noradrenaline [82] and serotonin [83, 84] from brain regions crucial for the maintenance of alertness or the storage of information [3, 49]. The cholinergic hypothesis has provided the rationale for the current treatments of cognitive impairments, such as Alzheimer's disease, mainly based on acetylcholinesterase inhibitors. However, substantial data support the multivariate nature of cognitive disorders pathology and suggest the involvement of other neurotransmitters such as serotonin, noradrenalin, dopamine, histamine, excitatory amino acids and neuropeptides among others [85–89]. Furthermore, region-selective decreases in dopaminergic, noradrenergic or serotonergic contents are associated with the level of age-related learning and memory impairments [90, 91]. Consequently, compounds designed specifically to act on multiple neural and biochemical targets may prove more suitable as cognitive enhancers. In this regard, H_3 receptor ligands, with their ability to modulate the synaptic availability of host neurotransmitters, should not be underestimated.

H_3 receptors regulate the cholinergic tone in the cortex, and influence animal performances in related cognitive tests

In rats, as in humans, projections from the nucleus basalis magnocellularis (NBM) provide the majority of cholinergic innervation to the cortex [92, 93]. Local stimulation of H_3 receptors decreases the cholinergic tone in the cortex [79, 80, 94]. This effect may have functional relevance, as systemic administration of H_3 receptor agonists impairs rat performance in object recognition and in a passive avoidance response at the same doses that moderate ACh release from the cortex of freely moving rats [80]. These tasks require an intact cortical cholinergic system [95]. ACh inhibition caused by H_3 receptor agonists is tetrodotoxin-sensitive [80], thus strongly suggesting that these receptors are located postsynaptically on intrinsic perikarya [80, 94]. As immepip, an H_3 receptor agonist, increases GABA release from the cortex of freely moving rats [96], it is conceivable that stimulation of cortical H_3 heteroreceptors releases GABA, which, in turn, inhibits ACh release. Histamine terminals in the cortex are relatively sparse, yet they may exert a powerful effect on cortical activity, since cortical GABAergic interneurons have extensive axon arborizations and control the tone of large populations of principal cells [97]. Decreased cholinergic neurotransmission, widely believed to underlie cognitive deficits, could account for H_3-receptor-agonists-elicited impairments observed in cognitive tests [80], and, by inference, explain H_3-receptor antagonists/inverse agonists procognitive effects (see above). In this regard another observation may be relevant. Local administrations of either clobenpropit or thioperamide into the NBM, which provides the cholinergic innervation to the cortex, increase cortical ACh release measured with dual-probe microdialysis in rats [35]. H_3-autoreceptors, and not heteroreceptors, are presumably involved, because NBM perfusion with the same compounds increases local histamine release as well [50]. Triprolidine, an H_1-receptor antagonist, fully antagonizes thioperamide-elicited ACh release [35], thus implicating postsynaptic H_{HI}-receptors, which are known to increase the tonic firing of NBM cholinergic neurons [98]. These results fit well with reports that intra-NBM injections of thioperamide improve place recognition memory [99], and i.c.v. administrations of a selective H_1-receptor agonist (2-(3-(trifluoromethyl)-phenyl-histamine) ameliorate the performance of rats in object recognition tasks [100]. Additional evidence supports that H_1 receptors function as a postsynaptic target for histamine to improve cognition [51, 101], therefore the action of H_3-receptor antagonists to augment NBM histamine levels might be one of the drivers of cognitive enhancement. These interactions may have implications for the cognitive decline associated with aging and Alzheimer's disease. The characteristic cortical cholinergic dysfunction may result from cell degeneration of both cholinergic and non-cholinergic neurons of the NBM [102], and from reduction in impulse flow from the NBM to the cortex [103]. Loss of cholinergic neurons would reduce the cortical cholinergic activity directly and degeneration of non cholinergic neurons may contribute to cholinergic hypofunction. Interestingly, binding of H_1 receptors, assessed by positron emission tomography, is significantly decreased in the brain of Alzheimer's disease patients compared to those of normal subjects [104]. These observations can be readily integrated: loss of NBM GABA neurons [105] that project primarily to cortical GABA interneurons [97] would increase the cortical GABAergic inhibitory tone on ACh release [106]. Also the decrease of excitatory inputs to the NBM cholinergic neurons, because of the reduction of H_1 receptors, may contribute to the cortical cholinergic hypofunction, although we cannot exclude that reduction of H_1 receptors is a consequence of the loss of cholinergic neurons.

Region-specific nature of the response to H_3 receptor ligands: the amygdala paradox

There is extensive evidence that crucial neural changes mediating emotional memory occur in the basolateral amygdala (BLA) [107–110]. Emotional memory may be assessed with contextual fear conditioning in which experimental animals learn to associate a mild electrical foot-shock with the environment where they receive the punishment. A critical event for emotional memory consolidation is the stimulation of muscarinic receptors within the BLA [111–114]. In the BLA H_3 receptor ligands modulate ACh release in a bimodal fashion and modify the expression of fear memories accordingly. Indeed, H_3 receptor antagonists/inverse agonists administration locally into the BLA impairs memory consolidation in contextual fear conditioning, as did the infusion of scopolamine [113]. Conversely, H_3-receptor agonists or oxotremorine ameliorate expression of this memory [114]. Noteworthy, BLA perfusion with H_3-agonists increases whereas with H_3-antagonists/inverse agonists decrease ACh release from the BLA at concentrations comparable to those affecting fear memory [113, 114]. These drugs impact presumably on inhibitory H_3-autoreceptors, as in the BLA H_3 receptor binding is strictly associated with the presence of histaminergic fibers [115], and local perfusion with H_3receptor antagonists/inverse agonists increases endogenous histamine release [116]. The report of impairing effect on the acquisition of an avoidance task, another task with high emotional content, following histamine administration into the BLA [117], supports these conclusions. These results contrast with the findings in the cortex, thus, H_3-receptors modulate ACh release with modalities that differ according to tissue architectural constraints, and to their role as auto- or hetero-receptors. If H_3 receptor antagonists/inverse agonists are beneficial in some behavioral models of cognition, presumably by increasing ACh levels in the cortex [50, 118], the opposite seems to be true for fear conditioning and ACh release in the BLA.

In the hippocampus histamine affects cognition involving non-cholinergic mechanisms

Fear conditioning comprises two components, a cued and a contextual one. Both components depend upon the amygdala, whereas the latter involves the hippocampus as well [110, 119]. Bilateral post-training injections into the dorsal hippocampus of H_2- or H_3-receptor agonists improve memory consolidation after contextual fear conditioning [120]. Yet, reports that histamine-receptor-mediated modulation of ACh is not detectable in this region [121, 122] call for a different explanation than just the interactions between the cholinergic and the histaminergic systems. Increasing evidence implicates ERK2 in fear-dependent neuronal plasticity [123]. Upstream components of ERK2 pathway, such as neurotransmitters and neurotrophins, may act during the critical period of memory consolidation by modulating ERK2 activity, as suggested by NMDA receptor stimulation during fear conditioning [124], or by nerve growth factor-induced effects during inhibitory avoidance [125]. Noteworthy, stimulation of either H_2 or H_3 histaminergic receptors activates ERK2 in hippocampal CA3 pyramidal cells [120], that are involved in stress-mediated effects on memory [126]. Moreover, hippocampal administration of U0126, a selective inhibitor of ERK-kinase, prevents memory improvements exerted by H_2- or H_3-receptor agonists [120]. The observation that stimulation of H_2 and H_3 receptors activate the ERK2 pathway in CA3 pyramidal cells and improve memory consolidation after contextual fear conditioning provides major insight into histamine receptor regulation of hippocampal function and the physiological mechanisms underlying learning in the mammalian nervous system. H_2 receptors are likely localized on CA3 pyramidal cells [120, 127]. Conversely, experiments with tetrodotoxin strongly suggest that H_3 receptors are not located on the CA3 pyramidal cells, nor on nerve endings impinging on them [120]. In the simplest scenario, H_3 receptors may promote the release of an, as yet unidentified neurotransmitter, which in turn activates ERK cascade in CA3 pyramidal cells. It is unlikely that H_3 autoreceptors modulate ERK phosphorylation in the hippocampus, since antagonism of the H_3 receptor by thioperamide that should increase endogenous histamine release, has no effect [120]. Several neurotransmitters, such as dopamine, glutamate and norepinephrine, activate the ERK cascade in the hippocampus [128] and histamine may interact with these neurotransmitters to orchestrate ERK2 phosphorylation in CA3 pyramidal cells. It has been proposed that brain histamine is a danger response signal, triggered by a variety of aversive stimuli such as stress, dehydration, hypoglycemia [49]. The histaminergic system may mediate hypothalamic influences on the hippocampus and amygdala to achieve an adequate behavioral response through neural circuits activated by emotional arousal [3]. In addition, R-alpha-methylhistamine, an H_3receptor agonist, improves normal [129] and scopolamine-impaired [130] rat performance in the Morris water maze, a paradigm which requires an intact hippocampus. As these regions appear engaged in the development of memory disorders associated with extreme emotional traumas [131–133], the use of histaminergic compounds may be proposed to alleviate disturbances of brain mechanisms underlying emotional memory formation that contribute to mood disorders such as panic attacks, specific phobias and generalized anxiety.

Concluding remarks

A wide variety of studies agree that the neuronal histaminergic system regulates some forms of cognition, and, inevitably, reports that pharmacological blockade of central H_3-receptors exerted procognitive activity in several cognitive tasks has raised considerable interest. Interactions between the histaminergic and cholinergic systems serve as one of the physiological correlates of the ability of animals to learn and remember. As therapies with cholinesterase inhibitors or muscarinic agonists have been generally unproductive [134], histamine receptors could represent the target for compounds that potentiate cholinergic functions and may produce beneficial effects on disorders where the cholinergic function is compromised. Yet, ACh/histamine interactions are complex and multifaceted, and the results are often contradictory, as both facilitatory and inhibitory effects of histamine on memory have been described. As it turned out, histaminergic H_3 receptor activation, for instance, modulates ACh release and cognitive processes, apparently with modalities that differ according to their role as auto- or hetero-receptors, or the architectural constraints that separate groups of transmitters in particular brain structures. Thus, it will be necessary to develop drugs selective for the receptor subtypes and the particular brain region of interest. However, molecular pharmacology is uncovering the extraordinary complexity of the H_3receptor: it shows functional constitutive activity, polymorphisms in humans and rodents with a differential distribution of splice variants in the CNS, and potential coupling to different intracellular signal-transduction mechanisms (reviewed in [34]). Thus, there is increasing interest and great effort is being channeled into developing ever more selective agonists, inverse agonists, pure antagonists for the H_3 receptor, as well as ligands for its various isoforms. This will be a great challenge in the years to come. Obviously, new discoveries create tremendous expectations, as these receptors are involved in cognition, the sleep-wake cycle, obesity,and epilepsy, which are the most actively pursued pathological conditions for the therapeutic potentials of selective H_3-receptor ligands.

References

1. Kwiatkowski H (1943) Histamine in nervous tissue. *J Physiol* 102: 32–41
2. Green JP (1964) Histamine and the nervous system. *Fed Proc* 23: 1095–1102
3. Haas H, Panula P (2003) The role of histamine and the tuberomamillary nucleus in the nervous system. *Nat Rev Neurosci* 4: 121–130
4. Green JP (1987) Histamine receptors,In: HY Meltzer et al. (eds): *Psychopharmacology: The third generation of progress*. Raven, New York, 273–279
5. Wada H, Inagaki N, Yamatodani A, Watanabe T (1991) Is the histaminergic neuron system a regulatory center for whole-brain activity? *Trends Neurosci* 14: 415–418
6. Panula P, Yang HY, Costa E (1984) Histamine-containing neurons in the rat hypothalamus. *Proc Natl Acad Sci* 81: 25722576
7. Watanabe T, Taguchi Y, Hayashi H, Tanaka J, Shiosaka S, Tohyama M, Kubota H, Terano Y, Wada H (1983) Evidence for the presence of a histaminergic neuron system in the rat brain:an immunohistochemical analysis. *Neurosci Lett* 39: 249254
8. Ericson H, Watanabe T, Köhler C (1987) Morphological analysis of the tuberomammillary nucleus of the rat brain: delineation of subgroups with antibody against L-histidine decarboxylase as a marker. *J Comp Neurol* 263: 1–24
9. Green JP, Prell GD, Khandelwal JK, Blandina P (1987) Aspects of histamine metabolism. *Agents Actions* 22: 1–15
10. Kollonitsch J, Patchett AA, Marburg S, Maycock AL, Perkins LM, Doldouras GA, Duggan DE, Aster SD (1978) Selective inhibitors of biosynthesis of aminergic neurotransmitters. *Nature* 274: 906–908
11. Inagaki N, Toda K, Taniuchi I, Panula P, Yamatodani A, Tohyama M, Watanabe T, Wada H (1990) An analysis of histaminergic efferents of the tuberomammillary nucleus to the medial preoptic area and inferior colliculus of the rat. *Exp Brain Res* 80: 374–380
12. Inagaki N, Yamatodani A, Shinoda K, Shiotani Y, Tohyama M, Watanabe T, Wada H (1987) The histaminergic innervation of the mesencephalic nucleus of trigeminal nerve in the rat brain: A light and electron microscopic study. *Brain Res* 418: 388–391
13. Tohyama M, Tamiya R, Inagaki N, Tagaki H (1991) Morphology of histaminergic neurons with histidine decarboxylase as a marker. In: Watanabe T, Wada H (eds): *Histaminergic neurons: morphology and function.* CRC Press, Boca Roton, 107–126
14. Takagi H, Morishima Y, Matsuyama T, Hayashi H, Watanabe T, Wada H (1986) Histaminergic axons in the neostriatum and cerebral cortex of the rat: A correlated light and electron microscope immunocytochemical study using histidine decarboxylase as a marker. *Brain Res* 364: 114–123
15. Michelsen K, Panula P (2002) Subcellular distribution of histamine in mouse brain neurons. *Inflamm Res* 51, Suppl 1: S46–48
16. Inagaki N, Fukui H, Taguchi Y, Wang N, Yamatodani A, Wada H (1989) Characterization of histamine H1-receptors on astrocytes in primary culture: [3H]mepyramine binding studies. *Eur J Pharmacol* 173: 43–51
17. Carman-Krzan M, Lipnik-Stangelj M (2000) Molecular properties of central and peripheral histamine H1 and H2 receptors. *Pflugers Arch* 439: 131–132
18. Hill SJ, Ganellin CR, Timmerman H, Schwartz JC, Shankley NP, Young JM, Schunack W, Levi R, Haas HL (1997) International union of Pharmacology. XIII. Classification of histamine receptors. *Pharmacol Rev* 49: 253–278
19. Oda T, Morikawa N, Saito Y, Masuho Y, Matsumoto S (2000) Molecular cloning and characterization of novel type of histamine receptor preferentially expressed in leukocytes. *J Biol Chem* 275: 36781–36786

20. Zhu Y, Michalovich A, Wu H-L, Tang K, Dytko GM (2001) Cloning, expression, and pharmacological characterization of a novel human histamine receptor. *Mol Pharmacol* 59: 434–441
21. Liu C, Ma X-J, Wilson S, Hofstra C, Blevitt J, Pyati J, Li X, Chai W, Carruthers N, Lovenberg T (2001) Cloning and pharmacological characterization of of a fourth histamine receptor (H_4) expressed in bone marrow. *Mol Pharmacol* 59: 420–426
22. Liu C, Wilson S, Kuei C, Lovenberg T (2001) Comparison of human, mouse, rat, and guinea pig histamine H4 receptors reveals substantial pharmacological species variation. *J Pharmacol Exp Ther* 299: 121–130
23. Smit M, Leurs R, Alewijnse A, Blauw J, Amongen G, Vandevrede T, Roovers E, Timmerman H (1996) Inverse agonism of histamine H2 antagonists accounts for upregulation of spontaneously active histamine H2 receptors. *Proc Nat Acad Sci* 93: 6802–6807
24. Bakker R, Wieland K, Timmerman H, Leurs R (2000) Constitutive activity of the H(1) receptor reveals inverse agonism of histamine H(1) receptor antagonists. *Eur J Pharmacol* 387: R5–R7
25. Morisset S, Rouleau A, Ligneau X, Gbahou F, Tardivel-Lacombe J, Stark H, Schunack W, Ganellin CR, Schwartz J-C, Arrang J-M (2000) High constitutive activity of native H3 receptors regulates histamine neurons in brain. *Nature* 408: 860–864
26. Wieland K, Bongers G, Yamamoto Y, Hashimoto T, Yamatodani A, Menge W, Timmerman H, Lovenberg T, Leurs R (2001) Constitutive activity of histamine h(3) receptors stably expressed in SK-N-MC cells: display of agonism and inverse agonism by H(3) antagonists. *J Pharmacol Exp Ther* 299: 908–914
27. Rouleau A, Ligneau X, Tardive-Lacombe J, Morisset S, Gbahou F, Schwartz JC, Arrang JM (2002) Histamine H3 receptor mediated [35S] GTP gamma[s] binding: evidence for constitutive activity of the recombinant and native rat and human H3 receptors. *Br J Pharmacol* 135: 383–392
28. Crochet S, Sakai K (1999) Effects of microdialysis application of monoamines on the EEG and behavioural states in the cat mesopontine tegmentum. *Eur J Neurosci* 11: 3738–3752
29. Cahill L, McGaugh J (1995) A novel demonstration of enhanced memory associated with emotional arousal. *Conscious Cogn* 4: 410–421
30. Parmentier R, Ohtsu H, Djebarra-Hannas Z, Valtx J-L, Watanabe T, Lin J-S (2002) Anatomical, physiological and pharmacological characteristics of histidine decarboxylase knock-out mice: evidence for the role of brain histamine in behavioral and sleep-wake control. *J Neurosci* 22: 7695–7711
31. Nishino S, Fujiki N, Ripley B, Sakurai E, Kato M, Watanabe T, Mignot E, Yanai K (2001) Decreased brain histamine content in hypocretin/orexin receptor-2 mutated narcoleptic dogs. *Neurosci Lett* 313: 125–128
32. Sakai K, Mansari ME, Lin J, Zhang J, Mercier GV (1990) The posterior hypothalamus in the regulation of wakefulness and paradoxical sleep. In: Mancia M (ed): *The diencephalon and sleep.* Raven Press, New York, 171–198
33. Lin JS (2000) Brain structures and mechanisms involved in the control of cortical activation and wakefulness, with emphasis on the posterior hypothalamus and histaminergic neurons. *Sleep Med Rev* 4: 471–503
34. Passani MB, Lin J-S, Hancock A, Crochet S, Blandina P (2004) The histamine H3 receptor as a novel therapeutic target for cognitive and sleep disorders. *Trends Pharmacol Sci* 25: 618–625

35. Cecchi M, Passani MB, Bacciottini L, Mannaioni PF, Blandina P (2001) Cortical Acetylcholine release elicited by stimulation of histamine H1 receptors in the nucleus basalis magnocellularis: a dual probe microdialysis study in the freely moving rat. Eur J Neurosci 13: 68-78
36. Lin J, Sakai K, Jouvet M (1994) Hypothalamo-preoptic histaminergic projections in sleep-wake control in the cat. *Eur J Neurosci* 6: 618–625
37. Lin JS, Hou Y, Sakai K, Jouvet M (1996) Histaminergic descending inputs to the mesopontine tegmentum and their role in the control of cortical activation and wakefulness in the cat. *J Neurosci* 16: 1523–1537
38. Mignot E, Taheri S, Nishino S (2002) Sleeping with the hypothalamus: emerging therapeutic targets for sleep disorders. *Nat Neurosci* 5 Suppl: 1071–1075
39. Huang Z-L, Qu W-M, Li W-D, Mochizuki T, Eguchi N, Watanabe T, Urade Y, Hayaishi O (2001) Arousal effect of orexin A depends on activation of the histaminergic system. *Proc Natl Acad Sci* 98: 9965–9970
40. Eriksson K, Sergeeva O, Brown R, Haas H (2001) Orexin/hypocretin excites the histaminergic neurons of the tuberomammillary nucleus. *J Neurosci* 21: 9273–9279
41. Passani MB, Bacciottini L, Mannaioni PF, Blandina P (2000) Central histaminergic system and cognitive processes. *Neurosci Biobehav Rev* 24: 107–114
42. Passani M, Blandina P (2004) The neuronal histaminergic system in cognition. *Curr Med Chem* 4: 17–26
43. deAlmeida M, Izquierdo I (1986) Memory facilitation by histamine. *Arch Int Pharmacodyn Ther* 283: 193–198
44. Kamei C (1990) Influence of certain H1-blockers on the step-through active avoidance response in rats. *Psychopharmacology* 102: 312–318
45. Huston JP, Wagner U, Hasenöhrl RU (1997) The tuberomammilary nucleus projections in the control of learning, memory and reinforcement processes: evidence for an inhibitory role. *Behav Brain Res* 83: 97–105
46. Klapdor K, Hasenöhrl RU, Huston JP (1994) Facilitation of learning in adult and aged rats following bilateral lesions of the tuberomammilary nucleus region. *Behav Brain Res* 61: 113–116
47. Frisch C, Hasenöhrl RU, Haas HL, Weiler HT, Steinbusch HWM, Huston J (1998) Facilitation of learning after lesions of the tuberomammillary nucleus region in adult and aged rats. *Exp Brain Res* 118: 447–456
48. Frisch C, Hasenöhrl RU, Huston JP (1999) Memory improvement by post-trial injection of lidocaine into the tuberomammilary nucleus, the source of neuronal histamine. *Neurbiol Learning and Memory* 72: 69–77
49. Brown RE, Stevens DR, Haas HL (2001) The physiology of brain histamine. *Prog Neurobiol* 63: 637–672
50. Blandina P, Efoudebe M, Cenni G, Mannaioni PF, Passani MB (2004) Acetylcholine, histamine and cognition: two sides of the same coin. *Learn Mem* 11: 1–8
51. Witkin JM, Nelson DL (2004) Selective histamine H3 receptor antagonists for treatment of cognitive deficiencies and other disorders of central nervous system. *Pharmacol Ther* 103: 1–20
52. Hancock AH, Fox GB (2004) Perspective on cognitive domains, H3 receptor ligands and neurological disease. *Expert Opin Investig Drug* 13: 1237–1248
53. Fox GB, Pan JB, Esbenshade TA, Bennani YL, Black LA, Faghih R, Hancock AA, Decker MW (2002) Effects of histamine H3 receptor ligands GT-2331 and ciproxifan in a repeated acquisition avoidance response in the spontaneously hypertensive rat pup. *Behav Brain Res* 131: 151–161

54. Fox GB, Pan JB, Radek RJ, Lewis AM, Bitner RS, Esbenshade TA, Faghih R, Bennani YL, Williams M, Yao BB et al. (2003) Two novel and selective nonimidazole H_3 receptor antagonists A-304121 and A-317920: II. In vivo behavioral and neurophysiological characterization. *J Pharmacol Exp Ther* 305: 897–908
55. Prast H, Argyriou A, Philippu A (1996) Histaminergic neurons facilitate social memory in rats. *Brain Res* 734: 316–318
56. Ligneau X, Lin J-S, Vanni-Mercer G, Jouvet M, Muir JL, Ganellin CR, Stark H, Elz S, Schunack W, Schwartz JC (1998) Neurochemical and behavioral effects of ciproxifan, a potent histamine H_3-receptor antagonist. *J Pharmacol Exp Ther* 287: 658–666
57. Meguro K-I, Yanai K, Sakai N, Sakurai E, Maeyama K, Sasaki H, Watanabe T (1995) Effects of thioperamide, a histamine H3 antagonist, on the step-through passive avoidance response and histidine decarboxylase activity in senescence-accelerated mice. *Pharmacol Biochem Behav* 50: 321–325
58. Giovannini MG, Bartolini L, Bacciottini L, Greco L, Blandina P (1999) Effects of histamine H_3 receptor agonists and antagonists on cognitive performance and scopolamine-induced amnesia. *Behav Brain Res* 104: 147–155
59. Onodera K, Miyazaki S, Imaizumi M, Stark H, Schunack W (1998) Improvement by FUB 181, a novel histamine H3-receptor antagonist, of learning and memory in the elevated plus-maze test in mice. *Naunyn-Schmiedeberg's Arch Pharmacol* 357: 508–513
60. Huang YW, Hua WW, Chen Z, Zhang L-S, Shen H-Q, Timmerman H, Leurs R, Yanai K (2004) Effect of the histamine H3-antagonist clobenpropit on spatial memory deficits induced by MK-801 as evaluated by radial maze in Sprague-Dawley rats. *Behav Brain Res* 151: 287–293
61. Stark H (2003) Recent advances in histamine H3/H4 receptor ligands. *Expert Opin Ther Patents* 13: 851–865
62. Bongers G, Leurs R, Robertson J, Raber J (2004) Role of H_3-receptor-mediated signaling in anxiety and cognition in wild-type and Apoe-/- mice. *Neuropsychopharmacology* 29: 441–449
63. Lovenberg TW, Roland BL, Wilson SJ, Jiang A, Pyati J, Huvar A, Jackson MR, Erlander MG (1999) Cloning and functional expression of the human histamine H3 receptor. *Mol Pharmacol* 55: 1101–1107
64. Clark EA, Hill SJ (1996) Sensitivity of histamine H3 receptor agonist stimulated [35S]GTP-gamma[S] binding to pertussis toxin. *Eur J Pharmacol* 296: 223–225
65. Hill SJ, Straw RM (1988) $Alpha_2$-Adrenoceptor-mediated inhibition of histamine release from rat cerebral cortical slices. *Br J Pharmacol*: 95: 1213–1219
66. Drutel G, Peitsaro N, Karlstedt K, Wieland K, Smit M, Timmerman H, Panula P, Leurs R (2001) Identification of rat H3 receptor isoforms with different brain expression and signaling properties. *Mol Pharmacol* 59: 1–8
67. Arrang JM, Garbarg M, Schwartz JC (1983) Auto-inhibition of brain histamine release mediated by a novel class (H_3) of histamine receptors. *Nature* 302: 832–837
68. Arrang JM, Garbarg M, Lancelot JC, Lecont JM, Pollard H, Robba M, Schunack W, Schwartz JC (1987) Highly-potent and selective ligands for histamine-H_3 receptors. *Nature* 327: 117–123
69. Itoh Y, Oishi R, Nishibori M, Saeki K (1991) Characterization of histamine release from the rat hypothalamus as measured by in vivo microdialysis. *J Neurochem* 56: 769–774
70. Mochizuki T, Yamatodani A, Okakura K, Takemura M, Inagaki N, Wada H (1991) In vivo release of neuronal histamine in the hypothalamus of rats measured by microdialysis. *Naunyn-Schmiedeberg's Arch Pharmacol* 343: 190–195

71. Itoh Y, Oishi R, Adachi N, Saeki K (1992) A highly sensitive assay for histamine using ion-pair HPLC coupled with postcolumn fluorescent derivatization: its application to biological specimens. *J Neurochem* 58: 884–889
72. Prast H, Fisher HP, Prast M, Philippu A (1994) In vivo modulation of histamine release by autoreceptors and muscarinic acetylcholine receptors in the rat anterior hypothalamus. *Naunyn-Schmiedeberg's Arch Pharmacol* 350: 599–604
73. Arrang JM, Garbarg M, Schwartz J (1987) Autoinhibition of histamine synthesis mediated by presynaptic H_3-receptors. *Neuroscience* 23: 149–157
74. Garbarg M, TrungTuong MD, Gros C, Schwartz JC (1989) Effects of histamine H_3-receptor ligands on various biochemical indices of histaminergic neuron activity in rat brain. *Eur J Pharmacol* 164: 1–11
75. Oishi R, Itoh Y, Nishibori M, Saeki K (1989) Effects of histamine H_3-agonist (R)-a-methylhistamine and the antagonist thioperamide on histamine metabolism in the mouse and rat brain. *J Neurochem* 52: 1388–1392
76. Cumming P, Shaw C, Vincent SR (1991) High affinity histamine binding site is the H_3receptor: characterization and autoradiographic localization in rat brain. *Synapse* 8: 144–151
77. Pollard H, Moreau J, Arrang JM, Schwartz J-C (1993) A detailed autoradiographic mapping of histamine H_3 receptors in rat brain areas. *Neuroscience* 52: 169–189
78. Martinez-Mir MI, Pollard H, Arrang JM, Raut M, Traiffort E, Schwartz JC, Palacios J M (1990) Three histamine receptors (H_1, H_2 and H_3) visualized in the brain of human and non human primates. *Brain Res* 526: 322–327
79. Clapham J, Kilpatrick GJ (1992) Histamine H_3 receptors modulate the release of [^{3}H]-acetylcholine from slices of rat entorhinal cortex: evidence for the possible existence of H_3 receptor subtypes. *Br J Pharmacol* 107: 919–923
80. Blandina P, Giorgetti M, Bartolini L, Cecchi M, Timmerman H, Leurs R, Pepeu G, Giovannini MG (1996) Inhibition of cortical acetylcholine release and cognitive performance by histamine H_3 receptor activation in rats. *Br J Pharmacol* 119: 1656–1664
81. Schlicker E, Fink K, Detzner M, Göthert M (1993) Histamine inhibits dopamine release in the mouse striatum via presynaptic H_3 receptors. *J Neural Transm* 93: 1–10
82. Schlicker E, Fink K, Hinterhaner M, Göthert M (1989) Inhibition of noradrenaline release in the rat brain cortex via presynaptic H_3 receptors. *Naunyn-Schmiedeberg's Arch Pharmacol* 340: 633–638
83. Schlicker E, Betz R, Göthert M (1988) Histamine H_3 receptor-mediated inhibition of serotonin release in the rat brain cortex. *Naunyn-Schmiedeberg's Arch Pharmacol* 337: 588–590
84. Threlfell S, Cragg SJ, Kallo I, Turi GF, Coen CW, Greenfield SA (2004) Histamine H_3 receptors inhibit serotonin release in substantia nigra pars reticulata. *J Neurosci* 24: 8704–8710
85. Decker MW, McGaugh JL (1991) The role of interactions between the cholinergic system and other neuromodulatory systems in learning and memory. *Synapse* 7: 151–168
86. Buccafusco JJ, Terry Jr AV (2000) Multiple central nervous system targets for eliciting beneficial effects on memory and cognition. *J Pharmacol Exp Ther* 295: 438–446
87. Hardy J, Adolfsson R, Alafuzoff I, Bucht G, Marcusson J, Nyberg P, Perdahl E, Wester P, Winblad B (1985) Transmitter deficits in Alzheimer's disease. *Neurochem Int* 7: 545–563
88. Schneider C, Risser D, Kirchner L, Kitzmuller E, Cairns N, Prast H, Singewald N, Lubec G (1997) Similar deficits of central histaminergic system in patients with Down syndrome and Alzheimer disease. *Neurosci Lett* 222: 183–186
89. Panula P, Rinnie J, Kuokkanen K, Eriksson KS, Sallmen T, Kalimo H, Relja M (1997) Neuronal histamine deficit in Alzheimer's disease. *Neuroscience* 82: 993–997

90. Birthelmer A, Stemmelin J, Jackisch R, Cassel J-C (2003) Presynaptic modulation of acetylcholine, noradrenaline, and serotonin release in the hippocampus of aged rats with various levels of memory impairments. *Brain Res Bull* 60: 283–296
91. Stemmelin J, Lazarus C, Cassel S, Kelche C, Cassel J-C (2000) Immunohistochemical and neurochemical correlates of learning deficits in aged rats. *Neuroscience* 96: 275–289
92. Mesulam MM, Mufson EJ, Wainer BH, Levey AI (1983) Central cholinergic pathways in the rat: an overview based on an alternative nomenclature (Ch1-Ch6). *Neuroscience* 10: 1185–1201
93. Heckers S, Ohtake T, Wiley RG, Lappi DA, Geula G, Mesulam MM (1994) Complete and selective cholinergic denervation of rat neocortex and hippocampus but not amygdala by an immunotoxin against the p75 NGF receptor. *J Neurosci* 14: 1271–1289
94. Arrang JM, Drutel G, Schwartz JC (1995) Characterization of histamine H_3 receptors regulating acetylcholine release in rat enthorinal cortex. *Br J Pharmacol* 114: 1518–1522
95. Goldman-Rakic PS (1987) Circuitry of primate prefrontal cortex and regulation of behavior by representational memory. In: Plum F (ed): *Handbook of physiology*, Vol. 5 American Physiol Soc, Bethesda, Maryland, 373–417
96. Giorgetti M, Bacciottini L, Bianchi L, Giovannini MG, Cecchi M, Blandina P (1997) GABAergic mechanism in histamine H_3 receptor inhibition of K^+-evoked release of acetylcholine from rat cortex in vivo. *Inflamm Res* 46: S3–S4
97. Freund TF, Meskenaite V (1992) g-Aminobutyric acid-containing basal forebrain neurons innervate inhibitory interneurons in the neocortex. *Proc Natl Acad Sci* 89: 738–742
98. Khateb A, Fort P, Pegna A, Jones BE, Mühlertaler M (1995) Cholinergic nucleus basalis neurons are excited by histamine in vitro. *Neuroscience* 69: 495–506
99. Orsetti M, Ferretti C, Gamalero R, Ghi P (2002) Histamine H3-receptor blockade in the rat nucleus basalis magnocellularis improves place recognition memory. *Psychopharmacology* 159: 133–137
100. Malmberg-Aiello P, Ipponi A, Blandina P, Bartolini L, Schunack W (2003) Pro-cognitive effect of a selective H1 receptor agonist, 2-(3-trifluoromethylphenyl)histamine, in the rat object recognition test. *Inflam Res* 52: S33–S34
101. Okamura N, Yanai K, Higuchi M, Sakai J, Iwata R, Ido T, Sasaki H, Watanabe T, Itoh M (2000) Functional neuroimaging of cognition impaired by a classical antihistamine, d-chlorpheniramine. *Br J Pharmacol* 129: 115–123
102. Whitehouse PJ, Price DL, Strubble RG, Clark AW, Coyle JT, DeLong MR (1982) Alzheimer's disease and senile dementia; loss of neurons in the basal forebrain. *Science* 215: 1237–1239
103. Aston-Jones G, Rogers J, Shaver RD, Dinan TG, Moss DE (1985) Age-impaired impulse flow from nucleus basalis to cortex. *Nature* 318: 462–464
104. Higuchi M, Yanai K, Okamura N, Meguro H, Arai H, Itoh M, Iwata R, Ido T, Watanabe T, Sasaki H (2000) Histamine H1 receptors in patients with Alzheimer's disease assessed by positron emission tomography. *Neuroscience* 99: 721–729
105. Gritti I, Mainville L, Mancia M, Jones BE (1997) GABAergic and other noncholinergic basal forebrain neurons, together with cholinergic neurons, project to the mesocortex and isocortex in the rat. *J Comp Neurol* 383: 163–177
106. Giorgetti M, Bacciottini L, Giovannini MG, Colivicchi MA, Goldfarb J, Blandina P (2000) Local GABAergic inhibitory tone of acetylcholine release from the cortex of freely moving rats. *Eur J Neurosci* 12: 1941–1948
107. Davis M (1992) The role of the amygdala in fear and anxiety. *Ann Rev Neurosci* 15: 353–375
108. LeDoux JE (2000) Emotion circuits in the brain. *Annu Rev Neurosci* 23: 155–184

109. Dalmaz C, Introini-Collison IB, McGaugh JL (1993) Noradrenergic and cholinergic interactions in the amygdala and the modulation of memory storage. *Behav Brain Res* 58: 167–174
110. Sacchetti B, Ambrogi Lorenzini C, Baldi E, Tassoni G, Bucherelli C (1999) Auditory thalamus, dorsal hippocampus, basolateral amygdala, and perirhinal cortex role in the consolidation of conditioned freezing to context and to acoustic conditioned stimulus in the rat. *J Neurosci* 19: 9570–9578
111. Power AE (2004) Muscarinic cholinergic contribution to memory consolidation: with attention to involvement of the basolateral amygdala. *Curr Med Chem* 11: 987–986
112. Vazdarjanova A, McGaugh JL (1999) Basolateral amygdala is involved in modulating consolidation of memory for classical conditioning. *J Neurosci* 19: 6615–6622
113. Passani MB, Cangioli I, Baldi E, Bucherelli C, Mannaioni PF, Blandina P (2001) Histamine H_3 receptor-mediated impairment of contextual fear conditioning, and in-vivo inhibition of cholinergic transmission in the rat basolateral amygdala. *Eur J Neurosci* 14: 1522–1532
114. Cangioli I, Baldi E, Mannaioni PF, Bucherelli C, Blandina P, Passani MB (2002) Activation of histaminergic H3 receptors in the rat basolateral amygdala improves expression of fear memory and enhances acetylcholine release. *Eur J Neurosci* 16: 521–528
115. Anichtchik OV, Huotari M, Peitsaro N, Haycock JW, Männistö PT, Panula P (2000) Modulation of histamine H_3 receptors in the brain of 6-hydroxydopamine-lesioned rats. *Eur J Neurosci* 12: 3823–3832
116. Cenni G, Cangioli I, Yamatodani A, Passani MB, Mannaioni PF, DiFelice AM, Blandina P (2004) Thioperamide-elicited increase of histamine release from basolateral amygdala of freely moving rats and its therapeutic implications. *Inflamm Res* 53: S53–S54
117. Alvarez E, Ruarte M (2002) Histaminergic neurons of the ventral hippocampus and the baso-lateral amygdala of the rat: functional interaction on memory and learning mechanisms. *Behavioural Brain Research* 128: 81–90
118. Bacciottini L, Passani MB, Mannaioni PF, Blandina P (2001) Interactions between histaminergic and cholinergic systems in learning and memory. *Beh Brain Res* 124: 183–194
119. Phillips RG, LeDoux JE (1992) Differential contribution of amygdala and hippocampus to cued and contextual fear conditioning. *Behav Neurosci* 106: 274–285
120. Giovannini M, Efoudebe M, Passani M, Baldi E, Bucherelli C, Giachi F, Corradetti R, Blandina P (2003) Improvement in fear memory by histamine elicited erk2 activation in hippocampal CA3 cells. *J Neurosci* 23: 9016–9023
121. Alves-Rodrigues A, Timmerman H, Willems E, Lemstra S, Zuiderveld OP, Leurs R (1998) Pharmacological characterisation of the histamine H3 receptor in the rat hippocampus. *Brain Res* 788: 179–186
122. Bacciottini L, Passani M, Giovannelli L, Cangioli I, Mannaioni P, Schunack W, Blandina P (2002) Endogenous histamine in the medial septum-diagonal band complex increases the release of acetylcholine from the hippocampus: a dual-probe microdialysis study in the freely moving animal. *Eur J Neurosci* 15: 1669–1680
123. Schafe GE, Nader K, Blair HT, LeDoux JE (2001) Memory consolidation of Pavlovian fear conditioning: a cellular and molecular perspective. *TINS* 24: 540–546
124. Atkins CM, Selcher JC, Petraitis JJ, Trzaskos JM, Sweatt JD (1998) The MAPK cascade is required for mammalian associative learning. *Nature Neurosci* 1: 602–609
125. Walz R, Lenz G, Roesler R, Vianna M, Martins V, Brentani R, Rodnight R, Izquierdo I (2000) Time-dependent enhancement of inhibitory avoidance retention and MAPK activation by post-training infusion of nerve growth factor into CA1 region of hippocampus of adult rats. *Eur J Neurosci* 12: 2185–2189

126. Roozendaal B, Quirarte G, McGaugh J (2000) Glucocorticoids interact with the basolateral amygdala beta-adrenoceptor–cAMP/cAMP/PKA system in influencing memory consolidation. *Eur J Neurosci* 15: 553–60
127. Yanovsky Y, Haas HL (1998) Histamine increases the bursting activity of pyramidal cells in the CA3 region of mouse hippocampus. *Neurosci Lett* 240: 110–112
128. Adams JP, Sweatt JD (2002) Molecular physiology: roles for the ERK MAP kinase cascade in memory. *Annu Rev Pharmacol Toxicol* 42: 135–163
129. Smith CPS, Hunter AJ, Bennett GW (1994) Effects of (R)-a-methylhistamine and scopolamine on spatial learning in the rat assessed using a water maze. *Psychopharmacology* 114: 651–656
130. Rubio S, Begega A, Santin L, Arias J (2002) Improvement of spatial memory by (R)-alpha-methylhistamine, a histamine H(3)-receptor agonist, on the Morris water-maze in rat. *Behav Brain Res* 129: 77–82
131. Emery N, Amaral D (2000) *The cognitive neuroscience of emotion*. Oxford University Press, New York
132. Amaral D (2002) The primate amygdala and the neurobiology of social behavior: Implications for understanding social anxiety. *Biological Psychiatry* 51: 11–17
133. Schneider F, Weiss U, Kessler C, Muller-Gartner HW, Posse S, Salloum JB, Grodd W, Himmelmann F, Gaebel W, Birbaumer N (1999) Subcortical correlates of differential classical conditioning of aversive emotional reactions in social phobia. *Biol Psychiatry* 45: 863–871
134. Kelly JS (1999) Alzheimer's disease: the tacrine legacy. *Trends Pharmacol Sci* 20: 127–129

Neurotransmitter Interactions and Cognitive Function
Edited by Edward D. Levin
© 2006 Birkhäuser Verlag/Switzerland

Cholinergic, histaminergic, and noradrenergic regulation of LTP stability and induction threshold: cognitive implications

Hans C. Dringenberg and Min-Ching Kuo

Department of Psychology and Centre for Neuroscience Studies, Queen's University, Kingston, Ontario, K7L 3N6, Canada

Introduction

As noted by William James [1], memory systems exhibit an astonishing degree of selectivity with regard to the information that is encoded and maintained. Of the almost infinite number of stimuli and events detected by peripheral sensory organs and relayed to the brain, only a small sample is selected and stored in one of multiple memory systems [2]. After the initial storage, further selection occurs, resulting in decay of encoding for most information. Thus, only a minute fraction of the total processed information shows stable encoding for prolonged (and in some cases life-long) time periods [3].

Very little is known about the mechanisms that allow memory systems to perform these selection processes. However, recent work has provided insights into some of the synaptic and neurochemical processes that may determine if information is stored and influence the duration and decay functions of this encoding. In this chapter, we will discuss published and recent, unpublished experiments examining the neurochemical mechanisms that influence the initiation and maintenance of synaptic modifications, thought to mediate experience-dependent plasticity and memory encoding in neuronal circuits [4–6]. We will emphasize whole animal *in vivo* studies that allow investigations of the interactions of endogenous, synaptically released transmitters and modulators, though important *in vitro* studies will also be discussed. The evidence we summarize suggests that levels of cholinergic and monoaminergic transmission play important roles in allowing synapses to initiate and maintain long-term potentiation (LTP), a type of plasticity characterized by an increased strength of coupling of glutamatergic synapses. These types of neurochemical interactions result in LTP characteristics that parallel some of the effects of neuromodulators on performance in tests assessing memory functions in mammalian species.

Parameters determining LTP induction and maintenance

As mentioned, memory systems are highly selective in terms of initial encoding and temporal maintenance of information. LTP, as a hypothetical encoding mechanism, mimics these features. At most synapses, induction of LTP is achieved only when presynaptic glutamatergic release is paired with sufficient postsynaptic depolarization to open N-methyl-D-aspartate (NMDA)-receptor channels and permit sufficient influx of calcium (Ca^{2+}) into postsynaptic dendrites [4, 7–9]. This requirement for concurrent pre- and postsynaptic activity provides a type of "encoding filter" that prevents many signals from inducing lasting changes in postsynaptic neurons.

Experimentally, concurrent pre- and postsynaptic activity (resulting in LTP) is typically achieved by applying high frequency stimulation (e.g., 100 Hz) to afferent pathways, whereas lower stimulation frequencies (e.g., $\leq$ 10 Hz) are less effective in LTP induction, or may result in NMDA-receptor-dependent long-term depression (e.g., [10, 11]). Further, the number of stimuli applied to the afferent pathway can affect the probability of LTP induction, and small changes in stimulus parameters can have surprisingly clear effects on the induction of LTP. For example, at basal dendrites of hippocampal CA1 neurons, 10 stimulation trains consisting of 10 stimuli each, but not 10 trains of seven stimuli, reliably induce LTP *in vivo* [12]. Thus, relatively minor modifications of activity patterns in presynaptic fibers have pronounced effects on plastic responses of the postsynaptic membrane.

Manipulations of presynaptic activity also affect the temporal persistence of LTP. One of the first reports of LTP induction in the hippocampus *in vivo* demonstrated that the duration of LTP could be enhanced by applying repeated episodes of high frequency afferent stimulation [13], a finding that has been replicated in numerous laboratories. For example, three bursts of 15 stimulation pulses delivered to the perforant path in urethane-anesthetized rats can produce a transient (4–7 h) potentiation of dentate gyrus field potentials, whereas 20 bursts of 15 pulses induce stable, late-phase LTP that shows no signs of decay for at least 8 h after induction [14]. Recently, induction paradigms have been developed that can produce hippocampal LTP *in vivo* that is stable for periods of up to 1 year [15].

The large majority of work characterizing LTP induction parameters has focused on the hippocampal formation. Neocortical synapses may be expected to show plastic properties quite distinct from those of the hippocampus, depending on the experimental preparation and specific type of synapses (e.g., thalamocortical, intracortical) under investigation. Extensive *in vivo* studies by Racine et al. [16, 17] have shown the neocortex of chronically prepared rats to be highly resistant to induction protocols that reliably produce LTP in the hippocampal formation. Thus, neocortical synapses may be governed by a unique set of rules and constraints with regard to plasticity induction.

We have carried out *in vivo* experiments to characterize induction protocols for LTP at thalamocortical synapses in urethane-anesthetized rats. For these experiments, we use theta-burst stimulation (five pulse bursts repeated at theta frequency of 5 Hz) of the lateral geniculate nucleus (LGN) to induce LTP of the field excitatory postsynaptic potential (fEPSP) recorded in the primary visual cortex (V1; Fig. 1A). Heynen and

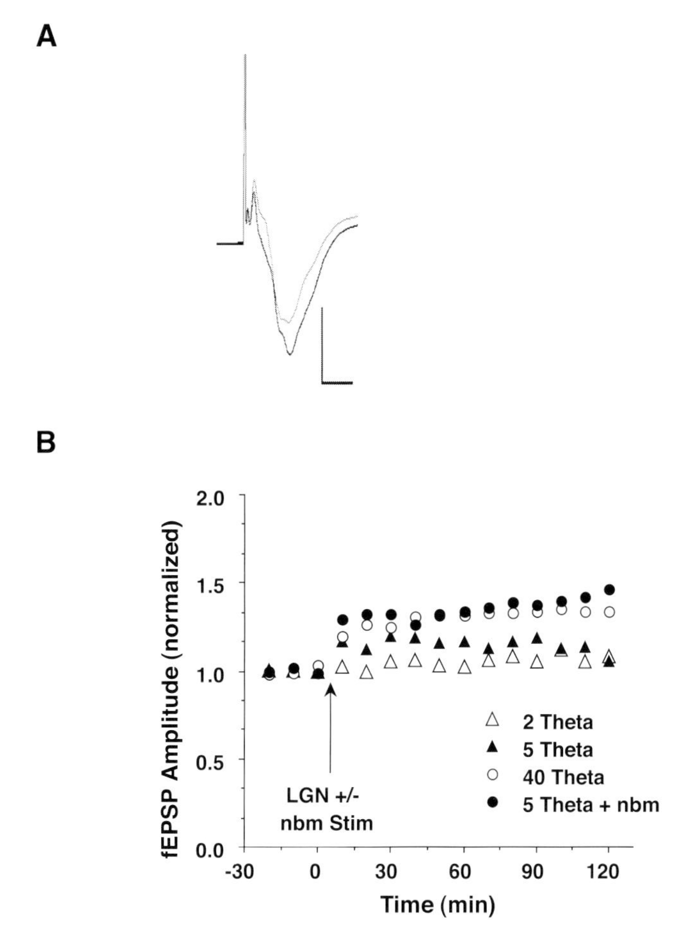

Figure 1. Long-term potentiation (LTP) between the lateral geniculate nucleus (LGN) and primary visual cortex (V1). (A) Excitatory postsynaptic field potentials (fEPSP, averages of 20 sweeps, calibration 10 ms and 0.5 mV) in V1 elicited before (gray) and 2 h after (black) thalamic theta-burst stimulation (5 theta cycles). (B) fEPSP amplitude before and after thalamic theta burst stimulation (2, 5, and 40 theta cycles) and after theta bursting (5 cycles) and subsequent stimulation of the nucleus basalis magnocellularis (nbm). LTP induction with 5 theta cycles plus nbm stimulation produced LTP equivalent to that induced with 40 theta cycles (n = 10–14/group).

Bear [18] have shown that these synapses remain highly plastic and express NMDA receptor-dependent LTP in the mature, adult brain *in vivo*. Thalamic stimulation with two theta cycles is insufficient to produce LTP, whereas five theta cycles produce a moderate ($\sim$ 20%), transient ($<$ 2 h) potentiation of fEPSP amplitude (Fig. 1B). Stronger induction protocols (10-40 theta cycles) produce robust ($\sim$ 40%), stable ($>$ 4 h) potentiation of thalamocortical fEPSPs (Fig. 1B; H. Dringenberg, B. Hamze, A. Wilson, M.-C. Kuo, *unpublished data*).

The studies summarized above serve to demonstrate that changes in presynaptic (glutamatergic) activity are sufficient to manipulate the probability of induction, as well as the duration of LTP at glutamatergic synapses in the forebrain. It is worthwhile mentioning that the LGN, like the hippocampus, shows endogenous oscillations at theta frequencies [19], suggesting that theta-burst stimulation constitutes a physiologically relevant induction regime.

Cholinergic modulation of LTP threshold and maintenance

In the intact brain, glutamatergic activity occurs in a complex, highly dynamic neurochemical environment that is fundamentally different from that of typical *in vitro* preparations. It is becoming increasingly clear that this neuromodulatory environment profoundly influences the characteristics of NMDA receptor-dependent LTP. Importantly, neuromodulatory factors can affect NMDA-dependent LTP without an apparent action on other types (e.g., AMPA receptor-mediated) of glutamatergic transmission. Thus, heterosynaptic interactions involving multiple transmitter systems likely play fundamental roles in optimizing synaptic plasticity and encoding mechanisms in cortical networks.

Cholinergic transmission in the forebrain has long been considered an important modulator of synaptic plasticity, memory consolidation, and other cognitive processes [20–22]. Detailed reviews of the role of acetylcholine (ACh) in plasticity of the sensory cortex are available [21, 23–26]. Less is known about the effects of ACh on LTP *in vivo*, especially for forebrain areas other than the hippocampal formation.

Muscarinic receptor effects

Elegant, early *in vivo* investigations by Krnjevic and Ropert [27] using anesthetized rats demonstrated that electrical stimulation of the medial septum, the main source of cholinergic innervation of the hippocampus [28], produced short-lasting (up to 300 ms) facilitation of CA1 synaptic responses elicited by commissural fiber inputs. This septal facilitation of hippocampal synapses could be reduced by muscarinic receptor antagonists, thus confirming the cholinergic nature of this effect. Interestingly, these investigators also raised the possibility of a small nicotinic component, an effect we discuss in more detail in a subsequent section. Subsequently, using *in vitro* slice preparations and iontophoretic ACh application, Markram and Segal [29] successfully demonstrated longer-lasting facilitatory effects of ACh on EPSPs in CA1 neurons. Importantly, in addition to an action on LTP induced by tetanic stimulation [30, 31], cholinergic-muscarinic receptor activation *in vitro* can induce LTP of the fEPSP in CA1 without high-frequency stimulation of glutamatergic fibers [32, 33]. This finding is of particular importance since it reveals that levels of cholinergic activity have profound effects on the type of afferent input required to elicit synaptic enhancement.

Recent *in vivo* investigations have shown that, in area CA1 of the hippocampus, endogenous ACh release induced by medial septal stimulation lowers the threshold for LTP induction, making a normally subthreshold tetanus effective in eliciting LTP [12]. Conversely, blocking muscarinic receptors inhibits LTP induced by moderate, but not strong, tetanic afferent stimulation [12] (see also [34]). In an intriguing experiment, Leung et al. [35] examined the effects of behaviorally stimulated ACh release on LTP in the CA1 field of freely moving rats. LTP was induced during different behavioral states known to correlate with high (active locomotion) and low levels (immobility, slow wave sleep) of hippocampal ACh release [36]. Induction during walking resulted in enhanced LTP measured 24 h later, an effect that was blocked by the muscarinic antagonist scopolamine or selective immunotoxic lesions of cholinergic cells in the medial septum [35]. Effects on induction thresholds were not assessed in this study. This experiment appears to be the first to demonstrate a naturalistic enhancement of glutamatergic synaptic coupling by ACh and highlights the importance of ongoing behavior, or related cognitive processes, as a factor influencing hippocampal plasticity.

Very few studies have examined the role of ACh in modulating LTP at neocortical synapses *in vivo*, even though pharmacological experiments employing systemic administration of cholinergic drugs support a role of ACh in modifying cortical LTP. Boyd et al. [37] have shown that cholinergic agonists and antagonists facilitate and block, respectively, LTP induction in the motor cortex of freely moving rats. In an interesting study using urethanc-anesthetized rats, Verdier and Dykes [38] showed that pairing of cutaneous electrical hindlimb and basal forebrain stimulation can result in long-lasting increases in the cutaneously elicited evoked potential recorded in the somatosensory cortex. This cholinergic facilitation of a sensory response occurs in the absence of high-frequency afferent stimulation, but several features of this enhancement are similar to those of typical, tetanus-induced LTP (e.g., time-course and NMDA dependence; [38]).

We have performed detailed examinations of the effects of endogenous, synaptically released ACh on the characteristics of NMDA-dependent LTP at thalamocortical synapses between the LGN and V1 in urethane-anesthetized rats. A weak theta-burst induction regime (five theta cycles) produces weak, early-phase LTP at these synapses ($< 20\%$ maximal potentiation of fEPSP amplitude, decay to 6% potentiation 2 h after induction). The same stimulation regime is effective in inducing strong, long-lasting LTP ($\sim 40\%$ potentiation, no decay for > 4 h) when paired with basal forebrain stimulation (10×100 Hz trains of 0.5 s duration) delivered 5 min after LTP induction (Fig. 1B; H. C. Dringenberg, B. Hamze, A. Wilson and M.-C. Kuo, *unpublished data*). A similar effect to stabilize weak, early-phase LTP is seen when transsynaptic inputs to the basal forebrain are activated [14, 39]. Strong, persistent LTP ($\sim 40\%$ maximal potentiation, stable for > 4 h after induction) produced by stimulation with 40 theta cycles is further enhanced ($\sim 60\%$ maximal potentiation) by basal forebrain stimulation 5 min after LTP induction (data not shown). Both of these effects of basal forebrain stimulation are reduced by scopolamine treatment (5 mg/kg, i.p.; Fig. 2), confirming the involvement of muscarinic receptors. It is of interest to note that the magnitude and duration of LTP induced by five theta cycles plus basal forebrain stimulation is not significantly different from LTP induced by stimulation with 40 theta cycles (Fig. 1B). Thus, it appears that the basal fore-

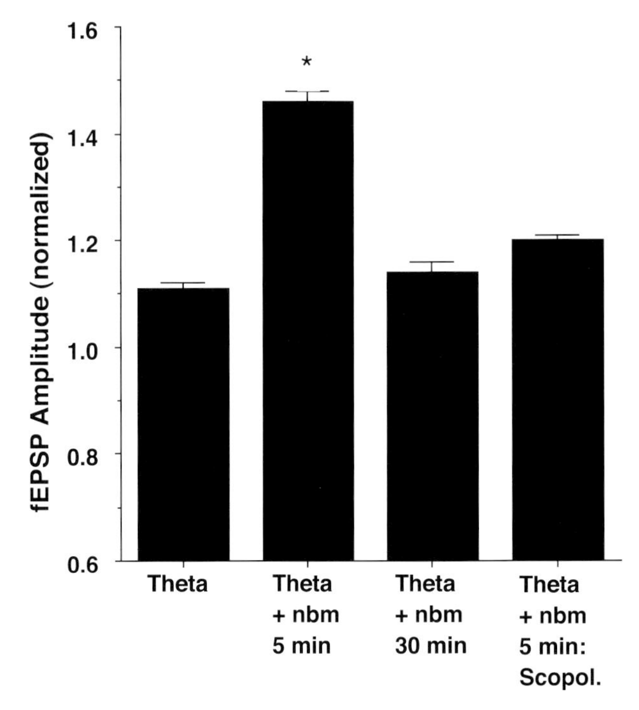

Figure 2. Thalamocortical fEPSP amplitude averaged for recordings taken from 3 to 4 hours after LTP induction (values shown are normalized to baseline). Induction of LTP with 5 theta cycles stimulation produced a small increase in fEPSP amplitude. Additional nbm stimulation 5 min, but not 30 min, following LTP induction resulted in robust LTP. This effect was reduced by scopolamine treatment (Scopol. 5 mg/kg, i.p.). $*$ indicates significant difference ($p < 0.05$) from all other groups ($n = 8$–12/group).

brain cholinergic system can substitute for strong glutamatergic excitation to produce long-lasting increases in synaptic strength at excitatory synapses.

Cholinergic reinforcement of thalamocortical LTP does not occur when basal forebrain stimulation is delivered delayed 30 min (rather than 5 min) following LTP induction (Fig. 2), indicating that a close temporal correlation between glutamatergic and cholinergic activity is required to facilitate plasticity. Further, consistent with other reports [12], these data suggest that ACh acts primarily on the molecular mechanisms of induction, rather than consolidation of glutamate-dependent synaptic strengthening. It is important to note that the effects of basal forebrain stimulation on LTP occur in the absence of significant changes in baseline (i.e., non-potentiated) thalamocortical transmission. Thus, cholinergic LTP enhancement likely is due to a synergistic interaction, rather than an additive effect of glutamatergic and cholinergic stimulation.

The results reviewed above indicate that the level of cholinergic-muscarinic activity exerts an important modulatory effect on glutamatergic, NMDA-dependent synaptic plasticity in the hippocampal formation and neocortex. Muscarinic receptor activation can act as a "filter" to set the threshold for the induction of synaptic modifications. Further, it can enhance the stability of potentiation over extended time periods. Thus, heterosynaptic interactions of cholinergic and glutamatergic inputs are critical in determining the precise characteristics of synaptic modifications in the forebrain.

Nicotinic receptor effects

The role of nicotinic receptors in the regulation of LTP has received less attention than that devoted to their muscarinic counterpart. This is, in part, related to the difficulty in investigating longer-term nicotinic effects due to rapid receptor desensitization [40], especially when exogenous ligands are used to stimulate the receptor. However, nicotinic receptors are abundant at both pre- and postsynaptic elements of glutamatergic synapses in the forebrain [41–44]. Thus, there is increasing interest in nicotinic-glutamatergic interactions and their effects on plasticity (see [45]), processes that could mediate the well-characterized cognitive effects of nicotinic receptor activation and blockade (e.g., [46, 47]).

Several *in vitro* studies have shown that both acute and chronic nicotinic receptor activation can facilitate LTP induction in the hippocampal slice preparation [48, 49]. For example, acute nicotine application *in vitro* lowers the threshold for tetanus-induced LTP in area CA1 without producing potentiation in the absence of tetanic stimulation [50]. Subsequent investigations revealed that the nicotinic modulation of hippocampal plasticity is complex and depends on both the timing and precise location of nicotinic activation (i.e., pyramidal cells versus interneurons; [51]).

It appears that only one study has examined nicotinic contributions to LTP *in vivo*. Matsuyama et al. [52] showed that in the dentate gyrus of anesthetized mice nicotine administration produces a dose-dependent potentiation of the dentate gyrus population spike in the absence of high-frequency stimulation of afferent fibers. This nicotine-induced LTP is long-lasting (> 2 h) and blocked by the nicotinic receptor antagonist mecamylamine given 10 min prior to, but not 1 h after, nicotine administration, which is indicative of a selective role of nicotinic activation in LTP induction [52].

Studies of endogenous, synaptically released ACh acting on nicotinic receptors are particularly important since they offer a way to minimize rapid receptor desensitization seen with continuous agonist exposure [40]. However, with regard to LTP, no such studies have been published to date, even though Krnjevic and Ropert [27] suspected that they observed a nicotinic component in their *in vivo* experiments on septal facilitation of hippocampal population spikes. Recently, Alan Fine and colleagues at the National Institute for Medical Research in London have used organotypic co-cultures to reconstruct the septo-hippocampal cholinergic pathway *in vitro*. This preparation mimics much of the synaptic organization of the cholinergic innervation of the hippocampus *in vivo* and allows examinations of the effects of synaptically released ACh on hippocampal glutamatergic transmission (A. Fine, personal communication). In unpublished experiments, Fine and co-workers have shown that stimulation of septal cholinergic cells can produce a robust, long-lasting (> 1 h) enhancement of hippocampal glutamatergic EPSPs. This effect is resistant to muscarinic receptor blockade, but reduced or abolished by the nicotinic receptor antagonist methyllycaconitine. Nicotinic enhancement of glutamatergic signaling occurs without high-frequency stimulation of glutamatergic fibers, but requires close temporal association between activity in cholinergic and glutamatergic fibers (< 2 s). The latter observation emphasizes the importance of using preparations that permit precisely timed, brief activation to assess the temporal dynamics of converging transmitter inputs, a requirement not met by exogenous drug applications.

In summary, like muscarinic receptors, nicotinic binding sites are in a position to exert powerful, modulatory influences over glutamatergic transmission and plasticity. It will be critical to determine whether the effects summarized above occur *in vivo* and by means of endogenous ACh release. Such studies will provide evidence to link nicotinic effects on plasticity to cognitive processes known to depend on nicotinic receptor activation [45–47].

Histamine

There is a growing consensus that histamine, a relative newcomer to the family of general neuromodulators, plays important roles in behavioral regulation and cognition (see reviews by [53–55]). Nevertheless, only very few studies have investigated the role of histamine in neocortical and hippocampal plasticity.

Early, pioneering investigations using *in vitro* preparations of hippocampal pyramidal cells showed that histamine can significantly potentiate NMDA-mediated currents [56, 57], suggesting that it may also affect the induction of long-lasting, NMDA-dependent plasticity. Histamine occupies a unique position in term of its ability to affect glutamate signaling since, in addition to binding to its own receptors, it can directly interact with the polyamine-binding site on the NMDA receptor complex [57]. Brown et al. [58] examined the action of bath-applied histamine on LTP induction in area CA1 of rat hippocampal slices. A weak tetanus to elicit short-lasting (< 1 h) potentiation of the fEPSP produced LTP (> 2 h) when given in the presence of 100 μM histamine. Histamine was washed out following tetanus, suggesting that its effect is on the initial induction of LTP, rather than processes of consolidation. The histaminergic LTP enhancement is resistant to histamine H1 or H2 receptor antagonists, suggesting that it involves a direct action of histamine on NMDA receptors [58].

We have examined histaminergic effects on *in vivo* LTP in thalamocortical pathways in the urethane-anesthetized rat. The use of electrical stimulation to elicit a relatively selective release of histamine is problematic, given the complex, heterogeneous neurochemical anatomy of the hypothalamic region. Thus, we use reverse microdialysis to apply histamine in close proximity (< 0.5 mm) of the cortical recording electrode. Histamine (0.01–10 mM) application itself does not produce consistent effects on fEPSP amplitude (Fig. 3A), but results in a dose-dependent enhancement of LTP induced by a strong (40 cycles) theta burst protocol (Figs. 3A and B; M.-C. Kuo and H.C. Dringenberg, unpublished data). To our knowledge, these are the first data to demonstrate a direct effect of histamine on neocortical synaptic plasticity *in vivo*. Preliminary data indicate that this effect cannot be blocked by histamine H1 and H2 receptor antagonists, consistent with the data obtained by Brown

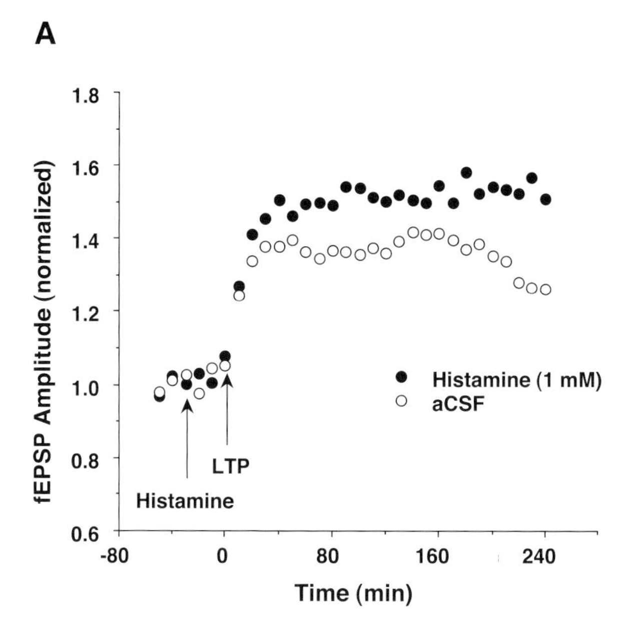

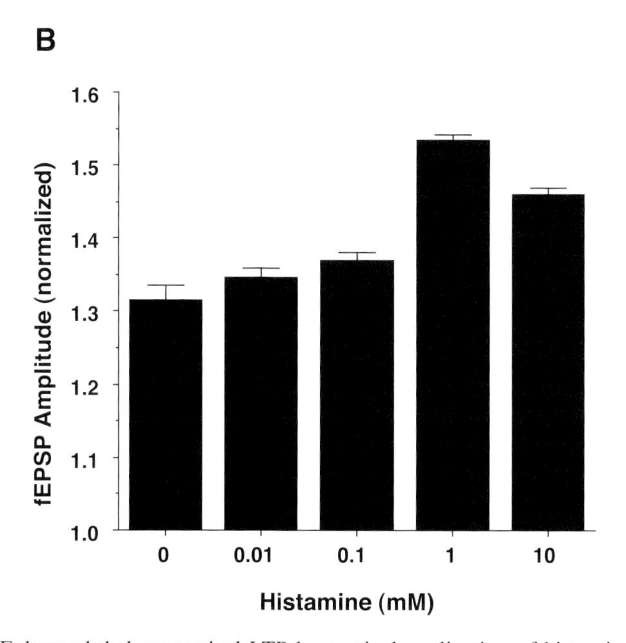

Figure 3. Enhanced thalamocortical LTP by cortical application of histamine by means of reverse microdialysis. (A) Histamine alone (first arrow indicates start on continuous application) did not change fEPSP amplitude but enhanced the effects of subsequent LTP induction (second arrow, 40 theta cycles, $n = 8$/group). (B) Dose-response data for histaminergic LTP enhancement. Values shown are averages of fEPSPs recorded from 3 to 4 h following LTP induction ($n = 8$/group).

et al. [58]. At present, it is not known whether histamine can lower the induction threshold for LTP at thalamocortical or intracortical synapses.

It is important to emphasize that effects of histamine on cortical plasticity and behavior may also be indirectly mediated by an action of histamine on the release of ACh in the forebrain [54]. Activation of histamine receptors in the medial septum and nucleus basalis produces hippocampal and cortical ACh release, respectively [59, 60]. In contrast, histamine acting locally in the neocortex exerts an inhibitory effect on (potassium-) stimulated, but not basal levels of ACh release, an effect mediated by histamine H3 receptors [61]. These observations further emphasize the complexity of mechanisms that can potentially mediate histaminergic effects on plasticity and cognition.

Noradrenaline

Relative to histamine, the modulation of hippocampal LTP by noradrenaline (NA) has received considerable attention and is well characterized for both *in vitro* and *in vivo* preparations. Early work by Lacaille and Harley [62] in hippocampal slices showed that application of NA produces an activity-independent enhancement of dentate gyrus fEPSPs and population spike amplitude, effects that can last beyond 30 min. Subsequently, in a series of elegant investigations, Harley and co-workers demonstrated that NA exerts similar effects in the hippocampus *in vivo*. Excitation of locus coeruleus neurons by means of local glutamate application facilitates dentate gyrus population spike responses in both anesthetized and freely moving rats [63–66]. This effect is blocked by propranolol, indicative of a critical role of beta-receptor activation [63, 64]. The majority of these earlier experiments suggested that the synaptic potentiation in the hippocampus induced by NA is relatively short-lasting (< 1 h). Nevertheless, these data pointed toward a role of endogenous NA release in enhancing hippocampal glutamatergic transmission in the absence of high frequency stimulation of hippocampal afferent fibers.

Recently, Walling and Harley [67] described a novel form of long-lasting, NA-induced plasticity in the rat dentate gyrus. In these experiments, glutamate infusions into the locus coeruleus of freely moving rats is without effect during the initial 3 h, but results in a pronounced increase in fEPSP amplitude in the dentate gyrus 24 h after the infusion. In other words, Walling and Harley observed a form of late-phase LTP without the initial, early-phase synaptic potentiation. Similar to the immediate, but short-lasting effects on population spike amplitude (a measure of cell excitability), this delayed effect of NA on synaptic strength is dependent on activation of beta-receptor [67]. Previously, a type of LTP that can occur in the absence of initial potentiation had been reported for the invertebrate *Aplysia* [68]. These data are of interest since they indicate that some forms of long-lasting synaptic facilitation are not necessarily a mere consequence of a strong, initial potentiation. Walling and Harley [67] speculate that separate short- and long-lasting effects of NA on glutamatergic transmission may mediate different cognitive processes, attention and memory, respectively.

Potent modulatory actions of NA on tetanus-induced LTP have also been characterized. Early-phase LTP in the dentate gyrus *in vivo* induced by weak tetanic stimulation can be converted into late-phase LTP by stimulation of the amygdala or medial septum [14, 69]. Both of these effects are blocked by propranolol, indicating that the amygdala and septum recruit NA- and beta-adrenoreceptor-dependent mechanisms to stabilize synaptic potentiation in the dentate gyrus. Noradrenergic mechanisms even appear to be engaged under experimental conditions when glutamatergic afferents to the dentate gyrus are stimulated without activation of additional, heterosynaptic inputs. Straube and Frey [70] used three LTP induction protocols consisting of weak, intermediate, and strong tetanic stimulation of the perforant path in freely moving rats, all of which produce long-lasting LTP (> 24 h). Blockade of beta-receptors at the time of induction abolishes and reduces, respectively, late-phase LTP elicited by the weak and intermediate stimulation protocol, while no effect is apparent for the strongest induction regime. Thus, complex, heterosynaptic mechanisms can be activated by stimulation of glutamatergic inputs. It will be of interest to establish whether the strongest induction protocol used by Straube and Frey [70] is truly independent of other heterosynaptic neuromodulators, or whether non-noradrenergic inputs are active with glutamate to produce this type of late-phase LTP resistant to NA-receptor blockade.

Recently, Straube et al. [71] demonstrated that exploration of a novel environment is effective in producing a beta-adrenoreceptor-dependent conversion of early- to late-phase LTP, similar to the effects of septal or amygdala stimulation described above [14, 69]. This effect occurs only with relatively short delays (15-30 min, but not 60 min) between the onset of exploration and LTP induction, suggesting that novelty is a critical factor in the observed effect. Importantly, the authors monitored for ongoing behavior to rule out contributions of behavioral state and locomotor activity known to affect hippocampal LTP maintenance [35]. Thus, like active locomotion, novelty exposure results in the release of neuromodulators that interact with weak (i.e., insufficient to produce late-phase LTP) glutamatergic signals to induce lasting synaptic modifications in the hippocampal formation.

Together, these results demonstrate that NA can exert a variety of effects on glutamatergic synapses in the hippocampal formation, including a potent enhancement of synaptic strength that occurs in the absence of tetanizing stimulation and is apparent only after relatively long delays (24 h) following the excitation of NA-containing neurons in the locus coeruleus *in vivo*.

Cognitive implications

How relevant are the data summarized above to behavior and cognitive functions? There is an extensive, controversial literature on the relationship of LTP to learning and memory formation, and insightful commentaries on various aspects of this debate have been published (e.g., [5, 6, 72–74]). A fundamental obstacle to answering this question lies in linking drastically different levels of analyses (experimentally manipulated synaptic strength measured at one or a few synapses; cognitive function in a behaving animal) in a causal manner. At present, it appears impossible to establish direct links between plastic events at isolated synapses and memory encoding in the intact nervous system (see [4, 5]). Nevertheless, the types of synaptic modifications that neurons are capable of expressing, under conditions of complex, dynamic neurochemical interactions, should bear some resemblance to characteristics of information encoding and memory formation in complete organisms. We will explore this point by discussing two instances of similarity between the effects of neuromodulators on LTP and on memory processes characterized in behavioral studies.

Muscarinic receptor involvement in memory stabilization

We have reviewed evidence indicating that cholinergic (muscarinic and nicotinic) receptor activation can transform transient potentiation of glutamatergic synapses into stable increases in synaptic strength. Interestingly, there is good evidence that the cholinergic system, and in particular muscarinic receptor activation, may play similar roles in the stabilization of memory encoding. Bartus et al. [75] used the eight-arm radial maze, requiring rats to remember which arms had been visited during an initial test session, to assess encoding strength in rats. Importantly, delay intervals ranging from 15 min to 8 h between the initial and a subsequent retention test were employed to detect temporally specific changes in memory decay. Control animals showed good performance across all intervals, with only slight decreases in performance at 4 and 8 h delays. Rats receiving neurotoxic lesions of the nucleus basalis (not selective to ACh-containing neurons) showed good performance at short retention intervals (15 min, 1 h), but revealed significant deficits at longer delays, with performance near chance level at 8 h [75]. Bartus and co-workers described similar, delay-dependent impairments of memory encoding in primates treated with scopolamine [76, 77], results that are also seen with selective cholinergic deafferentation of the rhinal cortex [78]. Unfortunately, much shorter retention intervals are used for this primate work (often < 60 s), making it difficult to relate the temporal dynamics of memory strength to LTP stabilization by ACh.

The data summarized above provide a clear demonstration that cholinergic inputs are not required for the initial storage, but play a critical role in the maintenance and consolidation of encoded information, a hypothesis confirmed by subsequent work (see [79] for review). Barros et al. [80] assessed the effects of intra-amygdaloid infusions of muscarinic drugs (given 4 min after training) on both short-term and long-term retention (1.5 and 24 h after training) of a simple inhibitory avoidance response in rats. Muscarinic stimulation and blockade enhanced and impaired, respectively, performance at 24 h without affecting performance 1.5 h after training. Similarly, cholinergic deafferentation of the amygdala results in a slight impairment in acquisition of an inhibitory avoidance response, which can be overcome by additional training trials. However, at 48 h after training, these lesions produce a clear deficit in retention performance [81]. It is unclear whether the amygdala itself is the site for memory encoding assessed in these tests, or whether muscarinic activation of the amygdala initiates signals that facilitate synaptic strength and encoding elsewhere in the brain; in fact, much that has been published speaks for the latter hypothesis [3, 79, 82]. The amygdala activates basal forebrain neurons and stimulates the cholinergic inputs to the neocortex and hippocampus [83, 84]. This mechanism may allow the amygdala to engage the cholinergic system to enhance plasticity and memory encoding elsewhere in the forebrain (see [85]).

These behavioral experiments demonstrate that, for some brain regions and behavioral tasks, ACh promotes long-term memory consolidation, with lesser involvement in short-term encoding (for further examples and discussion, see [79]). These data are reminiscent of the role of ACh to stabilize synaptic changes, for example by converting short-lasting into late-phase LTP by means of muscarinic receptor activation. It is worthwhile noting, however, that the available evidence points toward a broader role of ACh in cognition (e.g., working memory, attentional processes), in addition to the consolidation of long-term memory [22, 79, 80, 86]. It is tempting to speculate that some of these additional cognitive functions may relate to a cholinergic modulation of induction thresholds, rather than more delayed effects on synaptic stabilization.

Noradrenergic facilitation of long-term memory

It appears that NA can exert at least three different effects on glutamatergic transmission and plasticity in the hippocampal formation: a relatively short-lasting (< 1 h) facilitation of cell excitability, a reinforcement of tetanus induced LTP, and a slow-developing (over 24 h) increase in synaptic strength in the absence of high-frequency activation of glutamatergic afferents. The latter two observations predict that NA may exert a somewhat preferential effect on long-term (days), rather than intermediate-term memory (hours). Izquierdo et al. [87] examined the effects of NA infusions (administered immediately after training) into the hippocampus and entorhinal cortex on both intermediate- and long-term memory (1.5 and 24 h after training, respectively, assessed in the same animals) for an inhibitory avoidance response. Hippocampal infusions of NA enhanced 24 h retention without affecting performance assessed 1.5 h after training, whereas entorhinal infusions enhanced performance at both time points [87]. Thus, NA may indeed play a more important role in facilitating long-term encoding, at least with regard to an action in the hippocampus. Similarly, mice carrying a mutated tyrosine hydroxylase gene that show reduced central NA levels are impaired in long-term, but not short-term retention of several aversively conditioned responses [88]. Hippocampal tetanus-induced LTP (assessed for 1 h after induction) and water maze performance were normal in these mutants. Unfortunately, the stability of tetanus-induced late-phase LTP and the presence of the slow-developing, late potentiation described by Walling and Harley [67] were not assessed, making it impossible to compare synaptic potentiation to behavior at time points when performance was impaired in the mutant mice (24–48 h after training).

Recently, Quevedo et al. [89] extended these findings to humans by demonstrating that emotional arousal, known to facilitate memory by a β-adrenergic action [90], enhanced long-term memory (1 week) without affecting short-term memory (1 h) for verbal information. Thus, for both rodents and humans, it is possible to demonstrate noradrenergic facilitation of long-term memory in the absence of strengthened short-term encoding (see [87, 91]). These data bear some resemblance to the effect of NA to enhance LTP stability and induce delayed potentiation without necessarily exerting a more immediate action on glutamatergic synaptic coupling. Thus, converging behavioral and physiological evidence consistently suggests that short- and long-term plasticity are not merely different time points of a continuous, unitary phenomenon. Rather, they can occur as independent processes, each characterized by the involvement of distinct neurochemical mechanisms that control the strength of encoding (see [92]).

Conclusions

We started this chapter with William James' assertion that memory systems are highly selective in terms of the information that is encoded and maintained ("If we remembered everything, we should on most occasions be as ill off as if we remembered nothing." [1]). The data reviewed here suggest that regulatory effects exerted by neuromodulators can aid in the initiation and subsequent stability of plastic phenomena at glutamatergic synapses in forebrain areas important for memory encoding [2]. By setting a threshold for the induction of NMDA-dependent LTP, neuromodulators act as a filter to select signals for an initial storage process. A regionally selective release of modulators, as recently demonstrated for ACh [93], could account for the effects of selective attention on the initial encoding of incoming sensory signals. Neuromodulators can further refine the mnemonic landscape by stabilizing temporally limited synaptic enhancements, thereby determining the duration that encoded information is maintained. In this chapter, we have focused on the cholinergic, histaminergic, and noradrenergic systems, but other neuromodulators (especially dopamine; [94]) are known to play similar roles in plasticity regulation.

It is a tremendous challenge to relate synaptic changes directly to complex, temporally dynamic memory processes. Nevertheless, the neuromodulatory-glutamatergic interactions described here offer some mechanistic explanations of how nervous systems can perform the types of selection processes alluded to by James and others pioneers of modern memory research [3].

Acknowledgments

Research in the authors' laboratory and described in this chapter is supported by the Natural Sciences and Engineering Research Council of Canada (NSERC), whose support is gratefully acknowledged.

References

1. James W (1890) *Principles of psychology.* H. Holt & Co., New York
2. Squire LR (2004) Memory systems of the brain: a brief history and current perspective. *Neurobiol Learn Memory* 82: 171–177
3. McGaugh JL (2000) Memory – a century of consolidation. *Science* 287: 248–251
4. Malenka RC, Bear MF (2004) LTP and LTD: an embarrassment of riches. *Neuron* 44: 5–21
5. Martin SJ, Morris RGM (2002) New life in an old idea: the synaptic plasticity and memory hypothesis revisited. *Hippocampus* 12: 609–636
6. Roman FS, Truchet B, Marchetti E, Chaillan FA, Soumireu-Mourat B (1999) Correlations between electrophysiological observations of synaptic plasticity modifications and behavioral performance in mammals. *Progr Neurobiol* 58: 61–87
7. Abraham WC, Williams JM (2003) Properties and mechanisms of LTP maintenance. *Neuroscientist* 9: 463–474
8. Dineley KT, Weeber EJ, Atkins C, Adams JP, Anderson AE, Sweatt JD (2001) Leitmotifs in the biochemistry of LTP induction: amplification, integration and coordination. *J Neurochem* 77: 961–971
9. Sheng M, Kim MJ (2002) Postsynaptic signaling and plasticity mechanisms. *Science* 298: 776–780
10. Heynen AJ, Abraham WC, Bear MF (1996) Bidirectional modification of CA1 synapses in the adult hippocampus *in vivo*. *Nature* 381: 163–166
11. Shouval HZ, Bear MF, Cooper LN (2002) A unified model of NMDA receptor-dependent bidirectional synaptic plasticity. *Proc Natl Acad Sci USA* 99: 10831–10836
12. Ovsepian SV, Anwyl R, Rowan MJ (2004) Endogenous acetylcholine lowers the threshold for long-term potentiation induction in the CA1 area through muscarinic receptor activation: *in vivo* study. *Eur J Neurosci* 20: 1267–1275
13. Bliss TVP, Gardner-Medwin AR (1973) Long-lasting potentiation of synaptic transmission in the dentate gyrus of the unanesthetized rabbit following stimulation of the perforant path. *J Physiol* 232: 357–374
14. Frey S, Bergado-Rosado J, Seidenbecher T, Pape H-C, Frey JU (2001) Reinforcement of early long-term potentiation (early-LTP) in dentate gyrus by stimulation of the basolateral amygdala: heterosynaptic induction mechanisms of late-LTP. *J Neurosci* 21: 3697–3703
15. Abraham WC, Logan B, Greenwood JM, Dragunow M (2002) Induction and experience-dependent consolidation of stable long-term potentiation lasting months in the hippocampus. *J Neurosci* 22: 9626–9634
16. Racine RJ, Chapman CA, Trepel C, Teskey GC, Milgram NW (1995) Post- activation potentiation in the neocortex. IV. Multiple sessions required for induction of long-term potentiation in the chronic preparation. *Brain Res* 702: 87-93
17. Racine RJ, Teskey GC, Wilson D, Seidlitz E, Milgram NW (1994) Post- activation potentiation and depression in the neocortex of the rat: II. Chronic preparations. *Brain Res* 637: 83–96
18. Heynen AJ, Bear MF (2001) Long-term potentiation of thalamocortical transmission in the adult visual cortex *in vivo*. *J Neurosci* 21: 9801–9813
19. Hughes SW, Lörincz M, Cope DW, Blethyn KL, Kekesi KA, Parri HR, Juhasz G, Crunelli V (2004) Synchronized oscillations at α and θ frequencies in the lateral geniculate nucleus. *Neuron* 42: 253–268
20. Hasselmo ME (1999) Neuromodulation: acetylcholine and memory consolidation. *Trends Cog Sci* 3: 351–359

21. Rasmusson DD (2000) The role of acetylcholine in cortical synaptic plasticity. *Behav Brain Res* 115: 205–218
22. Sarter M, Bruno JP (1997) Cognitive functions of cortical acetylcholine: toward a unifying hypothesis. *Brain Res Rev* 23: 28–46
23. Edeline J-M (1999) Learning-induced physiological plasticity in the thalamocortical sensory systems: a critical evaluation of receptive field plasticity, map changes and their potential mechanisms. *Progr Neurobiol* 57: 165–224
24. Edeline J-M (2003) The thalamocortical auditory receptive fields: regulation by the states of vigilance, learning and the neuromodulatory systems. *Exp Brain Res* 153: 554–572
25. Gu Q (2003) Contributions of acetylcholine to visual cortex plasticity. *Neurobiol Learn Mem* 80: 291–301
26. Weinberger NM (2004) Specific long-term memory traces in primary auditory cortex. *Nat Rev Neurosci* 5: 279–290
27. Krnjevic K, Ropert N (1982) Electrophysiological and pharmacological characteristics of facilitation of hippocampal population spikes by the stimulation of the medial septum. *Neuroscience* 7: 2165–2183
28. Semba K (2000) Multiple output pathways of the basal forebrain: organization, chemical heterogeneity, and roles in vigilance. *Behav Brain Res* 115: 117–141
29. Markram H, Segal M (1990) Long-lasting facilitation of excitatory postsynaptic potentials in the rat hippocampus by acetylcholine. *J Physiol (Lond)* 427: 381–393
30. Brocher S, Artola A, Singer W (1992) Agonists of cholinergic and noradrenergic receptors facilitate synergistically the induction of long-term potentiation in slices of rat visual cortex. *Brain Res* 573: 27–36
31. Burgard EC, Sarvey JM (1990) Muscarinic receptor activation facilitates the induction of long-term potentiation (LTP) in the rat dentate gyrus. *Neurosci Lett* 116: 34–39
32. Auerbach JM, Segal M (1994) A novel cholinergic induction of long-term potentiation in rat hippocampus. *J Neurophysiol* 72: 2034–2040
33. Huerta PT, Lisman JE (1993) Heightened synaptic plasticity of hippocampal CA1 neurons during a cholinergically induced rhythmic state. *Nature* 364: 723–725
34. Markevich V, Scorsa AM, Dawe GS, Stephenson JD (1997) Cholinergic facilitation and inhibition of long-term potentiation of CA1 in the urethane-anaesthetized rats. *Brain Res* 754: 95–102
35. Leung LS, Shen B, Rajakumar N, Ma J (2003) Cholinergic activity enhances hippocampal long-term potentiation in CA1 during walking in rats. *J Neurosci* 23: 9297–9304
36. Dudar JD, Whishaw IQ, Szerb JC (1979) Release of acetylcholine from the hippocampus of freely moving rats during sensory stimulation and running. *Neuropharmacology* 18: 673–678
37. Boyd TE, Trepel C, Racine RJ (2000) Cholinergic modulation of neocortical long-term potentiation in the awake, freely moving rat. *Brain Res* 881: 28–36
38. Verdier D, Dykes RW (2001) Long-term cholinergic enhancement of evoked potentials in rat hindlimb somatosensory cortex displays characteristics of long-term potentiation. *Exp Brain Res* 137: 71–82
39. Dringenberg HC, Kuo M-C, Tomaszek S (2004) Stabilization of thalamocortical long-term potentiation by the amygdala: cholinergic and transcription-dependent mechanisms. *Eur J Neurosci* 20: 557–565
40. Zhang ZW, Vijayaraghaven S, Berg DK (1994) Neuronal acetylcholine receptors that bind alpha-bungarotoxin with high affinity function as ligand-gated ion channels. *Neuron* 12: 167–177

41. Fabian-Fine R, Skehel P, Errington ML, Davies HA, Sher E, Stewart MG, Fine A (2001) Ultrastructural distribution of the alpha7 nicotinic acetylcholine receptor subunit in rat hippocampus. *J Neurosci* 21: 7993–8003
42. Hunt S, Schmidt J (1979) The relationship of alpha-bungarotoxin binding activity and cholinergic termination within the rat hippocampus. *Neuroscience* 4: 585–592
43. Hunt SP, Schmidt J (1978) The electron microscopic autoradiographic localization of alpha-bungarotoxin binding sites within the central nervous system of the rat. *Brain Res* 142: 152–159
44. Levy RB, Aoki C (2002) Alpha7 nicotinic acetylcholine receptors occur at postsynaptic densities of AMPA receptor-positive and -negative excitatory synapses in rat sensory cortex. *J Neurosci* 22: 5001–5015
45. Jones S, Sudweeks S, Yakel JL (1999) Nicotinic receptors in the brain: correlating physiology with function. *Trends Neurosci* 22: 555–561
46. Barros DM, Ramirez MR, Dos Reis EA, Izquierdo I (2004) Participation of hippocampal nicotinic receptors in acquisition, consolidation and retrieval of memory for one trial inhibitory avoidance in rats. *Neuroscience* 126: 651–656
47. Levin ED, Bradley A, Addy N, Sigurani N (2002) Hippocampal $\alpha 7$ and $\alpha 4\beta 2$ nicotinic receptors and working memory. *Neuroscience* 109: 757–765
48. Hunter BE, de Fiebre CM, Papke RL, Kem WR, Meyer EM (1994) A novel nicotinic agonist facilitates induction of long-term potentiation in the rat hippocampus. *Neurosci Lett* 168: 130–134
49. Sawada S, Yamamoto C, Ohno-Shosaku T (1994) Long-term potentiation and depression in the dentate gyrus and effects of nicotine. *Neurosci Res* 20: 323–329
50. Fujii S, Ji Z, Morita N, Sumikawa K (1999) Acute and chronic nicotine exposure differentially facilitates the induction of LTP. *Brain Res* 846: 137–143
51. Ji D, Lape R, Dani JA (2001) Timing and location of nicotinic activity enhances or depresses hippocampal synaptic plasticity. *Neuron* 31: 131–141
52. Matsuyama S, Matsumoto A, Enomoto T, Nishizaki T (2000) Activation of nicotinic acetylcholine receptors induces long-term potentiation *in vivo* in the intact mouse dentate gyrus. *Eur J Neurosci* 12: 3741–3747
53. Bacciottini L, Passani MB, Mannaioni PF, Blandina P (2001) Interactions between histaminergic and cholinergic systems in learning and memory. *Behav Brain Res* 124: 183–194
54. Blandina P, Efoudebe M, Cenni G, Mannaioni P, Passani MB (2004) Acetylcholine, histamine, and cognition: two sides of the same coin. *Learn Mem* 11: 1–8
55. Haas H, Panula P (2003) The role of histamine and the tuberomammillary nucleus in the nervous system. *Nat Rev Neurosci* 4: 121–130
56. Bekkers JM (1993) Enhancement by histamine of NMDA-mediated synaptic transmission in the hippocampus. *Science* 261: 104–106
57. Vorobjev VS, Sharonova IN, Walsh IB, Haas HL (1993) Histamine potentiates N-methyl-D-aspartate responses in acutely isolated hippocampal neurons. *Neuron* 11: 837–844
58. Brown RE, Fedorov NB, Haas HL, Reymann KG (1995) Histaminergic modulation of synaptic plasticity in area CA1 of rat hippocampal slices. *Neuropharmacology* 34: 181–190
59. Bacciottini L, Passani MB, Giovannelli L, Cangiolo I, Mannaioni PF, Schunack W, Blandina P (2002) Endogenous histamine in the medial septum-diagonal band complex increases the release of acetylcholine from the hippocampus: a dual-probe microdialysis study in the freely moving rat. *Eur J Neurosci* 15: 1669–1680

60. Cecchi M, Passani MB, Bacciottini L, Mannaioni PF, Blandina P (2001) Cortical acetylcholine release elicited by stimulation of histamine H_1 receptors in the nucleus basalis magnocellularis: a dual-probe microdialysis study in the freely moving rat. *Eur J Neurosci* 13: 68–78
61. Blandina P, Giorgetti M, Bartolini L, Cecchi M, Timmerman H, Leurs R, Pepeu G, Giovannini MG (1996) Inhibition of cortical acetylcholine release and cognitive performance by histamine H_3receptor activation in rats. *Br J Pharmacol* 119: 1656–1664
62. Lacaille JC, Harley CW (1985) The action of norepinephrine in the dentate gyrus: beta-mediated facilitation of evoked potentials *in vitro*. *Brain Res* 358: 210–220
63. Harley CW, Milway JS (1986) Glutamate ejection in the locus coeruleus enhances the perforant path-evoked population spike in the dentate gyrus. *Exp Brain Res* 63: 143–150
64. Harley C, Milway JS, Lacaille J-C (1989) Locus coeruleus potentiation of dentate gyrus responses: evidence for two systems. *Brain Res Bull* 22: 643–650
65. Harley CW, Sara SJ (1992) Locus coeruleus bursts induced by glutamate trigger delayed perforant path spike amplitude potentiation in the dentate gyrus. *Exp Brain Res* 89: 581–587
66. Klukowski G, Harley CW (1994) Locus coeruleus activation induces perforant path-evoked population spike potentiation in the dentate gyrus of awake rat. *Exp Brain Res* 102: 165–170
67. Walling SG, Harley CW (2004) Locus ceruleus activation initiates delayed synaptic potentiation of perforant path input to the dentate gyrus in awake rats: a novel β-adrenergic- and protein synthesis-dependent mammalian plasticity mechanism. *J Neurosci* 24: 598–604
68. Emptage NJ, Carew TJ (1993) Long-term synaptic facilitation in the absence of short-term facilitation in *Aplysia* neurons. *Science* 262: 253–256
69. Frey S, Bergado JA, Frey JU (2003) Modulation of late phases of long-term potentiation in the rat dentate gyrus by stimulation of the medial septum. *Neuroscience* 118: 1055–1062
70. Straube T, Frey JU (2003) Involvement of beta-adrenergic receptors in protein synthesis-dependent late long-term potentiation (LTP) in the dentate gyrus of freely moving rats: the critical role of the LTP induction strength. *Neuroscience* 119: 473–479
71. Straube T, Korz V, Balschun D, Frey JU (2003) Requirement of β-adrenergic receptor activation and protein synthesis for LTP-reinforcement by novelty in rat dentate gyrus. *J Physiol (Lond)* 552.3: 953–960
72. Bliss TVP (1998) The saturation debate. *Science* 281: 1975–1976
73. Cain DP (1997) LTP, NMDA, genes and learning. *Curr Opin Neurobiol* 7: 235–242
74. Stevens CF (1998) A million dollar question: does LTP = memory? *Neuron* 20: 1–2
75. Bartus RT, Flicker C, Dean R, Pontecorvo M, Figueriedo J, Fisher S (1985) Selective memory loss following nucleus basalis lesions: long-term behavioral recovery despite persistent cholinergic deficiencies. *Pharmacol Biochem Behav* 23: 125–135
76. Bartus RT (1979) Evidence for a direct cholinergic involvement in the scopolamine-induced amnesia in monkeys: effects of concurrent administration of physostigmine and methylphenidate with scopolamine. *Pharmacol Biochem Behav* 9: 833–836
77. Bartus RT, Johnson HR (1976) Short-term memory in the rhesus monkey: disruption from the anti-cholinergic scopolamine. *Pharmacol Biochem Behav* 5: 39–46
78. Turchi J, Saunders RC, Mishkin M (2005) Effects of cholinergic deafferentation of the rhinal cortex on visual recognition memory in monkeys. *Ann Natl Acad Sci USA* 102: 2158–2161
79. Power AE, Vazdarjanova A, McGaugh JL (2003) Muscarinic cholinergic influences in memory consolidation. *Neurobiol Learn Memory* 80: 178–193

80. Barros DM, Pereira P, Medina JH, Izquierdo I (2002) Modulation of working memory and of long- but not short-term memory by cholinergic mechanisms in the basolateral amygdala. *Behav Pharmacol* 13: 163–167
81. Power AE, McGaugh JL (2002) Phthalic acid amygdalopetal lesions of the nucleus basalis magnocellularis induces reversible memory deficits in rats. *Neurobiol Learn Memory* 77: 372–388
82. Cahill L, Weinberger NM, Roozendaal B, McGaugh JL (1999) Is the amygdala a locus of "conditioned fear"? Some questions and caveats. *Neuron* 23: 227–228
83. Dringenberg HC, Vanderwolf CH (1996) Cholinergic activation of the electrocorticogram: an amygdaloid activating system. *Exp Brain Res* 108: 285–296
84. Dringenberg HC, Vanderwolf CH (1997) Neocortical activation: modulation by multiple pathways acting on central cholinergic and serotonergic systems. *Exp Brain Res* 116: 160–174
85. Power AE, Thal LJ, McGaugh JL (2002) Lesions of the nucleus basalis magnocellularis induced by 192 IgG-saporin block memory enhancement with posttraining norepinephrine in the basolateral amygdala. *Proc Natl Acad Sci USA* 99: 2315–2319
86. Bartus RT (2000) On neurodegenerative diseases, models, and treatment strategies: lessons learned and lessons forgotten a generation following the cholinergic hypothesis. *Exp Neurol* 163: 495–529
87. Izquierdo I, Medina JH, Izquierdo LA, Barros DM, de Souza MM, Souza T (1998) Short- and long-term memory are differentially regulated by monoaminergic systems in the rat brain. *Neurobiol Learn Mem* 69: 219–224
88. Kobayashi K, Noda Y, Matsushita N, Nishii K, Sawada H, Nagatsu T, Nakahara D, Fukabori R, Yasoshima Y, Yamamoto T et al. (2000): Modest neuropsychological deficits caused by a reduced noradrenaline metabolism in mice heterozygous for a mutated tyrosine hydroxylase gene. *J Neurosci* 20: 2418–2426
89. Quevedo J, Sant' Anna MK, Madruga M, Lovato I, de-Paris F, Kapczinski F, Izquierdo I, Cahill L (2003) Differential effects of emotional arousal in short- and long-term memory of healthy adults. *Neurobiol Learn Mem* 79: 132–135
90. Cahill L, Prins B, Weber M, McGaugh JL (1994) β-adrenergic activation and memory for emotional events. *Nature* 371: 702–704
91. Izquierdo LA, Barros DM, Vianna MRM, Coitinho A, Silva TD, Choi H, Moletta B, Medina JH, Izquierdo I (2002) Molecular pharmacological dissection of short- and long-term memory. *Cell Mol Neurobiol* 22: 269–287
92. Izquierdo I, Medina JH, Vianna MRM, Izquierdo LA, Barros DM (1999) Separate mechanisms for short- and long-term memory. *Behav Brain Res* 103: 1–11
93. Fournier GN, Semba K, Rasmusson DD (2004) Modality- and region-specific acetylcholine release in the rat neocortex. *Neuroscience* 126: 257–262
94. Li S, Cullen WK, Anwyl R, Rowan MJ (2003) Dopamine-dependent facilitation of LTP induction in hippocampal CA1 by exposure to spatial novelty. *Nat Neurosci* 6: 526–531

Neurotransmitter Interactions and Cognitive Function
Edited by Edward D. Levin
© 2006 Birkhäuser Verlag/Switzerland

Nicotinic-antipsychotic drug interactions and cognitive function

Edward D. Levin and Amir H. Rezvani

Department of Psychiatry and Behavioral Sciences, Duke University Medical Center, Durham, NC 27710, USA

Introduction

Neuronal nicotinic systems have been found to be important for a variety of cognitive functions including learning, memory and attention [1]. Nicotinic treatments hold promise for syndromes of cognitive dysfunction such as Alzheimer's disease, attention deficit hyperactivity disorder (ADHD) and schizophrenia [2–4]. The development of nicotinic treatment for cognitive dysfunction must take into account not only the mechanisms of nicotinic effects in both compromised and normal brains, but also interactions with other medications that are used to treat these disorders. Prime examples of these types of interactions are nicotinic-antipsychotic drug interactions in schizophrenia.

Schizophrenia is primarily considered to be a psychotic disorder, but it has become apparent that schizophrenia is also a syndrome of cognitive impairment [5]. Cognitive dysfunction is substantial in schizophrenia. This cognitive impairment ranges from impairment of sensory gating to attentional deficits. Deficits in attention, memory, learning and sensory modulation compromise the ability of people with schizophrenia to function adequately in everyday activities and to successfully reintegrate into society. Antipsychotic drugs can effectively combat hallucinations, but often have no effect or even exacerbate cognitive impairment. Clearly, better medications to improve cognitive function in schizophrenia are necessary. The NIH-sponsored MATRICS program has outlined the need and possible avenues for developing new therapeutic drugs for cognitive enhancement in schizophrenia [6]. A variety of pharmacological approaches for treating the cognitive impairments of schizophrenia have been tried; among those especially promising are nicotinic agonists, particularly nicotinic $\alpha7$ agonists [7]. For the development of novel nicotinic treatments for schizophrenia it would be quite advantageous to know the critical mechanisms of action for nicotinic involvement in cognitive function and interactions of nicotinic systems with actions of antipsychotic drugs for reversing or improving cognitive dysfunction.

Nicotinic involvement in cognitive function

Nicotine exerts its effects through multiple mechanisms. Some effects, like its promoting cigarette smoking, are adverse. Other effects, like nicotine-induced improvement in attention and memory (for review see [1]), are potentially beneficial and present novel therapeutic opportunities. Nicotine has primary effects on a variety of different receptor subtypes, including $\alpha 7$ and $\alpha 4\beta 2$ receptors, which are the best-characterized CNS nicotinic receptors in terms of behavioral function. In addition, nicotine has cascading effects via its action to release a variety of different neurotransmitters including acetylcholine, dopamine, norepinepherine, serotonin, GABA and glutamate [8–10].

Neuronal nicotinic acetylcholinergic (ACh) receptors play a critical role in memory function in both humans and experimental animals, with nicotine causing a significant improvement in attention learning and memory function [1, 11–16]. This provides the basis for its promise as a new treatment for cognitive disorders. In rats, nicotine or other nicotinic agonists significantly improves working memory performance in the radial-arm maze [1]. Nicotine also reverses haloperidol-induced memory impairments in rats [17]. To further this development and to provide a better understanding of the basic neural mechanisms of memory, it is important to determine the critical neural structures and nicotinic receptor subtypes necessary for nicotine-induced memory improvement. Our earlier studies have determined the involvement of nicotinic $\alpha 4\beta 2$ and $\alpha 7$ receptors in the ventral hippocampus as being particularly important for working memory function. Infusions of nicotinic $\alpha 4\beta 2$ and $\alpha 7$ nicotinic receptor antagonists in the ventral hippocampus cause working memory impairments in the radial-arm maze [18].

The $\alpha 7$ nicotinic agonist ARR-17779 caused a significant improvement in learning of the classic win-shift radial-arm maze and also caused a continuing improvement in learning on the repeated acquisition task in the radial-arm maze in which a new problem is presented each session [19]. On this same task, nicotine did not improve accuracy whereas the atypical nicotinic agonist lobeline did significantly improve accuracy [20]. The $\alpha 4\beta 2$ nicotinic agonist metanicotine (RJR 2403) significantly improved working memory function in rats on the eight-arm radial maze (more correct entries until the first error). Interestingly, this effect was evident both 1 h after perioral administration as well as 6 h after dosing, long after the compound had been catabolized, thus indicating a persistent effect of nicotinic stimulation [21]. Local infusion of nicotinic antagonists offers a good way to understand the role of nicotinic systems in memory function. The rapid assessment of cognitive function after local infusions of nicotinic antagonists enables one to determine the functional effect before chronic adaptation takes place. Then, the impact of chronic adaptation itself can be studied with the use of chronic slow infusion of selective nicotinic antagonists.

In addition to nicotine-induced memory improvement, there is evidence that nicotine can also improve attention in experimental animals [22–28]. Using an operant visual signal detection task, it has been demonstrated that a low dose range of nicotine (0.0125–0.05 mg/kg) caused an increase in percent correct rejection suggesting an improvement in attention as reflected in an increase in choice accuracy [26, 27, 29]. In the same procedure the nicotinic antagonist mecamylamine decreased choice accuracy by reducing both percent hit and percent correct rejection [26]. Mecamylamine has been shown also to impair attentional performance in another well-validated rodent model of attention, the five choice serial reaction time task [24, 30]. Using the same task, Ruotsalainen et al. only reported a decrement in reaction time, not accuracy following mecamylamine challenge in rats. The cognitive impairing effects of mecamylamine suggest the involvement of the neuronal nicotinic cholinergic system in normal cognitive functioning [31].

Nicotine agonist ABT-418 has also been shown to improve accuracy in the operant signal detection task [32]. Nicotinic analog treatment has also been shown to improve attention. Terry et al. [33] found that the nicotinic agonist SIB-1553A significantly improves performance of rats on a five-choice attentional task, but only when accuracy was reduced behaviorally with a distracting stimulus, or pharmacologically by the NMDA-sensitive glutamate receptor antagonist dizocilpine (MK-801).

Nicotine has also been shown to reverse attentional impairments in rats caused by basal forebrain lesions [24, 25, 28] or lesions of the septohippocampal pathways [34]. Interestingly, chronic nicotine infusion has been shown to significantly diminish the impairing effects of the typical antipsychotic drug haloperidol [35] and atypical antipsychotic drugs, clozapine and risperidone [35] on attentional performance in female rats using an operant visual signal detection task.

The issue concerning whether nicotine can improve attentiveness in normal nonsmokers who have no pre-existing attentional impairment has been addressed. Adult nonsmokers without ADHD symptoms were administered either 7 mg/kg/day nicotine patches or placebo for 4.5 h/day. It was found that the administration of nicotine significantly reduced the number of errors of omission on the continuous performance task (CPT task). No change in errors of commission was found. It was also found that the nicotine patch significantly decreased response time variability and increased a composite attention measure. Overall, this study demonstrated that nicotine given transdermally could improve attention in nonsmoking subjects who had no pre-existing attentional deficits [36].

Selective effects of nicotine on attentional processes have also been studied in smokers. Smokers who abstained from smoking for at least 10 h prior to testing were treated with 21 mg nicotine transdermal patches for either 3 or 6 h and tested for selective effects of nicotine on tests of attentional function as well as the Stroop test. It was shown that the 6 h, but not the 3 h nicotine patch enhanced the speed of number generation and the speed of processing in both the control and interference condition of the Stroop test. There were no effects on attentional switching of the Flexibility of Attention test. The authors suggest that nicotine mainly improves the intensity features of attention, rather than the selectivity features [37]. The nicotinic agonist ABT-418 was found to improve attentional symptoms of ADHD in adults with the syndrome [38]. Nicotine-induced attentional improvement has been found in MRI imaging studies to be accompanied by increased activation in the parietal cortex, thalamus and caudate [39]. Kumari et al. found that nicotine induced improvements in an N-back memory task were accompanied by increases in activity in several cortical regions of interest, the anterior cingulate cortex, superior frontal cortex and the superior parietal cortex, during N-back task performance [40].

Nicotinic systems and schizophrenia

Nicotinic receptor deficits in mainly $\alpha 7$ but also $\alpha 4\beta 2$ receptors are seen in the brains of people with schizophrenia [41, 42]. These nicotinic receptor deficits appear to play an important role in the manifestation of their cognitive impairment. The high rates of tobacco smoking in people with schizophrenia may be a form of self-medication, albeit a very dangerous form, to combat their cognitive impairment [43]. Nicotinic co-treatments may hold promise for reducing cognitive dysfunction in schizophrenia. Without the need for self-medication, effective nicotinic therapy could also help people with schizophrenia quit smoking thereby eliminating a substantial health risk as well as improving cognitive function.

Cognitive impairment has become recognized as a central component of schizophrenia and is an important negative side effect of antipsychotic drugs, which is a key reason why most schizophrenia patients do not successfully re-integrate into society [44–47]. Effective pharmacotherapy for schizophrenia must not only be antipsychotic; it must also ameliorate cognitive dysfunction. Unfortunately, classic antipsychotic drugs like haloperidol not only do not help reverse the cognitive impairment of schizophrenia; they cause further impairment [48, 49]. The newer "atypical" antipsychotics, like clozapine and risperidone, are better in that they have less severe cognitive impairing side effects and provide some improvement in attentional function. However, there is still a significant need for further improvements, particularly regarding memory-related functioning [49, 50]. There is a great need for pharmacotherapies that provide enhanced treatment of cognitive impairment in schizophrenia.

A critical clue to improved pharmacotherapy for the cognitive impairment of schizophrenia comes from the behaviors of people with schizophrenia. People with schizophrenia smoke cigarettes more heavily (88%) than almost any other group in the population [43]. A world-wide meta-analysis of smoking and schizophrenia demonstrated three times greater smoking in schizophrenia than in the general population and twice the incidence compared with other major mental illnesses [51], possibly as a form of self-medication [52]. It was shown that patients with schizophrenia also smoke cigarettes more intensively thereby increasing their nicotine intake [53]. Highly dependent smokers are in general those with more severe schizophrenic illness [54]. Smoking improves aspects of their cognitive impairment [55]. Interestingly, higher smoking rates appear to precede the onset of schizophrenia. Higher smoking rates are seen in those who later become schizophrenic [56]. This may be related to self-medication for cognitive deficits, which are present before the first break into schizophrenia, greater vulnerability to nicotine addiction in those prone to schizophrenia or possible role for nicotine in precipitating schizophrenia. There is a developing literature that people with schizophrenia may smoke heavily for self-medication to relieve cognitive impairments, which are a part of the syndrome of schizophrenia and potential adverse cognitive effects of antipsychotic drugs [57] although there are certainly other possibilities including enhanced vulnerability to the reinforcing properties of nicotine.

There is considerable evidence that people with schizophrenia appear to be self-medicating to counteract effects of schizophrenia and side effects of the antipsychotic drugs [58, 59]. For the development of new therapeutic avenues for treating the cognitive impairment of schizophrenia it is vital to determine the interactions of the candidate drugs with the antipsychotic drugs given to treat the hallucinations of schizophrenia. A series of studies have been conducted to assess the interactions of antipsychotic drugs with nicotinic systems underlying cognitive function. These studies have characterized nicotinic-antipsychotic drug interactions in cognitive functions including pre-pulse inhibition, working and reference memory and selective attention.

Nicotinic $\alpha7$ and $\alpha4\beta2$ receptor dysfunction may underlie both schizophrenia and smoking in schizophrenics [41, 60]. Patients with schizophrenia have a deficiency of $\alpha7$ nicotinic receptors in the hippocampus and frontal cortex [61, 62]. In particular, $\alpha7$ receptors in the hippocampus appear to be important for the cognitive impairment [63–65]. This may be due to decreased desensitization by hippocampal $\alpha7$ receptors in schizophrenics [66]. Hippocampal-based deficient auditory sensory gating in patients with schizophrenia can be normalized by nicotine administration via cigarette smoking [67]. Nicotine has been found in our studies and others to significantly improve cognitive function in patients with schizophrenia [48, 63]. Nicotinic co-treatment may provide significant improvement in the pharmacotherapy for the cognitive deficits of schizophrenia. Smoking improves sensory gating in patients with schizophrenia [68]. It has been shown that nicotine skin patch administration reduces attentional deficits of schizophrenia as well as attenuates cognitive deficits caused by classic neuroleptics [48]. Nicotine administered by a safer route than smoking such as nicotine skin patches could provide potential beneficial cognitive effects without the toxic effects of smoking. New nicotinic subtype selective agonists may be even safer and more effective. In research with rats, it has been found that the $\alpha7$ agonist ARR-17779 significantly improves learning in the win-shift and repeated acquisition procedures on the radial-arm maze. It also reverses the memory impairment caused by knife-cut lesions of the fimbria-fornix [19]. In the converse experiment, it was found that hippocampal infusion of the nicotinic $\alpha7$ antagonist MLA significantly impaired memory function in rats [18, 69–72]. This may be due to blockade of $\alpha7$ and $\alpha4\beta2$ nicotinic receptors on hippocampal interneurons in rats [73].

The involvement of brain $\alpha4\beta2$ as well as $\alpha7$ nicotinic receptors in schizophrenia-induced cognitive impairment is supported by the finding of decreased receptor levels in the brains of patients with schizophrenia [41]. Nicotine-induced stimulation of DA release in the frontal cortex is blocked by antagonism of $\alpha4\beta2$, but not $\alpha7$ receptors [74]. Thus, the relationship of low prefrontal DA activity to cognitive deficits in schizophrenia supports the involvement of $\alpha4\beta2$ nicotinic receptors in the cognitive impairment of schizophrenia. We have found that, like nicotine, the $\alpha4\beta2$ nicotinic agonist RJR 2403 significantly improves memory in rats on the radial-arm maze [21]. More specifically, it has been demonstrated that hippocampal infusion of the $\alpha4\beta2$ nicotinic antagonist (DHβE) causes significant memory impairments of rats on the radial-arm maze [18, 70, 71, 75]. Systemic nicotine administration was found to reverse the memory impairment caused by hippocampal $\alpha4\beta2$ nicotinic receptor blockade [75].

Potential of nicotinic treatment for schizophrenia

Nicotinic treatments may be useful for a variety of neuropsychiatric indications including schizophrenia [76–80]. Attentional improvement may be a key therapeutic effect [81, 82]. Smoking withdrawal induced deficits in attentional performance and spatial working memory in patients with schizophrenia [83]. This deficit was reversed with smoking and the smoking effect was blocked by the nicotinic antagonist mecamylamine. Nicotine normalizes smooth pursuit eye movements in people with schizophrenia, an effect which is accompanied by increased activity in the cingulate gyrus and lower activity in the hippocampus [84]. Nicotine improves antisaccade and eye tracking performance in patients with schizophrenia [85–87]. Nicotine skin patches improved N-back memory test performance in withdrawn smokers with schizophrenia [88]. This was accompanied by enhanced activation in the cingulate cortex and thalamic nuclei. In a complementary fashion, nicotinic blockade caused significant deficits in the N-back task [89]. Nicotine skin patch treatment in healthy volunteers significantly improved the speed of pre-attentive sensory processing as indexed by mismatch negativity to auditory stimuli in an oddball paradigm [90]. Nicotine nasal spray improved spatial organization and also improved memory in schizophrenia [91, 92]. Nicotine improves eye tracking, memory and attentional function in schizophrenia [82, 83, 86, 91].

Nicotinic $\alpha7$ receptors offer a promising avenue for novel drug development for treatment of the cognitive impairments of schizophrenia [7]. Abnormal $\alpha7$ genotype is significantly associated with schizophrenia, smoking in schizophrenia and deficient sensory gating [93–95]. There are abnormal $\alpha7$ receptors in schizophrenia [96] which are related to impaired sensory gating [97]. Because of the findings of $\alpha7$ receptor deficits in schizophrenia and the involvement of $\alpha7$ receptors in the cognitive impairment of schizophrenia, $\alpha7$ nicotinic receptor agonists are being developed for treatment of the cognitive impairments of schizophrenia. A variety of promising new $\alpha7$ agonists that penetrate the blood brain barrier and are bioavailable with oral administration have been developed [98, 99]. Tropisetron, a partial agonist at $\alpha7$ receptors, improves sensory gating (P50 inhibition) in patients with schizophrenia [100]. A selective $\alpha7$ nicotinic acetylcholine receptor agonist, PNU-282987, has been found to reverse sensory gating deficits caused by amphetamine in the rat and to stimulate whole cell currents in hippocampal cells [101]. The $\alpha7$ agonist DMXB enhances sensory gating likely through $\alpha7$ receptors in the hippocampus [102, 103]. Anabasine, an $\alpha7$ agonist, reversed the "popping" behavior, which is a model for schizophrenia in mice given dizocilpine [102]. The PPI deficit caused by isolation rearing is reversed by the $\alpha7$ agonist (R)-N-(1-Azabicyclo[2.2.2]oct 3-yl)(5-(2-pyridyl)thiophene-2-carboxamide [104]. Thus, nicotinic $\alpha7$ agonists appear to hold promise for further development to reverse cognitive impairment in schizophrenia.

Antipsychotic drug effects on cognitive function

Atypical antipsychotics such as clozapine and risperidone represent a great improvement over classic antipsychotics such as haloperidol in terms of cognitive sparing [49], that is an attenuation of the negative impact on cognitive function. Attentional function appears to be improved by these drugs [45, 50, 105–107]. However, effects on memory are more problematic. Clozapine has been found to have adverse effects on working memory [108–110]. In contrast, reference memory [111] was found to be improved by clozapine. Clozapine-induced working memory impairment has been identified in experimental animal models. In monkeys, clozapine impairs the accuracy of delayed response performance [112]. In rats, a similar effect has been seen with clozapine, causing a delayed response choice accuracy impairment [113]. Deficits in the delayed response memory task were also seen with haloperidol and risperidone [113]. Clozapine, haloperidol and risperidone were also found to impair memory performance in the Morris water maze [114]. Clozapine but not haloperidol improved PPI in DBA/2 mice, which have deficient $\alpha 7$ receptors [115]. This effect was blocked by the $\alpha 7$ antagonist α-bungarotoxin, but not the $\alpha 4\beta 2$ antagonist DHβE. Antipsychotics are often given for behavioral control and antipsychotic effects in the elderly [116]. With the elderly the involvement of $\alpha 4\beta 2$ receptors would be particularly important given the age-related decline in $\alpha 4\beta 2$ receptors [117]. With the advancement of the therapeutic goal beyond just antipsychotic activity, i.e. to improve cognitive function, it is imperative to determine the mechanisms of cognitive effects of the therapeutic drugs for schizophrenia. In this way, novel approaches for improving cognitive function in schizophrenics can be developed.

Antipsychotic drugs have been shown to cause working memory impairment in the radial-arm maze. The classic neuroleptic haloperidol has been shown in several studies to impair working memory [17, 118–120]. Olanzapine impairs accuracy, an action that is significantly attenuated by nicotine coadministration [121]. Risperidone attenuates nicotine-induced memory improvement [122]. Clozapine has been shown to significantly impair working memory function in normal rats [122]. The clozapine-induced memory impairment is significantly attenuated by nicotine coadministration.

The hallmark of the class of atypical antipsychotics is that they affect multiple receptor systems. Atypical antipsychotics have been called "MARTA" (multi-acting receptor targeted antipsychotics) drugs because they act on a variety of receptor systems [123]. This profile may be related to the affinity of clozapine for DA D_1, muscarinic ACh and serotonergic receptors and relative lack of affinity for DA D_2 receptors. Serotonin 5-HT_{2A}-blocking activity produces better cognitive function in patients with schizophrenia than drugs with predominantly dopamine D_2-blocking activity [124–126]. Antipsychotic drug actions blocking D_2 receptors have been found to be related to higher rates of smoking in patients with schizophrenia [127]. Given the efficacy of MARTA drugs, the next step may be to develop combinations of drugs to provide optimal therapy. With drug combinations additional targeted receptor actions can be achieved without having to devise a novel drug. The selection of the drugs for combination and their relative doses can be adjusted to achieve optimal results to fit the needs of individual patients. Drug combinations can be selected such that the therapeutic effects are complementary and the side effects are offset. Lower doses of each drug can be used in combination to further reduce the problem of unwanted side effects.

Antipsychotics have complex interactions with nicotinic systems. Nicotinic interactions with both DA and glutaminergic systems may be key for their efficacious cognitive effects in combination with antipsychotic drugs. Nicotine enhances DA and glutamate release in the frontal cortex [128, 129]. Nicotine also plays a protective role in attenuating D_2 receptor up regulation with chronic antipsychotic drug therapy [130]. One critical factor is the background activity of DA systems. Under normal levels of dopamine activity the antipsychotic clozapine increased firing rate and burst firing rate of VTA DA cells [131]. Dizocilpine reversed this effect, but MLA did not. But under conditions of high DA activity levels induced by $\alpha 7$ blockade, there was the reverse effect [132]. Nicotinic actions in the VTA may be critically important on DA involvement in frontal cortically based cognitive function.

Long-term haloperidol and fluphenazine administration disrupts the normal pattern of spatial reversal learning in monkeys [133]. Monkeys given several years of antipsychotic treatment were trained on position discrimination. They did not differ from controls on original learning, but failed to show either the normal increase in errors with the first reversal of reward contingencies or the normal improvement on subsequent reversals. It is as if each reversal of reward contingencies (left or right rewarded) was treated as an entirely new problem by the monkeys chronically treated with antipsychotic drugs [133]. Chronic haloperidol was found to significantly impair working memory function and spatial information processing speed in people with schizophrenia. These effects were found to be significantly reversed by nicotine administered via a skin patch [48].

Nicotinic interactions with antipsychotic drugs

Due to the serotonergic and dopaminergic properties of most antipsychotic drugs, it is not surprising to observe a functional interaction between the nicotinic system in the brain and antipsychotic drugs. Clozapine, an antipsychotic drug that blocks both dopaminergic and serotonergic receptors significantly impaired working memory performance, an effect which was significantly attenuated by acute nicotine (Fig. 1) [122]. Haloperidol and risperidone (Fig. 2) significantly attenuated the working memory improvement induced by nicotine in the eight-arm radial maze [122]. The antipsychotic drug olanzapine causes memory impairment in terms of impairing choice accuracy of rats in the radial-arm maze [121]. Nicotine co-treatment attenuated the memory impairment caused by olanzapine [121]. Clozapine and risperidone (Fig. 3) caused a dose-related impairment in selective attention (lower percent hit) on the visual signal detection operant task [35]. Haloperidol (Fig. 4) also caused a significant impairment in attentional performance, an effect that was attenuated by chronic nicotine (5 mg/kg/day) infusion for the first 2 weeks of treatment. Chronic nicotine infusion at the same dose level also attenuated the cognitive impairment of clozapine and risperidone [134].

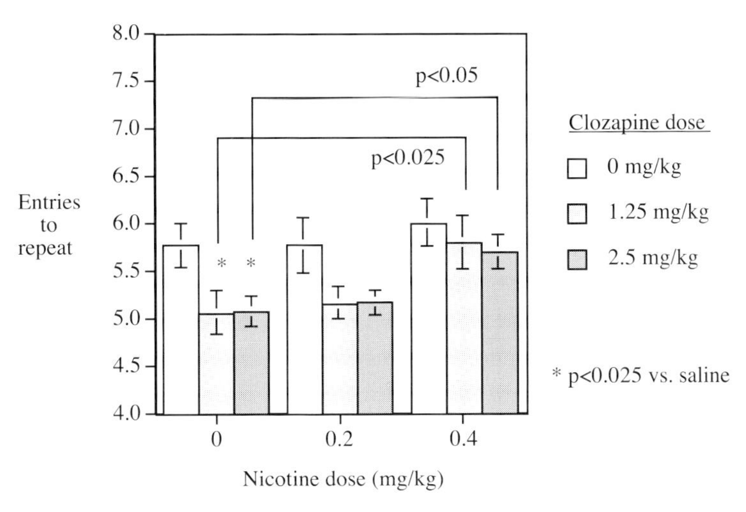

Figure 1. Nicotine interactions with clozapine and working memory performance on the radial-arm maze, reprinted from [122] with permission.

We have documented the interactions of nicotine with classic neuroleptic treatment. In parallel clinical and experimental animal studies, we have determined nicotine-haloperidol interactions with regard to cognitive function. In the clinical studies, we found that nicotine-induced cognitive improvement is not blocked by haloperidol, but rather nicotine is effective in reversing haloperidol-induced deficits. Nicotine administered via skin patches attenuated the working memory impairment caused by moderate and high doses of haloperidol [48]. The haloperidol-induced decrease in mental processing speed was also reduced by nicotine. Interestingly, the consistency of attentional response is improved by nicotine in a dose-related fashion regardless of the dose of haloperidol. In parallel basic studies in laboratory rats, we showed that nicotine-induced memory improvements were not blocked by haloperidol [17]. In mechanistic studies, we demonstrated the memory impairment is caused by intrahippocampal infusion of the dopamine D_2 antagonist raclopride [135], as well as memory impairments caused by intrahippocampal infusions of nicotinic antagonists [18].

There is little information concerning the interaction of nicotine with atypical neuroleptics. There is a decrease in cigarette smoking with clozapine administration [136]. This raises the possibility that clozapine may attenuate pharmacological effects of nicotine or that they have similar effects. Consistent with this idea is the preclinical finding of Brioni et al. that clozapine attenuates the discriminative stimulus effects of nicotine in rats [12].

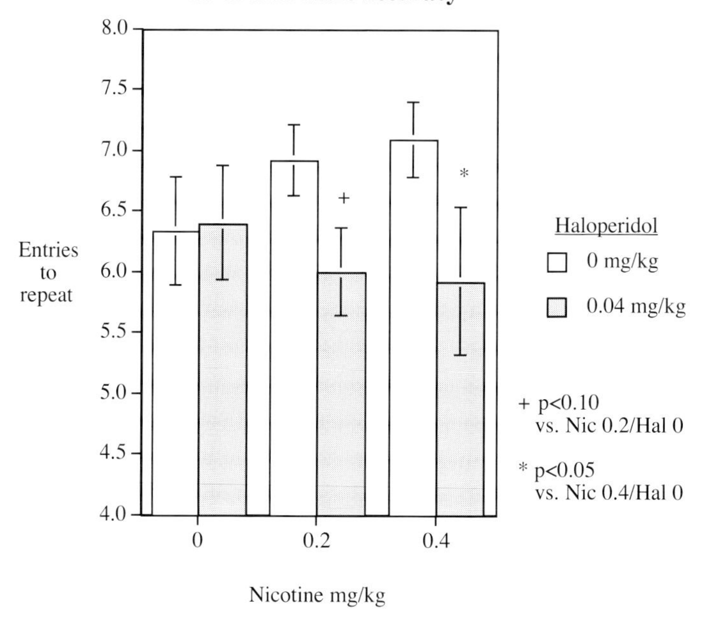

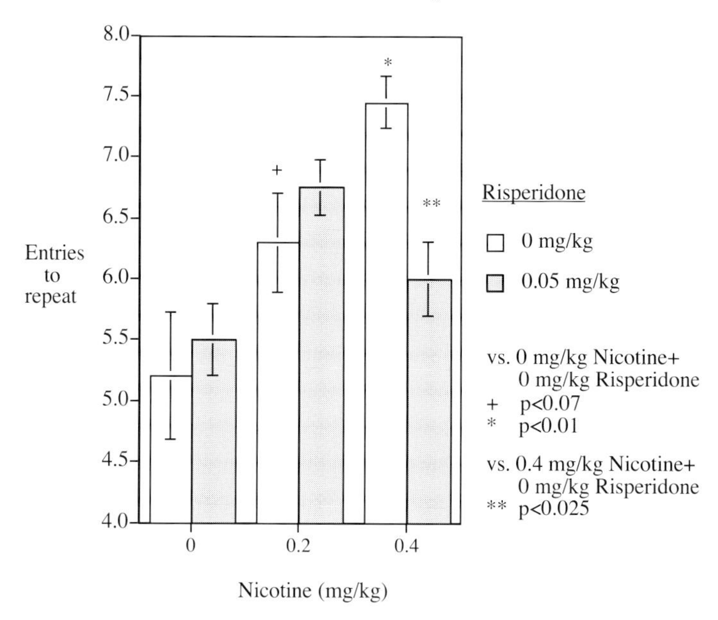

Figure 2. Nicotine interactions with haloperidol and risperidone and working memory performance on the radial-arm maze, reprinted from [122] with permission.

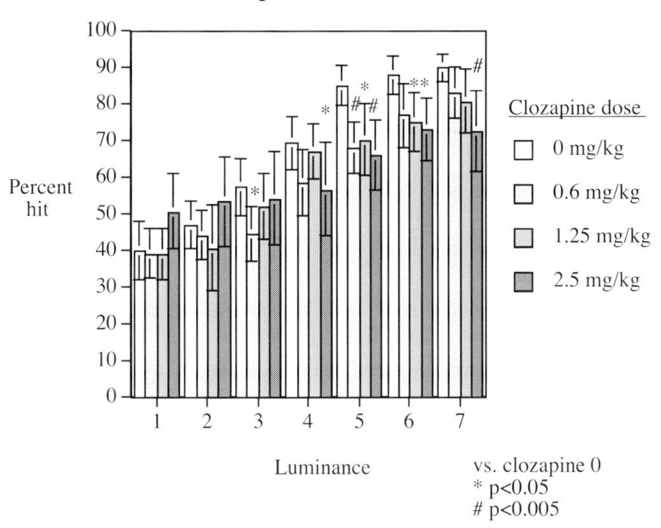

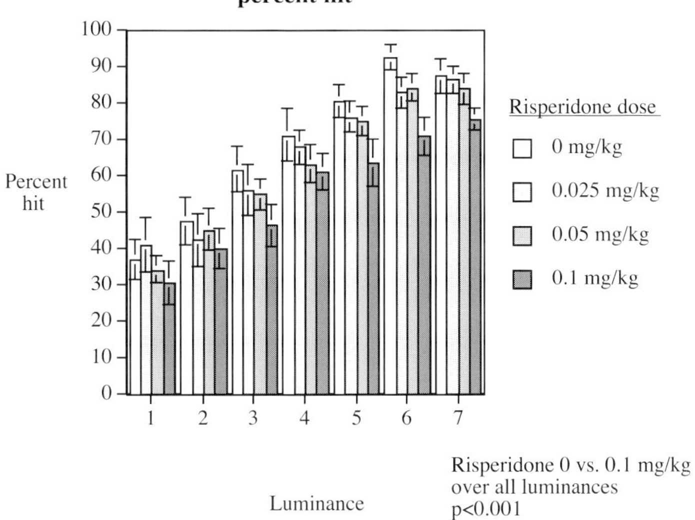

Figure 3. Clozapine and risperidone induced impairments on sustained attention, reprinted from [35] with permission from Elsevier.

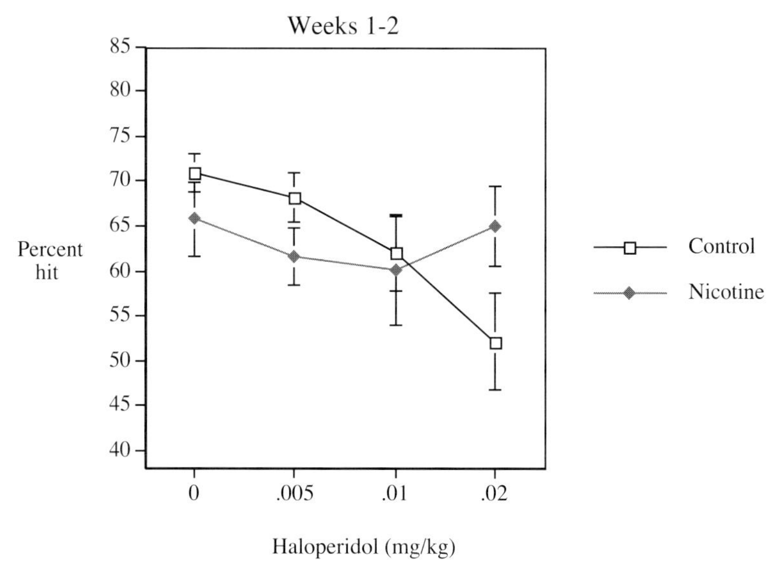

Figure 4. Chronic nicotine interactions with haloperidol and sustained attention, reprinted from [35] with permission from Elsevier.

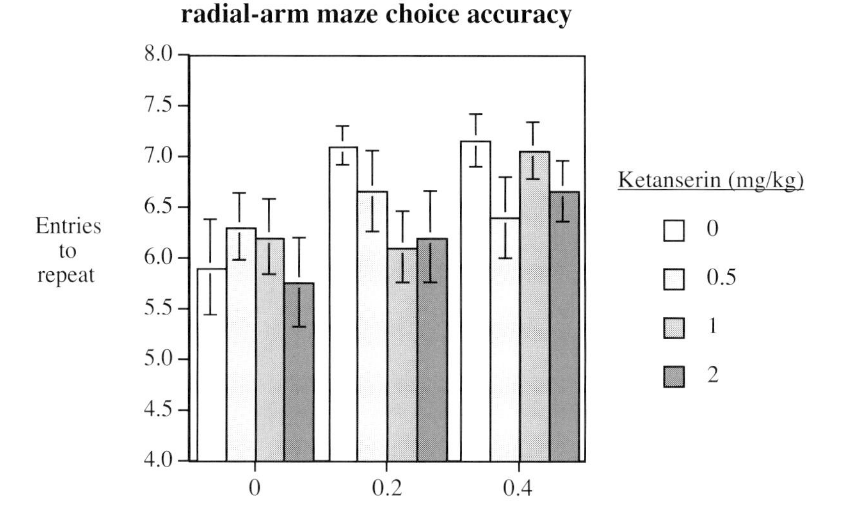

Figure 5. Nicotine interactions with ketanserin and working memory on the radial-arm maze, reprinted from [137] with permission from Elsevier.

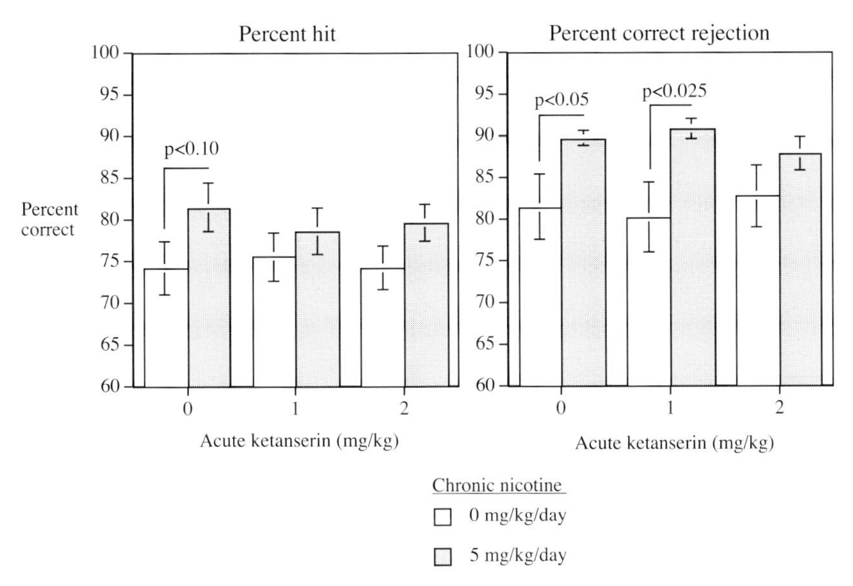

Figure 6. Chronic nicotine interactions with ketanserin and attentional performance, reprinted from [138] with permission.

The $5HT_2$ antagonist ketanserin attenuates nicotine-induced working memory improvement in rats (Fig. 5) [137]. Ketanserin also blocked the nicotine-induced improvement in attentional performance (Fig. 6) [138]. Since antipsychotic drugs, such as clozapine, act on a variety of different transmitter receptors it is important to determine which of these effects are key for the cognitive effects and for interactions with nicotinic effects. We have begun the pharmacological dissection of antipsychotic drug interactions with nicotine with the examination of the role of $5HT_2$ receptors.

Summary

In summary, neuronal nicotinic systems are important for a variety of aspects of cognitive function impacted by antipsychotic drugs. It has been demonstrated that antipsychotic drugs have memory and attentional impairing effects when given to unimpaired subjects. Nicotine can reduce some of these impairments, but antipsychotic drug administration can also attenuate nicotine effects. We have found that nicotinic agonists selective for $\alpha 7$ and $\alpha 4\beta 2$ receptor subtypes significantly improve learning and memory. Serotonergic actions of antipsychotic drugs may decrease efficacy of nicotinic co-treatments. When the antipsychotic drug clozapine and nicotine are administered to subjects with cognitive impairments caused by NMDA glutamate receptor blockade or hippocampal dysfunction they can significantly attenuate the attentional and memory impairments.

Nicotine has been shown in our studies to reverse the memory impairment caused by acute clozapine-induced memory improvement. Acute risperidone and haloperidol has been shown to attenuate nicotine-induced memory improvement. We have determined the role of hippocampal $\alpha 7$ and $\alpha 4\beta 2$ nicotinic receptors in the neural basis of nicotinic antipsychotic interactions. Local acute and chronic hippocampal infusion of either nicotinic $\alpha 7$ or $\alpha 4\beta 2$ antagonists cause significant spatial working memory impairment. Chronic hippocampal nicotinic antagonist infusions have served as a model of persistent decreases in nicotinic receptor level seen in schizophrenia and Alzheimer's disease. Clozapine attenuated the memory deficit caused by chronic suppression of hippocampal $\alpha 4\beta 2$ receptors while the amnestic effects of clozapine were potentiated by chronic suppression of hippocampal $\alpha 7$ receptors.

Nicotinic co-treatment may be a useful adjunct in the treatment of schizophrenia, to attenuate cognitive impairment of schizophrenia. Nicotine as well as selective nicotinic $\alpha 7$ and $\alpha 4\beta 2$ receptor agonists significantly improve working memory and attentional function. Nicotine treatment was found to be effective in attenuating the attentional and memory impairments caused by the psychototmimetic NMDA antagonist dizocilpine (MK-801), a model of the cognitive impairment of schizophrenia. Studies of the interactions of antipsychotic drugs with nicotinic agents provided quite useful information concerning possible co-treatment of people with schizophrenia with nicotinic therapy. Nicotine was found to significantly attenuate the memory impairments caused by the antipsychotic drugs clozapine and olanzapine. Interestingly, nicotine-induced cognitive improvement was significantly attenuated by the antipsychotic drug clozapine. One of the principal effects of clozapine is to block $5HT_2$ receptors. Ketanserin a $5HT_2$ antagonist significantly attenuated nicotine-induced improvements in attention and memory. Thus it appears that antipsychotic drugs with actions blocking $5HT_2$ receptors may limit the efficacy of nicotinic co-treatments for cognitive enhancement.

Acknowledgement

Research presented was supported by a grant from the National Institute of Mental Health grant MH64494.

References

1. Levin ED, McClernon FJ, Rezvani AH (2006) Nicotinic effects on cognitive function: Behavioral characterization, pharmacological specification and anatomic localization. *Psychopharmacology* 184: 523–539
2. Levin ED, Rezvani AH (2000) Development of nicotinic drug therapy for cognitive disorders. *Eur J Pharmacol* 393: 141–146
3. Newhouse PA, Potter A, Levin ED (1997) Nicotinic system involvement in Alzheimer's and Parkinson's diseases: Implications for therapeutics. *Drugs Aging* 11: 206–228
4. Rezvani AH, Levin ED (2001) Cognitive effects of nicotine. *Biol Psychiat* 49: 258–267
5. Tollefson GD (1996) Cognitive function in schizophrenic patients. *J Clin Psychiat* 57 Suppl 11: 31–39

6. Geyer MA, Tamminga CA (2004) Measurement and treatment research to improve cognition in schizophrenia: neuropharmacological aspects. *Psychopharmacology* 174: 1–2
7. Martin LF, Kem WR, Freedman R (2004) Alpha-7 nicotinic receptor agonists: potential new candidates for the treatment of schizophrenia. *Psychopharmacology* 174: 54–64
8. McGehee DS, Heath MJS, Gelber S, Devay P, Role LW (1995) Nicotine enhancement of fast excitatory synaptic transmission in CNS by presynaptic receptors. *Science* 269: 1692–1696
9. Wonnacott S, Irons J, Rapier C, Thorne B, Lunt GG (1989) Presynaptic modulation of transmitter release by nicotinic receptors. In: Nordberg A, Fuxe K, Holmstedt B, Sundwall A (eds): *Progress in Brain Research*. Elsevier Science Publishers B.V., 157–163
10. Wonnacott S, Kaiser S, Mogg A, Soliakov L, Jones IW (2000) Presynaptic nicotinic receptors modulating dopamine release in the rat striatum. *Eur J Pharmacol* 393: 51–58
11. Bartus RT, Dean RL, Flicker C (1987) Cholinergic psychopharmacology: An integration of human and animal research on memory. In: Meltzer HY (ed): *Psychopharmacology: the third generation of progress*. Raven Press, New York, 219–232
12. Brioni JD, Decker MW, Sullivan JP, Arneric SP (1997) The pharmacology of (-)-nicotine and novel cholinergic channel modulators. *Adv Pharmacol* 37: 153–214
13. Decker MW, Brioni JD, Bannon AW, Arneric SP (1995) Diversity of neuronal nicotinic acetylcholine receptors: Lessons from behavior and implications for CNS therapeutics – minireview. *Life Sci* 56: 545–570
14. Warburton DM (1992) Nicotine as a cognitive enhancer. *Prog Neuro-Psychopharmacol Biol Psychiatry* 16: 181–919
15. Levin ED (1992) Nicotinic systems and cognitive function. *Psychopharmacology* 108: 417–431
16. Levin ED, Simon BB (1998) Nicotinic acetylcholine involvement in cognitive function in animals. *Psychopharmacology* 138: 217–230
17. Levin ED (1997) Chronic haloperidol administration does not block acute nicotine-induced improvements in radial-arm maze performance in the rat. *Pharmacol Biochem Behav* 58: 899–902
18. Felix R, Levin ED (1997) Nicotinic antagonist administration into the ventral hippocampus and spatial working memory in rats. *Neuroscience* 81: 1009–1017
19. Levin ED, Bettegowda C, Blosser J, Gordon J (1999) AR-R17779, and alpha7 nicotinic agonist, improves learning and memory in rats. *Behav Neurosci* 10: 675–680
20. Levin ED, Christopher N (2003) Lobeline-induced learning improvements in rats in the radial-arm maze. *Pharmacol Biochem Behav* 76, 133–139
21. Levin ED, Christopher N (2002) Persistence of nicotinic agonist RJR 2403 induced working memory improvement in rats. *Drug Dev Res* 55: 97–103
22. Grilly DM (2000) A verification of psychostimulant-induced improvement in sustained attention in rats: effects of d-amphetamine, nicotine, and pemoline. *Exp Clin Psychopharmacol* 8: 14–21
23. Mirza NR, Bright JL (2001) Nicotine-induced enhancements in the five-choice serial reaction time task in rats are strain-dependent. *Psychopharmacology* 154: 8–12
24. Mirza NR, Stolerman IP (1998) Nicotine enhances sustained attention in the rat under specific task conditions. *Psychopharmacology* 138: 266–274
25. Muir JL, Everitt BJ, Robbins TW (1995) Reversal of visual attentional dysfunction following lesions of the cholinergic basal forebrain by physostigmine and nicotine but not by the 5-HT3 receptor antagonist, ondansetron. *Psychopharmacology* 118: 82–92
26. Rezvani AH, Bushnell P, Levin ED (2002) Nicotine and mecamylamine effects on choice accuracy in an operant signal detection task. *Psychopharmacology* 164: 369–375

27. Rezvani AH, Levin ED (2003) Nicotine interactions with the NMDA glutaminergic antagonist dizocilpine and attentional function. *Eur J Pharmacol* 465: 83–90
28. Stolerman IP, Mirza NR, Hahn B, Shoaib M (2000) Nicotine in an animal model of attention. *Eur J Pharmacol* 393: 147–154
29. Rezvani AH, Levin ED (2003) Nicotine-alcohol interactions and attentional performance on an operant visual signal detection task in female rats. *Pharmacol Biochem Behav* 76: 75–83
30. Grottick AJ, Higgins GA (2000) Effect of subtype selective nicotinic compounds on attention as assessed by the five-choice serial reaction time task. *Behav Brain Res* 117: 197–208
31. Ruotsalainen S, Miettinen R, MacDonald E, Koivisto E, Sirvio J (2000) Blockade of muscarinic, rather than nicotinic, receptors impairs attention, but does not interact with serotonin depletion. *Psychopharmacology* 148: 111–123
32. McGaughy J, Decker MW, Sarter M (1999) Enhancement of sustained attention performance by the nicotinic acetylcholine receptor agonist ABT-418 in intact but not basal forebrain-lesioned rats. *Psychopharmacology* 144: 175–182
33. Terry AVJ, Risbrough VB, Buccafusco JJ, Menzaghi F (2002) Effects of (+/−)-4-[[2-(1-methyl-2-pyrrolidinyl)ethyl]thio]phenol hydrochloride (SIB-1553A), a selective ligand for nicotinic acetylcholine receptors, in tests of visual attention and distractibility in rats and monkeys. *J Pharmacol Exp Ther* 301: 284–292
34. Levin ED, Christopher NC, Briggs SJ, Rose JE (1993) Chronic nicotine reverses working memory deficits caused by lesions of the fimbria or medial basalocortical projection. *Cognitive Brain Res* 1: 137–143
35. Rezvani AH, Caldwell D, Levin ED (2004) Nicotine-antipsychotic drug interactions and attentional performance. *Eur J Pharmacol* 486: 175–182
36. Levin ED, Simon BB, Conners CK (2000) Nicotine effects and attention deficit disorder. In: Newhouse P, Piasecki M (eds): *Nicotine in psychiatry: psychopathology and emerging therapeutics*. John Wiley, New York, 203–214
37. Mancuso G, Warburton DM, Melen M, Sherwood N, Tirelli E (1999) Selective effects of nicotine on attentional processes. *Psychopharmacology* 146: 199–204
38. Wilens TE, Biederman J, Spencer TJ, Bostic J, Prince J, Monuteaux MC, Soriano J, Fine C, Abrams A, Rater M, Polisner D (1999) A pilot controlled clinical trial of ABT-418, a cholinergic agonist, in the treatment of adults with attention deficit hyperactivity disorder. *Am J Psychiat* 156: 1931–1937
39. Lawrence NS, Ross TJ, Stein EA (2002) Cognitive mechanisms of nicotine on visual attention. *Neuron* 36: 539–548
40. Kumari V, Gray JA, Ffyche DH, Mitterschiffthalar MT, Das M, Zachariah E, Vythelingum GN, Williams SC, Simmons A, Sharma T (2003) Cognitive effects of nicotine in humans: An fMRI study. *Neuroimage* 19: 1002–1013
41. Durany N, Zochling R, Boissl KW, Paulus W, Ransmayr G, Tatschner T, Danielczyk W, Jellinger K, Deckert J, Riederer P (2000) Human post-mortem striatal alpha4beta2 nicotinic acetylcholine receptor density in schizophrenia and Parkinson's syndrome. *Neurosci Lett* 287: 109–112
42. Freedman R, Adams CE, Leonard S (2000) The alpha7-nicotinic acetylcholine receptor and the pathology of hippocampal interneurons in schizophrenia. *J Chem Neuroanat* 20: 299–306
43. Hughes JR, Hatsukami DK, Mitchell JE, Dahlgren LA (1986) Prevalence of smoking among psychiatric outpatients. *Am J Psychiatry* 143: 993–997
44. Cornblatt BA, Keilp JG (1994) Impaired attention, genetics, and the pathophysiology of schizophrenia. *Schiz Bull* 20: 31–46

45. Meltzer HY, Thompson PA, Lee MA, Ranjan R (1996) Neuropsychologic deficits in schizophrenia: relation to social function and effect of antipsychotic drug treatment. *Neuropsychopharmacology* 14: 27S–33S
46. Stip E (1996) Memory impairment in schizophrenia: perspectives from psychopathology and pharmacotherapy. *Can J Psychiatry* 41: S27–34
47. Alam DA, Janicak PG (2005) The role of psychopharmacotherapy in improving the long-term outcome of schizophrenia. *Essential Psychopharmacol* 6: 127–140
48. Levin ED, Wilson W, Rose JE, McEvoy J (1996) Nicotine-haloperidol interactions and cognitive performance in schizophrenics. *Neuropsychopharmacology* 15: 429–436
49. Mortimer AM (1997) Cognitive function in schizophrenia–do neuroleptics make a difference? *Pharmacol Biochem Behav* 56: 789–795
50. Lee MA, Jayathilake K, Meltzer HY (1999) A comparison of the effect of clozapine with typical neuroleptics on cognitive function in neuroleptic-responsive schizophrenia. *Schizophr Res* 37: 1–11
51. de Leon J, Diaz FJ (2005) A meta-analysis of worldwide studies demonstrates an association between schizophrenia and tobacco smoking behaviors. *Schizophr Res* 76: 135–157
52. Poirier MF, Canceil O, Bayle F, Millet B, Bourdel MC, Moatti C, Olie JP, Attar-Levy D (2002) Prevalence of smoking in psychiatric patients. *Prog Neuro-Psychopharmacol Biol Psychiat* 26: 529–537
53. Strand JE, Nyback H (2005) Tobacco use in schizophrenia: a study of cotinine concentrations in the saliva of patients and controls. *Eur Psychiat* 20: 50–54
54. Aguilar MC, Gurpegui M, Diaz F, de Leon J (2005) Nicotine dependence and symptoms in schizophrenia: naturalistic study of complex interactions. *Br J Psychiatry* 186: 215–221
55. George TP, Vessicchio JC, Termine A, Sahady DM, Head CA, Pepper WT, Kosten T R, Wexler BE (2002) Effects of smoking abstinence on visuospatial working memory function in schizophrenia. *Neuropsychopharmacology* 26: 75–85
56. Weiser M, Reichenberg A, Grotto I, Yasvitzky R, Rabinowitz J, Lubin G, Nahon D, Knobler HY, Davidson M (2004) Higher rates of cigarette smoking in male adolescents before the onset of schizophrenia: a historical-prospective cohort study. *Am J Psychiat* 161: 1219–1223
57. Kumari V, Postma P (2005) Nicotine use in schizophrenia: The self medication hypotheses. *Neurosci Biobehav Rev* 29: 1021–1034
58. Dalack GW, Healy DJ, Meador-Woodruff JH (1998) Nicotine dependence in schizophrenia: clinical phenomena and laboratory findings. *Am J Psychiat* 155: 1490–1501
59. Kosten TR, Ziedonis DM (1997) Substance abuse and schizophrenia: editors' introduction [see comments]. *Schizophr Bull* 23: 181–186
60. Stassen HH, Bridler R, Hagele S, Hergersberg M, Mehmann B, Schinzel A, Weisbrod M, Scharfetter C (2000) Schizophrenia and smoking: Evidence for a common neurobiological basis? *Am J Med Genet* 96: 173–177
61. Freedman R, Hall M, Adler LE, Leonard S (1995) Evidence in postmortem brain tissue for decreased numbers of hippocampal nicotinic receptors in schizophrenia. *Biol Psychiatry* 38: 22–33
62. Guan ZZ, Zhang X, Blennow K, Nordberg A (1999) Decreased protein level of nicotinic receptor alpha 7 subunit in the frontal cortex from schizophrenic brain. *Neuroreport* 10: 1779–1782
63. Adler LE, Olincy A, Waldo M, Harris JG, Griffith J, Stevens K, Flach K, Nagamoto H, Bickford P, Leonard S, Freedman R (1998) Schizophrenia, sensory gating, and nicotinic receptors. *Schizophrenia Bull* 24: 189–202

64. Leonard S, Breese C, Adams C, Benhammou K, Gault J, Stevens K, Lee M, Adler L, Olincy A, Ross R, Freedman R (2000) Smoking and schizophrenia: abnormal nicotinic receptor expression. *Eur J Pharmacol* 393: 237–242
65. Stevens KE, Freedman R, Collins AC, Hall M, Leonard S, Marks JM, Rose GM (1996) Genetic correlation of inhibitory gating of hippocampal auditory evoked response and alpha-bungarotoxin-binding nicotinic cholinergic receptors in inbred mouse strains. *Neuropsychopharmacology* 15: 152–162
66. Addington J (1998) Group treatment for smoking cessation among persons with schizophrenia. *Psychiatr Serv* 49: 925–928
67. Adler LE, Hoffer LD, Wiser A, Freedman R (1993) Normalization of auditory physiology by cigarette smoking in schizophrenic patients. *Am J Psychiatry* 150: 1856–1861
68. Olincy A, Johnson LL, Ross RG (2003) Differential effects of cigarette smoking on performance of a smooth pursuit and a saccadic eye movement task in schizophrenia. *Psychiat Res* 117: 223–236
69. Bettany JH, Levin ED (2001) Ventral hippocampal alpha7 nicotinic receptors and chronic nicotine effects on memory. *Pharmacol Biochem Behav* 70: 467–474
70. Levin ED, Addy N, Arthur D, Wagner Y, Stamm K (2001) *Acute and chronic a7 and a4b2 hippocampal nicotinic receptor blockade and memory function in rats*. Society for Research on Nicotine and Tobacco, Meeting March 23–25, Seattle, WA, USA
71. Levin ED, Bancroft A, Bettany J (2001) *Chronic systemic nicotine interaction with a7 and a4b2 hippocampal nicotinic receptors*. Society for Research on Nicotine and Tobacco, Meeting March 23–25, Seattle, WA, USA
72. Levin ED, Bradley A, Addy N, Sigurani N (2002) Hippocampal alpha 7 and alpha 4 beta 2 nicotinic receptors and working memory. *Neuroscience* 109: 757–765
73. McQuiston AR, Madison DV (1999) Nicotinic receptor activation excites distinct subtypes of interneurons in the rat hippocampus. *J Neurosci* 19: 2887–2896
74. Drew AE, Derbez AE, Werling LL (2000) Nicotinic receptor-mediated regulation of dopamine transporter activity in rat prefrontal cortex. *Synapse* 38: 10–16
75. Bancroft A, Levin ED (2000) Ventral hippocampal alpha4beta2 nicotinic receptors and chronic nicotine effects on memory. *Neuropharmacology* 39: 2770–2778
76. Singh A, Potter A, Newhouse P (2004) Nicotinic acetylcholine receptor system and neuropsychiatric disorders. *Idrugs* 7: 1096–1103
77. Sacco KA, Bannon KL, George TP (2004) Nicotinic receptor mechanisms and cognition in normal states and neuropsychiatric disorders. *Journal of Psychopharmacology* 18: 457–474
78. Ripoll N, Bronnec M, Bourin M (2004) Nicotinic receptors and schizophrenia. *Curr Med Res Opin* 20: 1057–1074
79. Newhouse P, Singh A, Potter A (2004) Nicotine and nicotinic receptor involvement in neuropsychiatric disorders. *Curr Topics in Medicinal Chem* 4: 267–282
80. Hogg RC, Bertrand D (2004) Nicotinic acetylcholine receptors as drug targets. *Curr Drug Targets – CNS & Neurological Disorders* 3: 123–130
81. Newhouse PA, Potter A, Singh A (2004) Effects of nicotinic stimulation on cognitive performance. *Curr Opinion in Pharmacol* 4: 36–46
82. Harris J, Kongs S, Allensworth D, Martin L, Tregallas J, Sullivan B, Zerbe G, Freedman R (2004) Effects of nicotine on cognitive deficits in schizophrenia. *Neuropsychopharmacology* 29: 1378–1385
83. Sacco KA, Termine A, Seyal A, Dudas MM, Vessicchio JC, Krishnan-Sarin S, Jatlow PI, Wexler BE, George TP (2005) Effects of cigarette smoking on spatial working memory and attentional deficits in schizophrenia: Involvement of nicotinic receptor mechanisms. *Arch Gen Psychiatry* 62: 649–659

84. Tregellas JR, Tanabe JL, Martin LF, Freedman R (2005) FMRI of response to nicotine during a smooth pursuit eye movement task in schizophrenia. *Am J Psychiatry* 162: 391–393
85. Avila MT, Sherr JD, Hong E, Myers CS, Thaker GK (2003) Effects of nicotine on leading saccades during smooth pursuit eye movements in smokers and nonsmokers with schizophrenia. *Neuropsychopharmacology* 28: 2184–2191
86. Larrison-Faucher AL, Matorin AA, Sereno AB (2004) Nicotine reduces antisaccade errors in task impaired schizophrenic subjects. *Prog Neuropsychopharmacol Biol Psychiatry* 28: 505–516
87. Sherr JD, Myers C, Avila MT, Elliott A, Blaxton TA, Thaker GK (2002) The effects of nicotine on specific eye tracking measures in schizophrenia. *Biol Psychiatry* 52: 721–728
88. Jacobsen LK, D'Souza DC, Mencl WE, Pugh KR, Skudlarski P, Krystal JH (2004) Nicotine effects on brain function and functional connectivity in schizophrenia. *Biol Psychiatry* 55: 850–858
89. Green A, Ellis KA, Ellis J, Bartholomeusz CF, Ilic S, Croft RJ, Phan KL, Nathan PJ (2005) Muscarinic and nicotinic receptor modulation of object and spatial N-back working memory in humans. *Pharmacol Biochem Behav* 81: 575–584
90. Inami R, Kirino E, Inoue R, Arai H (2005) Transdermal nicotine administration enhances automatic auditory processing reflected by mismatch negativity. *Pharmacol Biochem Behav* 80: 453–461
91. Myers CS, Robles O, Kakoyannis AN, Sherr JD, Avila MT, Blaxton TA, Thaker GK (2004) Nicotine improves delayed recognition in schizophrenic patients. *Psychopharmacology* 174: 334–340
92. Smith RC, Singh A, Infante M, Khandat A, Kloos A (2002) Effects of cigarette smoking and nicotine nasal spray on psychiatric symptoms and cognition in schizophrenia. *Neuropsychopharmacology* 27: 479–497
93. De Luca V, Wong AH, Muller DJ, Wong GW, Tyndale RF, Kennedy JL (2004) Evidence of association between smoking and alpha7 nicotinic receptor subunit gene in schizophrenia patients. *Neuropsychopharmacology* 29: 1522–1526
94. De Luca V, Wang H, Squassina A, Wong GW, Yeomans J, Kennedy JL (2004) Linkage of M5 muscarinic and alpha7-nicotinic receptor genes on 15q13 to schizophrenia. *Neuropsychobiology* 50: 124–127
95. Leonard S, Gault J, Hopkins J, Logel J, Vianzon R, Short M, Drebing C, Berger R, Venn D, Sirota P et al. (2002) Association of promoter variants in the alpha7 nicotinic acetylcholine receptor subunit gene with an inhibitory deficit found in schizophrenia. *Arch Gen Psychiatry* 59: 1085–1096
96. Gault J, Hopkins J, Berger R, Drebing C, Logel J, Walton C, Short M, Vianzon R, Olincy A, Ross RG, Adler LE, Freedman R, Leonard S (2003) Comparison of polymorphisms in the alpha7 nicotinic receptor gene and its partial duplication in schizophrenic and control subjects. *Am J Med Genet Part B, Neuropsychiatric Genetics* 123: 39–49
97. Freedman R, Olincy A, Ross RG, Waldo MC, Stevens KE, Adler LE, Leonard S (2003) The genetics of sensory gating deficits in schizophrenia. *Current Psychiatry Reports* 5: 155–161
98. Deutsch SI, Rosse RB, Schwartz BL, Weizman A, Chilton M, Arnold DS, Mastropaolo J (2005) Therapeutic implications of a selective alpha7 nicotinic receptor abnormality in schizophrenia. *Israel J Psychiat Related Sci* 42: 33–44
99. Tatsumi R, Fujio M, Satoh H, Katayama J, Takanashi S, Hashimoto K, Tanaka H (2005) Discovery of the alpha7 nicotinic acetylcholine receptor agonists. (R)-3′-(5-Chlorothiophen-2-yl)spiro-1-azabicyclo[2.2.2]octane-3,5′-[1′,3′]oxazolidin-2′-one as a novel, potent, selective, and orally bioavailable ligand. *J Med Chem* 48: 2678–2686

100. Koike K, Hashimoto K, Takai N, Shimizu E, Komatsu N, Watanabe H, Nakazato M, Okamura N, Stevens KE, Freedman R, Iyo M (2005) Tropisetron improves deficits in auditory P50 suppression in schizophrenia. *Schizophr Res* 76: 67–72
101. Hajos M, Hurst RS, Hoffmann WE, Krause M, Wall TM, Higdon NR, Groppi VE (2005) The selective alpha7 nicotinic acetylcholine receptor agonist PNU-282987 [N-[(3R)-1-Azabicyclo[2.2.2]oct-3-yl]-4-chlorobenzamide hydrochloride] enhances GABAergic synaptic activity in brain slices and restores auditory gating deficits in anesthetized rats. *J Pharmacol Exp Ther* 312: 1213–1222
102. Mastropaolo J, Rosse RB, Deutsch SI (2004) Anabasine, a selective nicotinic acetylcholine receptor agonist, antagonizes MK-801-elicited mouse popping behavior, an animal model of schizophrenia. *Behav Brain Res* 153: 419–422
103. O'Neill HC, Rieger K, Kem WR, Stevens KE (2003) DMXB, an alpha7 nicotinic agonist, normalizes auditory gating in isolation-reared rats. *Psychopharmacology* 169: 332–339
104. Cilia J, Cluderay JE, Robbins MJ, Reavill C, Southam E, Kew JN, Jones DN (2005) Reversal of isolation-rearing-induced PPI deficits by an alpha7 nicotinic receptor agonist. *Psychopharmacology* 182: 214–219
105. Goldberg TE, Weinberger DR (1996) Effects of neuroleptic medications on the cognition of patients with schizophrenia: a review of recent studies. *J Clin Psychiatry* 57 Suppl 9: 62–65
106. McGurk SR (1999) The effects of clozapine on cognitive functioning in schizophrenia. *J Clin Psychiatry* 60 Suppl 12: 24–29
107. Meltzer HY, McGurk SR (1999) The effects of clozapine, risperidone, and olanzapine on cognitive function in schizophrenia. *Schizophr Bull* 25: 233–255
108. Goldberg TE, Weinberger DR (1994) The effects of clozapine on neurocognition: an overview. *J Clin Psychiatry* 55 Suppl B: 88–90
109. Hoff AL, Faustman WO, Wieneke M, Espinoza S, Costa M, Wolkowitz O, Csernansky JG (1996) The effects of clozapine on symptom reduction, neurocognitive function, and clinical management in treatment-refractory state hospital schizophrenic inpatients. *Neuropsychopharmacology* 15: 361–369
110. Perez-Gomez M, Junque C (1999) Clozapina: estudios neuropsicologicos y de resonancia magnetica. *Actas Esp Psiquiatr* 27: 341–346
111. Hagger C, Buckley P, Kenny JT, Friedman L, Ubogy D, Meltzer HY (1993) Improvement in cognitive functions and psychiatric symptoms in treatment-refractory schizophrenic patients receiving clozapine. *Biol Psychiat* 34: 702–712
112. Murphy BL, Roth RH, Arnsten AF (1997) Clozapine reverses the spatial working memory deficits induced by FG7142 in monkeys. *Neuropsychopharmacology* 16: 433–437
113. Didriksen M (1995) Effects of antipsychotics on cognitive behaviour in rats using the delayed non-match to position paradigm. *Eur J Pharmacol* 281: 241–250
114. Skarsfeldt T (1996) Differential effect of antipsychotics on place navigation of rats in the Morris water maze. A comparative study between novel and reference antipsychotics. *Psychopharmacology* 124: 126–133
115. Simosky JK, Stevens KE, Adler LE, Freedman R (2003) Clozapine improves deficient inhibitory auditory processing in DBA/2 mice, via a nicotinic cholinergic mechanism. *Psychopharmacology* 165: 386–396
116. Neil W, Curran S, Wattis J (2003) Antipsychotic prescribing in older people [see comment]. *Age & Ageing* 32: 475–483
117. Hellstrom-Lindahl E, Court J (2000) Nicotinic acetylcholine receptors during prenatal development and brain pathology in human aging. *Behav Brain Res* 113: 159–168
118. Levin ED, Galen DM, Ellison GD (1987) Chronic haloperidol effects on oral movements and radial-arm maze performance in rats. *Pharmacol Biochem Behav* 26: 1–6

119. Ellison GD, Johansson P, Levin ED, See RE, Gunne L (1988) Chronic neuroleptics alter the effects of the D_1 agonist SK&F 38393 and the D_2 agonist LY 171555 on oral movements in rats. *Psychopharmacology* 96: 253–257
120. McGurk S, Levin ED, Butcher LL (1988) Cholinergic-dopaminergic interactions in radial-arm maze performance. *Behav Neural Biol* 49: 234–239
121. Levin ED, Petro A, Beatty A (2005) Olanzapine interactions with nicotine and mecamylamine in rats: effects on memory. *Neurotoxicol Teratol* 27: 459–464
122. Addy N, Levin ED (2002) Nicotine interactions with haloperidol, clozapine and risperidone and working memory function in rats. *Neuropsychopharmacology* 27: 534–541
123. Zhang W, Bymaster FP (1999) The in vivo effects of olanzapine and other antipsychotic agents on receptor occupancy and antagonism of dopamine D1, D2, D3, 5HT2A and muscarinic receptors. *Psychopharmacology* 141: 267–278
124. Green MF, Marshall BD, Jr., Wirshing WC, Ames D, Marder SR, McGurk S, Kern R S, Mintz J (1997) Does risperidone improve verbal working memory in treatment-resistant schizophrenia? *Am J Psychiat* 154: 799–804
125. Sharma T, Mockler D (1998) The cognitive efficacy of atypical antipsychotics in schizophrenia. *J Clin Psychopharmacol* 18: 12S–19S
126. Weinberger DR, Gallhofer B (1997) Cognitive function in schizophrenia. *Int Clin Psychopharmacol* 12 Suppl 4: S29–36
127. de Haan L, Booij J, Lavalaye J, van Amelsvoort T, Linszen D (2006) Occupancy of dopamine D2 receptors by antipsychotic drugs is related to nicotine addiction in young patients with schizophrenia. *Psychopharmacology* 183: 500–505
128. Drew AE, Werling LL (2003) Nicotinic receptor-mediated regulation of the dopamine transporter in rat prefrontocortical slices following chronic in vivo administration of nicotine. *Schizophr Res* 65: 47–55
129. Lambe EK, Picciotto MR, Aghajanian GK (2003) Nicotine induces glutamate release from thalamocortical terminals in prefrontal cortex. *Neuropsychopharmacology* 28: 216–225
130. Silvestri S, Negrete JC, Seeman MV, Shammi CM, Seeman P (2004) Does nicotine affect D2 receptor upregulation? A case-control study. *Acta Psychiatrica Scandinavica* 109: 313–317; discussion 317–318
131. Schwieler L, Engberg G, Erhardt S (2004) Clozapine modulates midbrain dopamine neuron firing via interaction with the NMDA receptor complex. *Synapse* 52: 114–122
132. Schwieler L, Erhardt S (2003) Inhibitory action of clozapine on rat ventral tegmental area dopamine neurons following increased levels of endogenous kynurenic acid. *Neuropsychopharmacology* 28: 1770–1777
133. Levin ED, Gunne LM (1989) Chronic neuroleptic effects on spatial reversal learning in monkeys. *Psychopharmacology* 97: 496–500
134. Rezvani AH, Caldwell D, Levin ED (2006) Chronic nicotine interactions with clozapine and risperidone and attentional function in rats. *Prog Neuropsychopharmacol Biol Psychiatry* 30: 190–197
135. Wilkerson A, Levin ED (1999) Ventral hippocampal dopamine D1 and D2 systems and spatial working memory in rats. *Neuroscience* 89: 743–749
136. McEvoy J, Freudenreich O, McGee M, Vanderzwaag C, Levin ED, Rose J (1995) Clozapine decreases smoking in patients with chronic schizophrenia. *Biol Psychiat* 37: 550–552
137. Levin ED, Icenogle L, Farzad A (2005) Ketanserin attenuates nicotine-induced working memory improvement in rats. *Pharmacol Biochem Behav* 82: 289–292
138. Rezvani AH, Caldwell DP, Levin ED (2005) Nicotine-serotonergic drug interactions and attentional performance in rats. *Psychopharmacology* 179: 521–528

Neurotransmitter Interactions and Cognitive Function
Edited by Edward D. Levin
© 2006 Birkhäuser Verlag/Switzerland

Function and dysfunction of monoamine interactions in children and adolescents with AD/HD

Robert D. Oades

Biopsychology Research Group, University Clinic for Child and Adolescent Psychiatry, Virchowstr. 174, 45147 Essen, Germany

Introduction

A consideration of how unusual function of the monoaminergic transmitters can contribute to the clinical picture of childhood attention-deficit/hyperactivity disorder (AD/HD) involves an understanding of three concepts: What are the main features of AD/HD, how does normal brain anatomy and function develop, and how do the monoaminergic pathways interact? With this context one is equipped to look at the evidence for unusual monoamine activity and interactions in contributing to the problems found in children with AD/HD.

This chapter proposes a way to integrate the features that these concepts have in common. The first part is concerned with a description of how childhood AD/HD appears in the clinic, at home or at school. This picture then acquires structure with specific features defined by laboratory testing. To understand what might be "dis-ordered" supposes knowledge of the organization in normal brain structure and in particular, how the organization of stimulus and response develops in the child and the adolescent. Important here is that much of the functional order is orchestrated by the monoamines. The third part sketches out where and how the long axon monoaminergic pathways reach out across brain structures and exert (normally) an adaptive modulation of function under changing circumstances. Further details are provided in other chapters.

I shall emphasize childhood AD/HD with modest reference to its manifestation in adults I shall concentrate on the three main monoamines (dopamine, DA; noradrenaline, NA and serotonin, 5-HT) with only minor reference to adrenaline. Nonetheless this material has implications for the origin and course of AD/HD outside the early developmental period. Further, it will become apparent that the full consequences of changed monoamine activity can only be fully appraised within the context of the interactions with other amine- (e.g., acetylcholine) and amino-acid transmitters (e.g., GABA and Glutamate).

AD/HD – a clinical picture

The diagnosis of AD/HD usually concerns young people between the ages of 7 and 18 years. The manual of the American Psychiatric Association (APA: DSM-IV [1]) requires the presence of 6/9 features for the inattentive type, a separate 6/9 features for the type with hyperactivity and impulsivity, or both for the more usual combined type. The decision is based on longer structured or semi-structured interviews that ask 60–80 questions (or more) from two informants (usually a parent and a teacher) in order to show that the reported problems can occur independently of the situation. These features, impairing the function of the child, must have been present before the seventh birthday.

The health professional will get an image of motor restlessness (chair rotation, alternately sit or stand, move from toy to toy/task-to-task, fidgeting). Fine motor control can appear clumsy. Movement is often led by impulsivity. From observation alone it is often difficult to distinguish impulsiveness driven by a distracter, changing desires/motivations or an inability to withhold prepotent tendencies. Concentration is difficult unless the situation is novel. Social abilities are poorly developed (e.g., few friends, interruption of discourse), self-esteem is often low and the ability to organize or plan deficient. The latter can incur poor judgment and risk-taking. Changes in the quality of motivational features (e.g., the need to drink, assess reinforcement), stress- and emotional control (e.g. temper tantrums) often complete the clinical picture (review [2]).

AD/HD – neuropsychological features

It must be emphasized that there is no function typical of normal child development that is completely absent in those with AD/HD. Lesions are not implicated. The patient is sometimes "normal", but the problems persist in different contexts. A child appearing for an MR- or electrophysiological investigation can appear remarkably "cool," for the time being. There have been innumerable disagreements over what constitutes a classical or "core" phenotype. Of course, a way out is to define sub-groups by one or by another feature (e.g., referrals *vs.* non-referrals [3], inattentive *vs.* hyperactive-combined subtypes [4], with/without different comorbid disorders [5] internalisers (fearful anxious types)/externalisers (fearless impulsive types [6], more or fewer than seven repeats on the dopamine D4 receptor gene [7] those with high theta/low beta EEG ratios *vs.* those with high beta EEG power [8], medication responders/non-responders [9, 10] and more. It is ironic that the feature with the most widespread applicability appears to be that of intra-individual variability [11] – where it is the variance of response time that is usually considered.

Yet it is possible that the difficulties of AD/HD children can be both differentiated and reduced to a few conventional fields of ability. Thus, variance in the speed of performance relates to motor abilities in general, in the sense of neuromuscular development [12], but also to poorly controlled supplemental motor activity and physiological state control [13]. Similarly the variance in accuracy can be explained by inattentiveness [12], in the sense that distracters can delay [14], focused attention/non-target detection is slow [15], and indeed signal-detection indices of perceptual sensitivity (e.g., d-prime) are low [16, 17]. The errors that so often result do not incur the usual slowing of the next response, implying the impaired processing of feedback and contingent executive control [18, 19]. There are two major processes here, the top-down control of information processing, and the short-term sensitivity to reinforcement. If these are abnormal, one consequence is that children with AD/HD often express an aversion to delays in event-rates. In other words there are two separate features (dual pathway, [20], executive dysfunction and delay aversion) that each make significant, independent contributions to predictions of AD/HD symptoms.

A number, if not all, of these features of AD/HD could be summarised under the rubric of a "disorder of impulsivity" [7]. There is some truth in this. The term "impulsivity" has three components – acting on the spur of the moment (motor), not focusing on the task in hand (attentional), and not planning ahead (executive [21]) that can all lead to ill-considered action. But it would be wise when attributing unusual neurochemistry to non-adaptive function to separate the control systems for cognitive and behavioural impulsivity [22]. The alternative to lumping is to split the disorder into numerous sub-types. This will always have some explanatory value for specific features, but it is worth considering, for example, the experience of Nigg and colleagues [23]. They examined executive function, motor abilities and flexibility of cognitive set, and found that the similarities between diagnostically inattentive and combined subgroups were much more striking than the differences (cf. also [24]).

Unusual brain functions in children with AD/HD are associated with inattention (perception and selection), poorly controlled (executive) decision processing (conflict management), non-adaptive evaluation of reinforcement contingencies and situationally inappropriate motor activity. These impairments are reflected in each of the successive stages of information processing that are so clearly and precisely represented by scalp electrophysiological records (event-related potentials, ERPs) in the first half second after an event: Stimulus-elicited cortical excitation (N1 reduced [25]) interference control (P2 larger [26]) stimulus categorization (N2 reduced [27]) effortful updating of short-term memories (P3 reduced [28]) assessment of stimulus "target-ness" (processing negativity reduced [29]), assessment of mistakes (error-related negativity/Ne/Pe reduced [30]), and motor organization (LRP reduced [31]).

Normal brain development

With an interest in AD/HD in mind, interest in normal anatomical and cognitive development centers on the classical peripubertal age for referral (8–14 years) with curiosity extending to earlier features (potentially relating to causality) or how matters progress or disappear in young adults.

Myelination, white matter development, begins in the second trimester, develops linearly from 4 years and continues through (and beyond) the third decade. In the meanwhile frontal lobe gray-matter develops slowly and gradually to 8 years of age when prefrontal development (rostral to the precentral sulcus) takes off and develops rapidly until about 14 years. Having peaked prior to adolescence, the grey matter volume then declines [32]. This process is attributed to the pruning of connections [33], and may start as early as 7 to 10 years of age in sensory and in frontal association cortices, respectively. The thickness of the cortex decreases across the whole period from 8–20 years [34]. The peripubertal age also sees the rise of hemispheric differences (e.g., around the inferior frontal sulcus: cf. language development on the left). Some of these differences are gender specific [35].

Brain, especially white-matter-volumes, increase continually over three decades: overall increases of volume are found in many parts of the frontal, parietal and mid temporal (limbic) lobes, while more definite decreases occur in the lateral cortices, basal ganglia and thalamic nuclei [36–38]. These studies have shown that maturation progresses in waves, rostrally in the frontal and laterally in the temporal lobes. Interestingly these separate developmental axes are reflected in a functional study showing the "migration" along these axes of the sources of activity underlying the detection, registration and response to changes of auditory stimulation [39]. Such maturational processes continue into the frontal and temporal poles throughout the third decade. Indeed, frontal grey/white matter ratios continue to decrease (linearly) even beyond that age [40].

Normal neuropsychological development

Linear increases in the rate of development of postural and sensorimotor coordination peak around 6 and 10 years of age, respectively. Continued development, particularly of the latter, depends increasingly on experience and its consequences, described as "enhanced programming resources" and online feedback processing [41, 42]. Tapping into such problems may reflect the core problems of AD/HD children in cognition, on which this chapter concentrates. Thus, it should be borne in mind that motor coordination does not become mature until relatively late (in the second decade), alongside attentional and executive functions [38]. In contrast, sensory functions, orientation and speech-related abilities develop earlier in the first decade.

In late childhood (around 7 years ± 1 year) children make a qualitative leap in their cognitive abilities, allowing measures to be made of tests that have a qualitative if not a quantitative similarity to those used in the neuropsychological testing of adolescents and adults. In particular they are able to orient between cues and master conflicting stimuli about as well as older children [43]. However, the speed and accuracy of switching attention continues to improve with age.

As would be expected from anatomical developments briefly described above, the transition of puberty (around 12 years ± 1 year) coincides with the maturation of many abilities associated with the function of the frontal, or especially the prefrontal lobes. These include abstract reasoning, use of goals in making plans, inhibitory control, verbal fluency, verbal delayed recall, novelty-seeking, even finding a degree of independence from the family [35, 44].

But fine grain analyses of development have been rare. A series of studies by Luna and colleagues [45] on speeds of processing, the ability to inhibit voluntary responses and working memory use were all based on variations of an oculomotor task, thereby controlling for the comparison of qualitatively different task requirements. They reported that adult levels of response inhibition were not achieved before the age of 14 years[1], independent of speeds of processing that matured a year later. Working memory performance, which depended modestly on the other two variables considered, did not attain adult levels until 19 years of age.

The development of the stages of information processing is illustrated in an exemplary way with ERP measures. The arrival of sensory information in the thalamus and sensory cortices is marked by the P1/P50. Maturation to adult levels involves a decrease of amplitude and latency by about a third between 5 and 15 years [48]. The gating of the ERP response to a second stimulus (as marked by P50 in a paired click paradigm) is extremely variable at puberty [49], and may not achieve adult expression until the end of the teens [50]. The development of excitation elicited by a salient stimulus (N1), along with the suppression of processing of other stimuli (P2), as a preliminary to its being further processed, has been described for subjects aged from 5 to 30 years [51, 52] .The N1/P2 adult waveform only becomes evident at 13-14 years of age. The decreases of the latency and amplitude characteristics of the peak and the dipoles do not mature until after 16 years. Around puberty the topographic distribution of the P50 peaks across the scalp move posterior and N1 peaks lose their rightward asymmetry However, P2 peaks do not move rostrally to their central adult locations until the end of adolescence. The categorization of stimuli (marked by N2) and context-updating (marked by P3) attain their bilateral frontal and parietal topography by around 17 years of age. The amplitudes of these components show a linear and curvilinear development with age, respectively, and mature around 15 years of age with latency attaining adult levels some 3 years later [53, 54]. Indicators of automatic selective processes (mismatch negativity, MMN) develop about 3 years earlier than controlled attention-related processes (negative-difference, Nd). While MMN topography becomes bilaterally distributed after puberty, the latency reaches adult levels around 17 years, but the dipoles continue to migrate along with normal frontal and temporal lobe expansion through the third decade [39, 51].

The monoamine pathways

As their names suggest there are three major dopaminergic (DA) innervation systems in the forebrain, with their mesencephalic origins in the ventral tegmental area (VTA) and substantia nigra (SN) in the brainstem – the mesocortical, mesolimbic and nigro-striatal projections [55]. The density of mesocortical DA pathways in primates increases rostrally across the cortices. For example, the increase in the rostral auditory association cortices is already markedly higher than in the more caudal temporal lobe. A moderate then higher innervation is found moving from somatosensory over motor to prefrontal association areas. The axons are especially dense in layers I and

[1] The emphasis is on adult levels of performance. In the preceding peripubertal phase children can execute such tasks (e.g. Go/no-go), but they recruit much larger areas in the frontal lobes [46] and the amplitudes of the ERPs show that their categorization of stimuli and evaluation of errors made on these and conflict tasks are in general remarkably small [47].

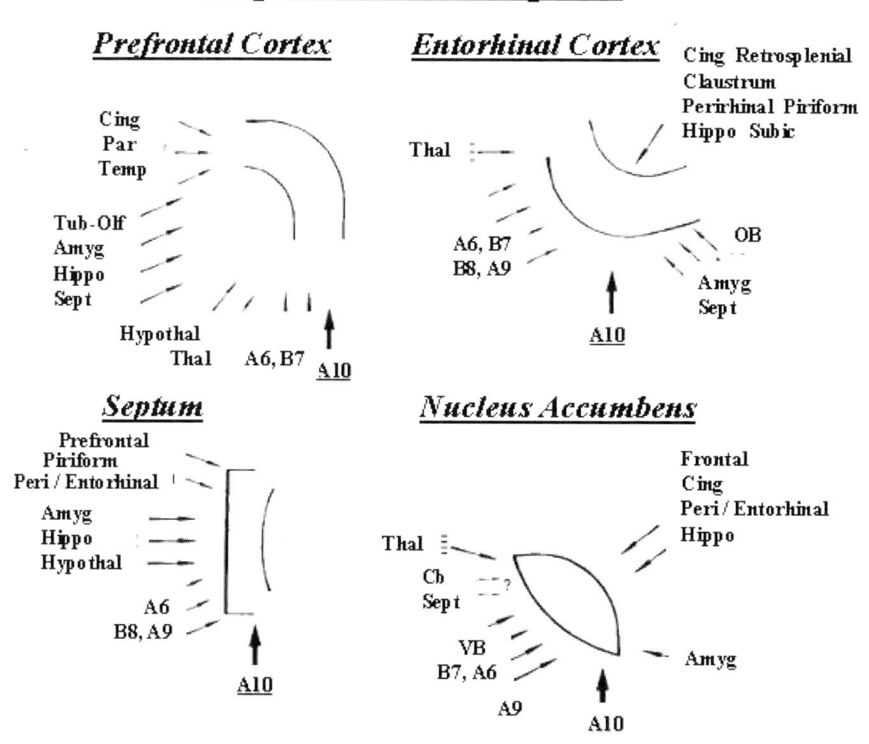

Figure 1. Nodes for the convergence of afferent fiber input on two mesocortical and two mesolimbic DA projection regions (prefrontal and entorhinal cortices, the nucleus accumbens and septum). Reproduced from [55] with permission from Elsevier.
Amygdala (Amyg), Cerebellum (Cb), Cingulate cortex (Cing), Claustrum, Entorhinal cortex, Frontal cortex, Hippocampus (Hippo), Hypothalamus (Hypothal), Infero-temporal cortex (Temp), Olfactory bulbs (OB), Parietal cortex (Par), Prefrontal Perirhinal, Piriform and Retrosplenial cortex, Septum (Sept), Thalamic nuclei (Thal), Tuberculum-olfactorium (Tub-Olf), Ventral noradrenergic bundle (VB): Monoaminergic nuclei (A/B 6-10).

II and again in V and VI [56]. DA D1 receptors (dense in I-IIIa, moderate in V and VI) are present at one to two orders of magnitude more than those of the D2-family, but in this D2-family the D4 type of receptors are more evident in the neocortices (e.g. layer V), and the D2 types in the limbic and temporal regions. Important recipients of mesolimbic innervation include the entorhinal and cingulate cortices (transitional and archicortices), parts of the hippocampus and amygdala, and the ventral striatum (nucleus accumbens and septum). Oades and Halliday [55] pointed out that these regions are "nodes of convergence" of input from very many brain regions and represent excellent opportunities for DA activity to influence the shifting of the control of their efferent output between different afferent sources (Fig. 1).

The main noradrenergic (NA) projections to the limbic and cortical brain regions of concern here arise in the locus coeruleus (LC) of the pontine brainstem. NA fibers project throughout the forebrain, to the phylogenetically older archicortices (hippocampus and amygdala), the neocortical mantle, but also the cerebellum. This more dorsal pathway along with a more ventral one from the nucleus tractus solitarius also innervate several subcortical regions including the thalamus and hypothalamus [57]. Innervation in the neocortices increases from layers I–V with highest densities in II and IV with greater densities of the alpha and beta receptors in the more superficial layers [56]. Alpha-2a sites, prominent in frontal regions, may be pre- or post-synaptic in location, while alpha-1 sites more often exert effects presynaptically, the former inhibiting, and the latter enhancing monoamine release [58].

Relevant to forebrain function, serotonergic (5-HT) projections originate in the median and dorsal raphe on the border of the pons (containing the LC) and midbrain (containing the VTA). There is some overlap between the areas innervated, but the dorsal raphe projects more anteriorly, to the frontal cortices and basal ganglia, and the median raphe somewhat more to limbic structures and the diencephalon. The sensory and motor cortices display a decidedly patchy distribution of low and high levels of innervation [59]. Much of the input arrives in layers III and IV [60]. Two of the most studied 5-HT binding sites in the CNS are the 5-HT1a and 5-HT2a receptors. The former is often characterised as an autoreceptor, and the latter postsynaptic, but this is not an exclusive compartmentalization (e.g. 5-HT1a sites are active postynaptically on cholinergic neurons). Stimulation of either site can lead to increased catecholamine outflow[2] [61–64].

Monoamines – development

DA neurons enter the cortical plate early in the second trimester. DA has a trophic role at this early stage, whereby impairments can have consequences on the later thickness and connectivity of the cortex [65]. From birth to puberty the number of axons can increase six-fold before pruning processes set in. Numbers of DA receptors peak in mid-childhood, already decreasing well before puberty (D1 earlier than D2: [66]. Across adolescence to adulthood the number of D1 sites falls by nearly 50% and D2 sites by nearly 60% [67]: thereafter numbers of D1 sites decrease by a few percent per year. The implication that the D1/D2 ratio falls with age is noteable. In studies of rodents the peak for D2 receptors seems to be larger in males, and despite the ensuing reductions, levels are still higher than in females through adolescence [68]. (The same study also described more D1 sites in right than left sided subcortical regions that lasted from the post-pubertal period into adulthood: this is reflected by measures of DA and its metabolite DOPAC that showed a lower turnover in the left hemisphere until inter-hemispheric coupling matured in young adulthood [69]. Such findings are yet to be confirmed for humans.) The DA transporter system follows a different

[2] This generalization glosses over the variation with brain region, receptor sub-type (e.g., 5-HT2c, 5-HT1b), the mechanism (through an effect on release or synthesis) and whether the catecholamine neuron is in a tonic- or burst-firing state.

course, peaking at puberty and gradually decreasing right on through to 50 or 60 years of age (postmortem study [66]. This matches the inverse changes for the synthesis of DA (by tyrosine hydroxylase) that in non-human primates continues to develop right through into adulthood [70]The gradual decrease of transport mechanisms may accurately reflect functional activity and are directly reflected by the gradual decrease of DA turnover seen in urinary measures taken between 10 and 20 years of age [71].

NA development in the human fetus follows, but at first lags a little behind that for DA in the perinatal period [72, 73]; but if data from animal studies pertain then it soon speeds up and overtakes that for DA [74]. In studies of primates and other animals alpha-2 and alpha-1 types of receptor also follow each other in developmental waves, with the alpha-2 ahead at birth. But levels fall off after birth as numbers of alpha 1 sites increase. Yet by puberty alpha-1 sites are decreasing more rapidly than the alpha-2 sites. Transport mechanisms are gradually reduced following puberty but increase again by the end of adolescence (review [59]). This post-pubertal decrease followed by an increase across the teenage period is reflected in urinary indicators of NA turnover [71].

5-HT development reflects first a prenatal neurotrophic role, and second a postnatal expansion of neural innervation and function. A study of Rhesus monkeys from 2 weeks to 10 years of age [70] showed that while the development of catecholamine-containing appositions on cortical pyramidal cells reached half adult levels by 6 months of age, 5-HT appositions had already attained adult levels by 2 weeks. Prepubertal development, though considerable, appears paradoxically to be functionally slower than that for DA, such that CSF measures suggest a near doubling of the ratio of DA to 5-HT metabolites over the prepubertal period (review [59]). Post-mortem tissue [75] and urinary measures [71] suggest that rather like the situation with NA, 5-HT turnover decreases initially post-pubertally, but then rises again at the end of the second decade. If studies of rodent development are any guide considerable lateralized differences are to be expected. Neddens and colleagues [76] reported a rightward emphasis of fiber density in the neocortices and a leftward emphasis in the limbic cortices.

Clearly there remains a lot of detail on the development of the various features of monoamine systems to be described: the near absence of knowledge of the relative abundance of the different receptor subtypes is striking and only partly explained by the fairly recent availability of suitable ligands. The results reported in this section show that there is no simple way to say that the functional activity of one or the other monoamine (let alone their interactions) is more or less than adult levels at a given age. First the baseline of adult levels is continually changing with age. Secondly it remains unfortunately equivocal whether any specific function considered is more accurately represented by turnover, synthesis rates, transport mechanisms, or the development of synaptic appositions on innervated pyramidal or non-pyramidal cells. Each of these features develops at different non-linear rates.

Monoamines interactions pertaining to normal cognition

Brain-damage or insults to the monoamine systems alone do not allow unequivocal conclusions to be drawn about hypo- or hyper-function in the affected system. But they do provide some insight into the normal situation by seeing in what domains there are dysfunctions. Preclinical studies (e.g. reviews [77–79]) suggest that damage impairing NA function increases distractibility. NA tunes the influences of the inputs competing to control the output of an NA innervated region. Low to high tonic firing rates are associated with inattention, and low arousal to agitation and stressed states. In contrast phasic firing occurs when stimulation is relevant, other activity should be tuned down [80]. Impaired 5-HT function is associated with impulsivity, whereby decreased function may relate to outbursts of aggression, while increases are associated with cognitive impulsiveness [81–83, 22]. By analogy with the role of NA in tuning, studies of stimulus control suggest that 5-HT very often appears to influence transmission by exerting a volume-control or gain function [59, 84]. By contrast, the role of (increasing) DA activity has been described as one of facilitating the likelihood of a switch occurring between one of two inputs controlling the output of a given brain region [79]. Reducing DA function thus leads to the slowed switching of a particular cued response [85]. This can be advantageous in initial learning. In contrast, high activity enhances switching as in divided attention, or between attentional and task sets (e.g. trail making, or discrimination reversal [86, 87]). While low and high levels of DA and NA activity respectively demonstrate the different roles of tuning and switching in initial learning, there are other situations in the control of ongoing behaviour when their function can appear rather similar as a result of the presence of different receptor subtypes[3].

There are numerous complications that make for difficulties in the interpretation of the results of the manipulation of any one of the monoamines. I shall mention a few. NA neurons have sites that will transport NA and DA, and others that can release NA or DA [89]. This makes it very difficult to determine precisely the mechanism by which, say, psychostimulants achieve a specific cognitive effect. Questions are not limited to the role of DA. NA is known not only for its high affinity for the alpha-2 and low affinity for the alpha-1 binding site, but is a relatively good ligand at the DA D4 site [90]. Interactions between the two catecholamines are also documented. For example, NA receptors have even been hypothesised to "gate" DA release [91].

It has long been realised that 5-HT input frequently inhibits DA activity Now a better understanding of the HT2a binding site has shown that this effect must also extend to the NA system [64]. However, opposite effects on catecholamine release are attributed to 5-HT1b, 5-HT1d and 5-HT3 binding sites. The fact that both alpha-NA

[3] Arnsten [77] provides an example of NA involvement in switching between channels of activity. Information may be faithfully transmitted from the thalamus to the cortex under conditions of sufficient NA release to engage $\alpha 1$ and β NA receptors. But when low levels of NA are released $\alpha 2$ receptors are engaged. Then, thalamic neurons enter a burst mode which prevents information transfer [88]. In this way, the varying affinities of NA for $\alpha 2$ vs. $\alpha 1$ or β NA receptors acts rather like a "switch to alter neuronal, and the ensuing behavioral state."

and 5-HT1 sites may be found in pre- and post-synaptic locations warns against generalizing about a transmitter's activity being associated with unidimensional changes of any one cognitive ability [59].

AD/HD: (1) Indicators of monoamine metabolism – theory

Let us take a "top-down" approach from the viewpoint of theories currently advanced to explain AD/HD problems. There are two to three broad explanations, that nonetheless do not acount for all features, and two to three that account for a domain of dysfunction, but extension beyond these domains remains controversial.

First, there is the dual pathway theory [93] and the cognitive energetic model [93]. The former directly invokes monoaminergic involvement and provides the background to the rest of this chapter. The latter is pitched at the psychological level of state regulation with physiological underpinnings, but elaborates little on the monoaminergic contribution. A related account [13] explicitly accounts for a range of AD/HD problems (variability and maturation) at the level of energy availability in CNS function, but only indirectly invokes modulation by the monoamines.

Other theories aim at generalizing from specific domains of performance such as response inhibition [94, 95] to executive function and affect control, and the "dynamic developmental theory" [96] that concentrates on the registration of reinforcement and related motivational consequences (see also reviews in [5, 97]). All these theories depend on functions modulated by DA (*prima unter pares*). They tend to overlook the role of NA and 5-HT, but do admit dependence on the interactions with excitatory and inhibitory transmitters (Glutamate, GABA and acetylcholine), without much elaboration.

Most of these theories also do not pay adequate attention to explanations that could account for rates of comorbidity, maturation lag, impulsivity, stress-responsivity and sleep-wake patterns, to name a few other abnormal features associated with the phenomenon of AD/HD.

AD/HD: (2) Indicators of monoamine metabolism – a dual pathway

This theory invokes a role for the mesocortical DA system in modulating (deficient) dorsal fronto-striatal glutamatergic mediation of some executive functions. It also envisions a role for the mesolimbic DA system in the anomalously functioning reward and motivation-influencing circuits of the more ventral frontal-accumbens glutamatergic system[92].

Mesocortical pathway

Direct evidence for the involvement of the mesocortical pathway is rather recent. Neuroimaging evidence from subjects with AD/HD suggests less activity in the right prefrontal regions and parts of the basal ganglia (the caudate nucleus and pallidum) during a continuous performance test of sustained attention (in children [98]), but also in these areas (inferior frontal) and in the cingulate region during stop-signal and Go/no-go tests of impaired response inhibition and impulsivity (in adolescents [99, 101]. Indeed, no significant increase was found in AD/HD children on interference suppression (as exhibited during performance of a flanker task [102]) where the activity recorded in normal children in the mid- and inferior frontal regions correlates with success [103]. The emphasis on right inferior frontal regions is warranted by a detailed study relating the location of brain damage to stop-task performance in brain-damaged adult subjects [104]. But we should also note with regard to the fMRI studies that blood oxygenation (BOLD) signals are low across many brain regions, even in the cingulate gyrus during Stroop tasks when performance in the interference condition was actually unimpaired [105].

In general, MR-anatomical studies of AD/HD subjects give little clue as to whether any particular region, such as those just mentioned, is altered in size or development. A small reduction is recorded as widespread through the cerebral and cerebellar lobes [106]. However, grey matter reduction in the right prefrontal [107], as well as in the caudate regions [108] in these studies is noteworthy.

The prefrontal and cingulate regions discussed receive a mesocortical DA innervation. But is DA involved? Relevant to this point are further studies on the ability to switch attentional set. The ability as tested by the trail-making test has been identified as potentially belonging to the core cognitive endophenotype of AD/HD [23]. In a task where the subject had to map words/symbols to response hand under changing conditions, switching proved especially inefficient for those with brain damage to mid- and the already described right inferior frontal region [109]. Such switches have been related to DA activity [79], and in accord with expectations methylphenidate enhances performance of AD/HD children in the stop-task [110] and reduces the cost of switching between letter/number sets [111, 112].

As one of the striking features of prefrontal blood flow activation during cognitive challenge is that these are absent or reduced in adolescent and adult subjects with AD/HD [fMRI above, also PET studies [113, 114], it is important to note that behavioral responses and brain activity in these regions are altered by methylphenidate treatment. However, while thalamic or cerebellar activity may increase, that in the relevant frontal regions *decreases* [115]. This must in part be a reflection of the marked increase of synaptic DA (and blockade of DA reuptake, 50% at therapeutic doses) known to follow treatment with methylphenidate in healthy subjects [116]. In turn such changes have been directly and quantitatively linked to the interest, motivation and success in subjects who completed simple maths tests [117]. However, two further findings provide a clue of how, with care, these results should be interpreted. Firstly, in cocaine-addicts methylphenidate actually increases metabolism in BA11 and BA25 (orbitofrontal cortex) regions registering salience, motivational and emotional reactivity [118]. Secondly increases of PET metabolic measures were recorded after double dosing [119]. In both situations increases of DA D2 binding are expected, and it is binding in the DA D2 family of receptors that correlates with metabolism across a whole range of frontal cortical regions [120]. Indeed, the variability of biochemical or behavioral response depends on the individual baseline for DA D2-like binding.

So one may entertain the hypothesis that the AD/HD deficit may be related to an unexpected low or a relatively low level of DA binding in the individual, and his or her baseline binding status. However, if an increased chance of binding is to be therapeutic, it should probably reflect the rapid on/off (high k_{off}) type (i.e. impulse related). The reasoning is first that synthetic activity marked by PET studies of DOPA decarboxylase are lower in frontal regions of adult AD/HD patients [121]. (Higher levels seen in the midbrain of younger patients [122] may reflect the mesolimbic pathway (see below). This would lead to a low availability of DA, especially when there is impulse activity. Secondly, a faster clearance of DA (by catecholomethyl-transferase, COMT) is associated with improved performance in tests of sustained attention and time estimation – [123, 124] especially in the inattentive type of AD/HD patient. Faster clearance is achieved by those with the valine variant of a functional polymorphism (Val158Met) of the COMT gene than by those with the methionine variant.

Now, we should add the complication that in the frontal cortices the binding site referred to may be the DA D4 site that is the more abundant member of the D2-family present. The type of rapid binding referred to above may well be influenced by the number of transmembrane repeated elements to be found in the molecular structure of the receptor. The D4 gene with seven (or two) repeats may be the form showing biased transmission in Occidental and Asian samples of AD/HD [125, 126]. Currently, the contrast of groups with or without the seven repeats shows relevant but rather minor cognitive problems. Those without the seven repeats showed more variable responses, longer response times and were mildly inattentive [7, 127]. Those with seven repeats were without problems on a color-word, cued detection or rapid choice reaction time task [127], yet more impulsive on a Go/no-go task [7]. A third laboratory has reported that homozygotes for the four repeat form tended to be those with a reduced brain volume [128, 129]. Our understanding of the mechanisms at work here is clearly in a process of evolution, but the evidence points to important variability in DA D4 function in AD/HD.

Cortical NA

With the, as yet, modest effects noted to be associated with several (but not all) forms of the D4 binding site, one should consider the interaction of the mesocortical DA system with other monoamines. The intimate interactions of NA with DA processes cannot be overlooked. The NA transporter (NET) can take up both NA and DA [130]. Such neurons can also release both NA and DA [89, 131]. Further NA is a high affinity ligand for the DA D4 binding site [78, 90]. NA receptors may even control the cortical release of DA, for with the alpha-1b site knocked out animals showed no extracellular release of DA in response to amphetamine treatment [91]. The role of NA must be considered in view of the well documented therapeutic effects of the newer (atomox-etine), as well as the older uptake inhibitors (desipramine, imipramine), the alpha-2 agonists (clonidine, guanfacine), as well as the psychostimulants methylphenidate and amphetamine that affect both catecholamines similarly [132].

The role of NET in the function of the "mesocortical pathway" is prominent in the response to methylphenidate, as it is far more abundant than the DA transporter [133]. Indeed, some changes in the NET genotype (G1287A, NET1) have already been reported to be associated with AD/HD [134] and in particular the symptoms of hyperactivity and impulsivity [135] (pace negative results for *other* polymorphisms in three studies [136, 138]). These symptoms are improved by atomoxetine treatment [139]. Tantalizing but as yet equivocal evidence has been reported for associations of polymorphisms of the synthetic enzyme and alpha-2 receptor sites with inattentive symptoms [140, 142].

Effects of NA associated with cognition probably occur through one of the three forms of the alpha-2 receptor located largely postsynaptically and with a high affinity for NA. (Alpha-1 and beta sites have a lower affinity for NA and may come into action in stress situations associated with high levels of NA [77]. In the monkey model infusion of guanfacine into the ventralateral PRF strengthened associative learning and impulse control [143, 144]. In dorso-lateral regions an alpha-2 antagonist induced some behavioral hyperactivity, more errors of commission on sustained attention tasks and no-go errors on Go/no-go tasks [77, 145, 146], reminiscent of the features of AD/HD children. These effects are consistent with what we know about the normal role of NA. The locus coeruleus, the pontine nucleus of origin of the cortical NA fibers, shows tonic slow firing rates in the waking state: the appearance of stimuli relevant to the ongoing situation elicits clear phasic increases of neuronal firing, thereby also suppressing responses to irrelevant stimuli [80]. This role is consistent with a "tuning" function for NA activity [79].

While published descriptions of neuroimaging studies relevant to the role of NA in AD/HD are still awaited, there are some data from electrophysiological studies. The sort of AD/HD subject that profits from imipramine treatment (that may affect NA and 5-HT systems) is one who shows EEG characteristics of a maturational lag [147]; these subjects show a widespread increase of theta power, expected to decrease with development, but reduced power in the beta and alpha bands posteriorly). The theta power also tends to normalise following methylphenidate treatment, especially over right frontal regions [148]. Robust clinical responders to psychostimulant medication show an anterior/posterior ratio of the P300 ERP amplitude exceding 0.5; just over half of the subjects tested on atomoxetine also showed this characteristic [149]. In a visual or auditory oddball paradigm methylphenidate treatment is associated with increasing the small P3a and P3b characteristic of unmedicated patients [148, 150, 151]. Indeed, sometimes both latency and the amplitude variability across subjects is reduced by methylphenidate treatment [152]. The enhancing effect on P3 (and processing negativity) is largely seen with target processing, consistent with an NA facilitated tuning effect [153, 154]. Probably reflecting both the NA and DA effects of methylphenidate, psychostimulant treatment also normalises early stages of information processing (a reduction of the large N1 and P2 amplitude, and increases of the size of the N2 in Go/no-go tasks [155, 156]).

Cortical 5-HT

It is not widely appreciated that changes in the 5-HT system may contribute to the clinical picture in AD/HD. This view arises out of the lack of an effect of the major pharmacotherapeutic agents on 5-HT activity[4]. Hence there have been few studies of direct relevance to this chapter. Genetic, biochemical and neuropsychological evidence has recently been reviewed [59].

One must first bear in mind that in brain regions where there is a common innervation from DA and 5-HT fibers, 5-HT activity modulates that of DA. Receptors are found on mesocortical DA fibers where 5-HT2c sites modulate tonic DA outflow, while HT2a sites affect active DA transmission [68, 158][5]. Thus it is not surprising that CSF measures of the metabolites of both monoamines are often inter-correlated, and were reported to decrease in AD/HD subjects responding to methylphenidate treatment [162].

From a functional point of view shifts of attention facilitated by methylphenidate are impaired by reducing 5-HT synthesis in healthy young adult subjects [163]. Let us take the example of the cognitive challenge of conditioned blocking. Healthy children switch out the influence of superfluously related stimuli while learning a conditioned association [164]. This is associated positively with levels of DA metabolites (HVA) excreted, but negatively in AD/HD children experiencing difficulties with conditioned blocking. Additionally the AD/HD children showed a positive association with the removal of 5-HT metabolites (5-HIAA). This is consistent with the AD/HD children removing high levels of 5-HIAA and showing low HVA/5-HIAA ratios of relative metabolic activity. This result contributed to the author' suggestion that with respect to 5-HT activity AD/HD children show hypodopaminergic activity [165]. This is also consistent with the authors' report of correlations between cognitive impulsivity measured on the stop-task and decreasing affinity of the 5-HT transporter that would lead to higher levels of 5-HT in the synapse and correspondingly more metabolism [22]. Rubia and colleagues [166] also report fMRI evidence from young adults of cognitive control by the 5-HT system. Decreased 5-HT synthesis induced by an amino acid drink related to more left/righthand choice errors on a go/no-go task using arrow-cues. The change in 5-HT levels was associated with decreased BOLD signal from the inferior and orbital frontal cortices, but an increased signal in the temporal lobe. (The former regions were noted above to be of special interest in explaining function in AD/HD.)

In continuous performance tests, perceptual sensitivity (d-prime) falls with an increased excretion of 5-HT metabolites [16]. The relationship of DA to 5-HT activity (HVA/5-HIAA) is depressed in some samples of AD/HD children [165], although increases of this ratio may reflect motor activity [167]. Let us consider some direct

[4] It is also not widely appreciated that atomoxetine binds to the 5-HT transporter with an affinity, very approximately, only an order of magnitude less than for the NET. For comparison it binds to DAT with an affinity three orders of magnitude less, and methylphenidate has an affinity for the 5-HT transporter well over four orders of magnitude less [157].

[5] The HT2a effects are better documented from the mesocortical projection and the HT2c effect on tonic DA outflow from mesolimbic projections [159, 161]

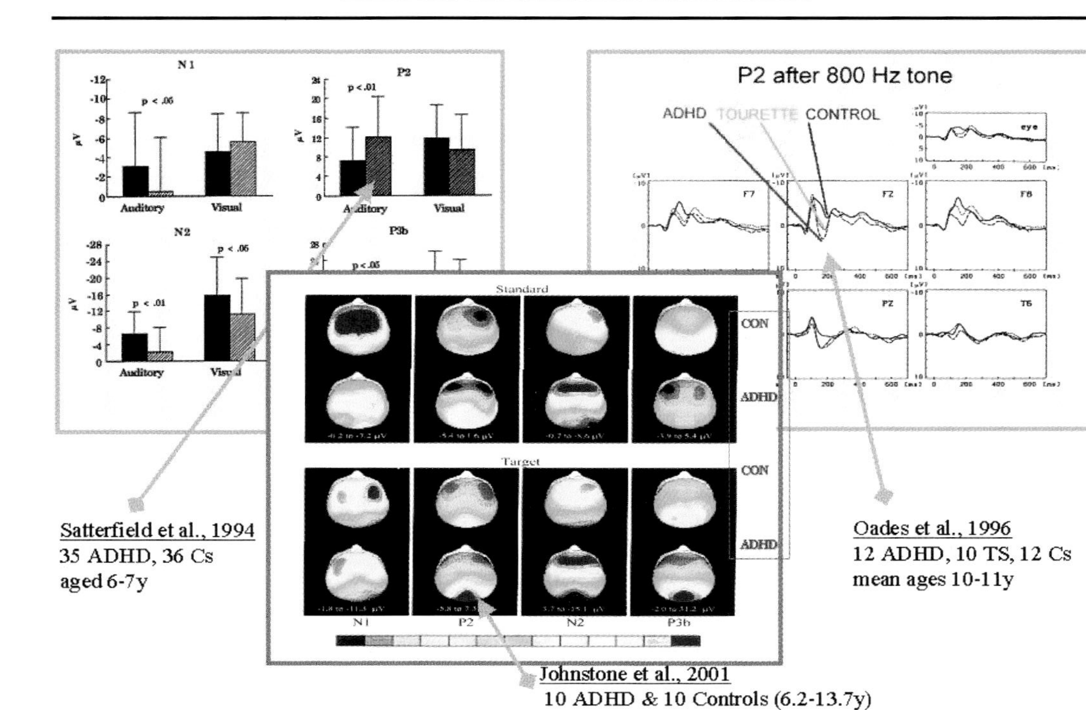

Figure 2. Three ERP studies of AD/HD children showing a P2 component of large amplitude that may reflect anomalous serotonergic activity. (The figures are modified after [26, 249, 250] and reproduced with the permission of Elsevier, Blackwells and the author, respectively.)

measures of the role of 5-HT in the processing of salient stimuli in the sensory and association cortices.

The amplitude of the N1 to P2 ERP elicited by auditory stimuli can depend on their loudness. These two components reflect the excitatory response to salient stimuli and the allocation of resources for further processing. The augmenting response reflects 5-HT neurotransmission and has been used to predict clinical responses to 5-HT agonists in affect disorders [168]. The slope is decreased following 5-HT uptake inhibition [169]. Although the activity of other transmitters (e.g. DA and acetylcholine) can also influence responsiveness [169, 170] the P2 component can be viewed as a marker of the role of 5-HT in the interplay with the catecholamines in the auditory cortices [171]. Long ago it was noticed that the response of autistic children to fenfluramine and AD/HD children to methylphenidate could be predicted by the augmenting response [172, 173]. More recently, numerous studies describe the frequent occurrence of unusually large P2 amplitudes in AD/HD children – three are illustrated in Fig. 2. The 5-HT influence may be more widespread. 5-HT suppression through amino acid drinks increases mismatch negativity (that marks the detection of deviant stimulation) – so increased activity may impair. The impairment of right frontal MMN in AD/HD children may reflect this [26]. The MMN sources known to include the right inferior frontal region are also those noted in fMRI studies (discussed above) to be sensitive to AD/HD impulsivity and 5-HT activity [99, 166]. One of the other sources of mismatch negativity is located in the cingulate cortex [174], alongside dipoles for the event-related responses recorded after error commission. One of these components (the Pe) may be reduced in AD/HD children [19]. Responses to error commission are sensitive to the activity of the 5-HT transporter. Variations in the transcriptional control region of the gene (5HTTLPR) come in short and long versions. The low activity short variant is associated with larger error responses in healthy subjects [175] – so that one would predict that the long variant may be associated with reduced Pe. Indeed biased transmission of the long allele has been reported recently for AD/HD [176]. Associations of the one or the other form with the 7-repeat DA D4 allele have been related to opposite extremes of temperament and anxiety in infants [177], and together with those for 5-HT may represent significant markers for AD/HD [178]. Lastly, supporting the thesis of over-activity in the 5-HT system, reductions of the 5-HT metabolite have been noted for hyperactive children responding to medication [179].

Against this background, it may be borne in mind that there are several mechanisms that could mediate the 5-HT/DA interactions in AD/HD. Thus, the nature of the 5-HT transporter (5-HTTLPR) will affect the expression of 5-HT binding sites, for example, the short allele is associated with a lower binding potential of the HT1a site [180]. Agonism here is associated with reducing 5-HT activity that inhibits DA release in terminal regions [181]. This could be one mechanism to combat hyperserotonemia. In contrast, agonism at DA D2 sites has been shown in microdialysis investigations directed at the dorsal raphe origin of 5-HT projections to increase 5-HT release [182, 183]. This would suggest caution in the exploration of useful DA agonists. With regard to ongoing treatment with methylphenidate, 5-HT agonism (quipazine) in animals can interact to enhance the down regulation of the DA

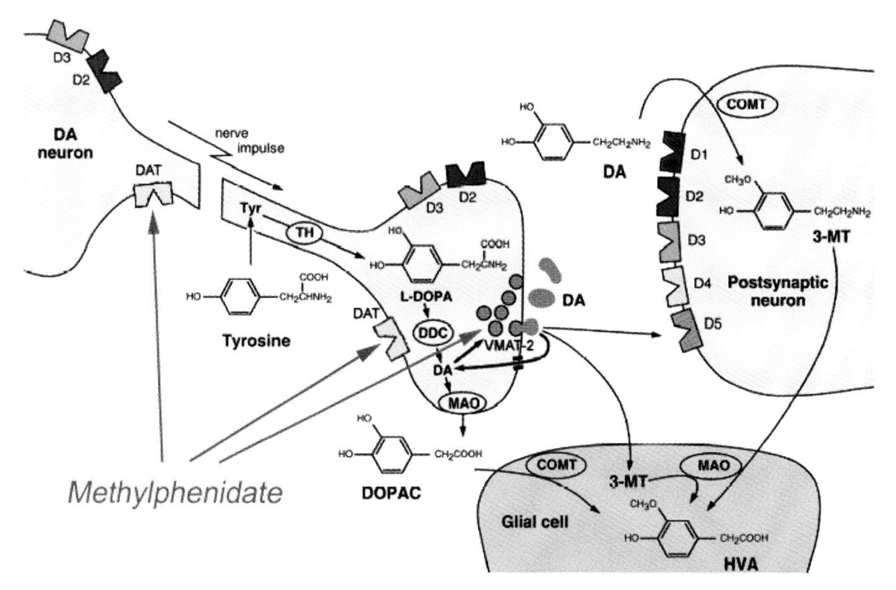

Figure 3. A scheme illustrating the synapse of a dopaminergic neuron, with the presynaptic bouton on the left at the end of an axon leading from the cell body, and the post-synaptic element on the right. The five types of DA receptor that may occur post-synaptically are illustrated although they would not all be found in the same synapse. The contribution from an astrocyte is symbolised by the glial cell below. The synthetic pathway for DA is illustrated pre-synaptically. The points for the potential action of medication (methylphenidate) are illustrated as a) the DA transporter on the cell body and on the bouton, and b) the vesicle monoamine transporter (VMAT-2) where newly synthesised DA is taken up prior to exocytosis in the cleft (Modified after [96] and reproduced with the permission of Cambridge University Press.)

transporter [184]. On the presynaptic bouton stimulation of both the D2 autoreceptor and the DA uptake site can change the sequestering by the vesicular monoamine transporter (VMAT-2) of transmitter be it DA or 5-HT [185, 186] (Fig. 3).

Mesolimbic pathway (DA)

Leading animal models have shown that the DA transporter (DAT) appears both to work inefficiently and be over-expressed in the mesocortical pathway. By contrast, these models disagree on the nature of the different situation in the mesolimbic system [187]. Mesocortical function is dominated by the NET control of both DA and NA clearance and release, exacerbated by disorder in the relatively sparsely distributed DAT control. NET is barely present in most of the regions modulated by the mesolimbic projections, but DAT is prominently represented.

The major targets of the mesolimbic DA pathway ascending from the mesencephalic VTA are the nucleus accumbens, amygdala and the hippocampal complex [55]. These regions receive topographically distributed glutamatergic input from dorsal and orbital frontal cortices, and provide feedback via GABAergic and glutamatergic pathways over several thalamic nuclei. Unusual activity in these constituent circuits modulated by the mesolimbic afferents are postulated to account for the aversion of many AD/HD children to delays. They can wait, but usually prefer a small reinforcement over waiting for a larger one (reward discounting [92]). Support for this being a prominent determinant of AD/HD behavior comes from many studies [188–191]. This characteristic is interpreted as an inefficient coupling between current responses and future rewards. The result is a reduced control by future salient events on current events The gradient between the two is short and steep [96]. The difficulty lies not in arguing whether there are problems in processing delays and discounting rewards in children with AD/HD, but in refining our understanding of what are the components of this phenomenon. For example, animals with lesions of the amygdala also prefer immediate over later, larger rewards. However damage to the input from the orbital frontal cortex has the reverse effect [191]. This could be described as a system that controls "impulsivity" [193]. Do meso-accumbens DA pathways mediate incentive motivation and reward [194], or do they (more parsimoniously) enhance a switch between circuits influencing the processing of more or less salient information [196]. It should not be overlooked that communication about reward (via some DA pathways) has much to do with its mediation by the orexin/hypocretin output from the lateral hypothalamus and amygdala [196].

At the behavioural level there is an apparent choice of AD/HD children to respond to immediate events over other possibilities. How does DA availability affect this? The answer here requires an understanding of what may be happening at the synapse of an AD/HD patient with/without medication (Fig. 3). Normally in the basal ganglia (in contrast to mesocortical regions) the ratios of DA, DAT and receptor densities are similar and the function of DAT is likely to be a major contributor to DA signaling [133]. Efficient DAT limits the duration of DA induced synaptic activity – at low DA levels it stimulates DA release, at higher levels the DA D2 autoreceptor attenuates release [133]. One would presume that psychostimulants are efficacious, as the first of these two processes is impaired. But this need not mean that the DA system is hypoactive. The increase could activate the D2 autoreceptors to reduce the (over-)release of DA, especially that associated with the neural impulse. Indeed methylphenidate also reduces the rate of spontaneous firing in mesolimbic neurons [197]. Thus the overall effect of treatment could be to increase tonic, but to decrease phasic DA release [198]. This would seem to fit the data from Schultz's monkeys [194]. He related a fast phasic component of the neural response to reward prediction. This may be too strong in AD/HD and should be attenuated to allow delayed behavioural response. Grace [198] suggested that through delayed development the reduced cortical glutamatergic input to the accumbens would lead to a hypoactive DA system. This proposal has been incorporated in the dynamic developmental model of Sagvolden [96].

In adult subjects with AD/HD striatal DAT binding was reported to be unusually high (a SPECT study) and was reduced by nearly 30% after a month of methylphenidate treatment [199]. This supports the notion (above) that tonic levels of DA would increase, as confirmed for normal adults [200]. Interestingly, in animals, co-administration of methylphenidate with nicotine (there are presynaptic acetylcholine receptors on mesolimbic neurons) increased DA levels in an additive manner [201]. This may provide a basis for apparent attempts at self-medication through cigarette smoking. Important for the distinction between the function of tonic and phasic activity, and its behavioral effect, Volkow's PET studies in humans show that methylphenidate-induced increases in DA are associated with an enhanced perception of a stimulus as salient [202]. While such perception is clearly relevant for the interest in and motivation generated by such stimuli, it relativises the emphasis placed on mesolimbic reinforcement processes in the direction of the attentional mechanisms I have emphasized.

There is evidence for genetic variation in the production of more and less efficient DAT. The 10-repeat allele for DAT (3′ variable number tandem repeat polymorphic site in 3′ region of the gene SLC6A3) is reportedly over-active. To obtain this beneficial behavioural, attentional and biochemical response to methylphenidate it is advantageous **not** to be homozygous for the 10/10 repeat allele of DAT [203–208] – even though the EEG of homozygotes is somewhat normalized after treatment[6] [206]. Although there is modest reason for suggesting a biased transmission of the 10/10 variant in AD/HD [210, 211], many studies do not find this – implying that we should be looking for other types of DAT variant.

As suggested above there is evidence for the involvement of the ventral striatum, thalamus and orbital-frontal cortex in discriminating reinforcement contingencies (or their saliency) in normal subjects [212] and that the 10/10 allele is associated with size reduction of the nearby caudate nucleus [128]. However, there is sparse evidence that methylphenidate is associated with changes of the aversion to delays. Yet, we have long known that the steep reinforcement gradient shown by the spontaneously hypertensive rat model of AD/HD is improved after methylphenidate treatment [213]. Immediate reinforcement was less effective and responses for delayed reinforcement were strengthened. The same effect of treatment was reported from a study of adults with a history of criminal behavior [214]. One presumes that the weak signal provided by a cued delay of reinforcement is amplified by the drug's effect on DA release. This seems to be supported by another PET study of normal adults from the Volkow team [215] showing that while the sight of food elicited no change in the dynamics of DA activity, there was a major response if the subjects had received a prior dose of methylphenidate. However, the apparent support from animal work is a bit difficult to reconcile with other rodent studies showing that chronic treatment in the pre- and peri-adolescent period resulted in less interest in natural rewards (e.g. sucrose, novelty and sex: [216]. This qualification and the interpretation of Volkow's data would seem to put emphasis on the processing of the "signal" rather than on incentive and motivation.

[6] The opposite effect (increased theta power) on the magnetic form of the EEG after methylphenidate treatment was reported for a group of ADHD patients who had not been genotyped [209].

Mesolimbic pathway (5-HT)

The previous section introduced the interactions of 5-HT with DA in regions innervated by the mesocortical projections. Such interactions are relevant in areas innervated by the mesolimbic system, and do concern the questions about impulsivity, of reinforcement mechanisms and motivation just addressed.

In AD/HD children cognitive impulsivity measured by a reduced probability of inhibition in the stop-task, is associated with decreased affinity (increased Kd in platelets) of the 5-HT transporter [22] (Fig. 4)[7]. With regard to the reinforcement mechanisms, stimulants like amphetamine (therapeutic in AD/HD) and cocaine act presynaptically on DA transport. Both alter 5-HT dynamics. Indeed, if the DA transporter is knocked out in rodents reinforcement measured by cocaine administration [217] or conditioned place preference to amphetamine [218] remains until a 5-HT_{1a} antagonist is administered. Further, the sensitivity to reinforcement administered by intracranial self-stimulation to the hypothalamus is increased by treating the median raphe nucleus with a 5-HT1a agonist [219]. Interactions between 5-HT and DA systems are central to considerations of cognitive impulsivity and the associated evaluation of reinforcement.

There is a large body of animal research that clearly shows the involvement of 5-HT interactions with DA in the mediation of the mechanisms underlying the preferred choice of AD/HD children for receiving immediate rather than delayed rewards. Measures taken with a dozen agents blocking NA and 5-HT uptake (but not DA uptake) show that there is an increased efficiency for obtaining water presented on a schedule of differential reinforcement at low rates of response (DRL [220, 221]). A similar effect was seen in young adult criminals given paroxetine while performing a task where a short delay resulted in a small reward, but a longer delay gave more reinforcement [222]. It may be noted that sub-chronic paroxetine down regulates pre- and post-synaptic 5-HT1a sites in normal young adults [223]. In confirmation, enhancing activity at the HT1a sites in animals leads to problems with delaying response for reinforcement [224, 225]. Enhancing activity at HT1b sites attenuates the effects of psychostimulants like amphetamine in decreasing impulsivity and promoting responses to targets [226] while HT2 *antagonism* may also lead to impulsive responding [227]. Comparison between animals bred for high or low sensitivity to 5-HT1a stimulation showed the latter with high response rates, and low reward rates on a DRL schedule [228]: these effects were improved with reuptake inhibitors. Reduced 5-HT activity promoted the selection of the delayed but larger reward [229, 230]. Recent thinking (and experiment) about these mechanisms led to the suggestion that while DA systems should be active during behavioral decisions requiring effort and concerning delay, 5-HT systems were *needed* for the latter [231].

[7] Cognitive impulsivity should not be confused with poor control of aggressive responses, often seen in ADHD children, especially those with comorbid conduct disorder. For disruptive behavior the association with the affinity of the transporter was the opposite (Fig. 4), consistent with a significant literature on the role of 5-HT in aggression [22].

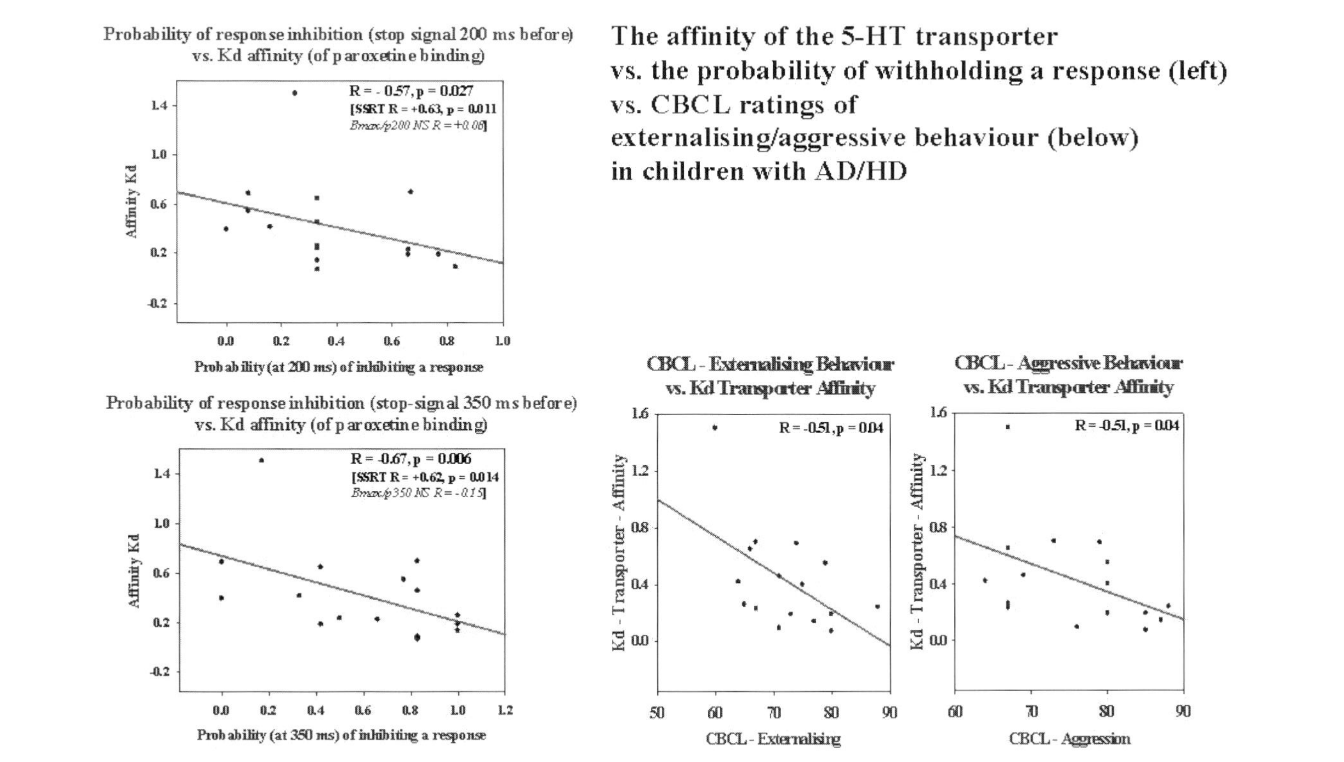

Figure 4. The relationship of the affinity (Kd) of the 5-HT transporter on platelets sampled from children with AD/HD with (left) their ability to withhold response if required on the stop-task (stop-signal reaction time SSRT) - the lower the probability of inhibiting a response (i.e., the more impulsive) the higher the Kd (lower affinity: Bmax was unrelated). On the right the reverse relationship between increasing Kd and more aggressive behavior is shown. (Modified after [22] and reproduced with the permission of Taylor/Francis.)

Thus, overall, there is reason to believe that 5-HT plays a marked role in the sensory, reinforcement, inhibitory and motor processes that are disturbed in AD/HD. At least in relation to 5-HT activity, the DA system seems to be hypoactive.

The status of peripheral and central nervous monoamine systems

Measures of the elimination of monoamine metabolites are indirect indicators of transmitter activity. It is difficult to identify the sources of these metabolites. But it is of both basic and clinical interest that there is some broad support for the relative activities between the monoamines, and some associations for these ratios with measures of symptoms or cognitive activity in young subjects with or without AD/HD.

NA metabolism

Levels of the metabolite MHPG (3-methoxy-4-hydroxyphenyl glycol), possibly an indicator of resting NA metabolism, are reported to be unusually low in AD/HD in 8/13 studies [59]. Raised levels of other metabolites such as NMN (normetanephrine) have been reported, possibly reflecting increased sympathetic actvity [17, 179], as associated with the stress of a cognitive task [17, 232]. Sub-chronic treatment with methylphenidate often results in further decreases of MHPG in peripheral catchments [233, 238] that correlate with improvements in symptom ratings [237, 239]. Speculatively, this may reflect a reduction of NA overflow resulting in the better control of DA/5-HT interactions via the high affinity alpha-2 rather than the alpha-1 site that is more closely related to activity in stressful situations.

DA metabolism

Pharmacological blocking of peripheral catecholamine breakdown shows that 15-20% of HVA may have a central origin. As a group levels are reported as normal, sometimes a bit low in CSF [240], plasma [241] and urine [235, 242]. Psychostimulant treatment tends to lower HVA excretion (in urine, plasma and CSF), if not quite to the same extent as the effect on MHPG [179, 233, 242, 243]. Shekim et al. [235, 236] reported a rate-dependent effect with high levels being lowered and low levels raised. Down-regulation has been reported to relate to decreases of symptoms, more especially for measures of hyperactivity than of attention [162, 240, 241, 244]. Together these data suggest that in comparison with NA metabolism the DA system is relatively hyperactive [165], even if some indicators suggest that DA metabolic activity is lower than normal. For example, Konrad [17] reported that impulsive errors of commission on a CPT-ax task related to rates of eye-blinking, and hence indirectly DA activity. Further, signal detection measures on a test of sustained attention (CPTax) were inversely related to HVA in normal children; no such relationship was found in age-matched children with AD/HD [16].

5-HT metabolism

A markedly lower ratio of DA to 5-HT metabolites (HVA/5-HIAA) reported in AD/HD subjects would be consistent with slightly lower DA and higher 5-HT metabolism [165]. But this result has not been supported in all samples [167, 245]. However, the increased 5-HIAA levels reported were shown to correlate closely and inversely with two quite separate measures of attentional ability, namely conditioned blocking and sensitivity (d-prime) on the CPT-ax task [16, 164]. These results along with those for the stop-task (see Fig. 3 are consistent with an over-availability of 5-HT in the synapses of children with AD/HD.

Could there be a simple explanation for the proposed relatively hyper-serotonergic (*vs.* DA) situation? Uzbekov [179] proposed one possibility. His laboratory found that while stimulant treatment (sydnocarb) reduced the high levels of 5-HIAA, N-methyl-nicotinamide (N-mna) levels rose. N-mna is the end product of the alternative metabolic pathway for the 5-HT precursor L-tryptophan. One may entertain the possibility that over activity of the indoleamine was pharmacologically diverted to an alternative metabolic route. This would be consistent with a psychostimulant induced reduction of 5-HT levels [246]. The hypothesis is open to test.

Conclusions

The diagnosic manuals maintain that AD/HD incurs differentially a broad range of cognitive (inattention), motor (hyperactive) and impulsive (response inhibition) problems. The core of this was described some 50 years ago [247]. The bases for these and related problems lie along a cerebellar – pontine/mesencephalic – cerebro-cortical axis (cf. patho-physiological findings, [248]). Recent experimental and pharmacological work points to a large contribution from the monoaminergic pathways originating in the mid/hind brain to the dysfunctions in the target areas innervated by dopamine (DA), noradrenaline (NA) and serotonin (5-HT). A significant proportion of these (dys)functions can be attributed to executive processes, the evaluation of stimuli and the reinforcement potentially associated with these events. Monoamine activity is discussed within the context of a dual-pathway theory of AD/HD function [92]. In this context mesocortical contributions to neuropsychological performance are described here for NA (with respect to DA) and mesolimbic contributions to reinforcement-related processes are described for 5-HT (with respect to DA). To divide the roles of the pathways in this way is useful but does tend to over simplify. Thus, different forms of impulsivity depend on mesolimbic and on mescortical interactions. To summarise in terms of DA activity being proportionately higher than that for NA or lower than that for 5-HT has a degree of validity but is a generalization masking some of the details of the mechanisms involved. The realization of cognitive process in the form of adaptive behavior necessarily incurs additional local GABAergic feedback, glutamatergic cortico-striatal integration and moderation by cholinergic input.

References

1. American Psychiatric Association (2000) *Diagnostic and statistical manual of mental disorders: DSM-IV-TR*. Washington, D.C.
2. Swanson JM, Sergeant JA, Taylor EA, Sonuga-Barke EJS, Jensen PS, Cantwell DP (1998) Attention-deficit hyperactivity disorder and hyperkinetic disorder. *Lancet* 351: 429–432
3. Smith AB, Taylor EA (2006) Response inhibition and hyperactivity in clinical and non-clinical populations: a meta-analysis using the stop task. In: Oades RD (ed): *Attention-deficit/hyperactivity disorder and the hyperkinetic syndrome: current ideas and ways forward*. Hauppauge, New York: Nova Science Publishing Inc., 203–225
4. Johnstone SJ, Barry RJ, Anderson JW (2001) Topographic distribution and developmental timecourse of auditory event-related potentials in two subtypes of attention-deficit hyperactivity disorder. *Int J Psychophysiol* 42: 73–94
5. Levy F (2004) Synaptic gating and ADHD: a biological theory of comorbidity of ADHD and anxiety. *Neuropsychopharmacol* 29: 1589–1596
6. Banaschewski T, Brandeis D, Heinrich H, Albrecht B, Brunner E, Rothenberger A (2003) Association of ADHD and conduct disorder - brain electrical evidence for the existence of a distinct subtype. *J Child Psychol Psychiat* 44: 356–376
7. Rubia K, Asherson P, Taylor EA, Curran S (2006) Association between the 7-repeat allele of the dopamine D4 receptor gene and specific impulsivity measures in attention deficit hyperactivity disorder (ADHD). In: Oades RD (ed): *Attention-deficit/hyperactivity disorder and the hyperkinetic syndrome: current ideas and ways forward*. Hauppauge, New York: Nova Science Publishing, Inc.. 187-201
8. Clarke AR, Barry RJ, McCarthy R, Selikowitz M (2002) EEG-defined subtypes of children with attention-deficit/hyperactivity disorder. *Clin Neurophysiol* 112: 2098–2105
9. Clarke AR, Barry RJ, McCarthy R, Selikowitz M, Croft RJ (2002) EEG differences between good and poor responders to methylphenidate in boys with the inattentive type of attention-deficit/hyperactivity disorder. *Clin Neurophysiol* 113: 1191–1198
10. Konrad K, Günther T, Hanisch C, Herpertz-Dahlmann B (2004) Differential effects of methylphenidate on attentional functions in children with attention-deficit/hyperactivity disorder. *J Am Acad Child Adolesc Psychiat* 43: 191–198
11. Castellanos FX, Tannock R (2002) Neuroscience of attention-deficit/hyperactivity disorder: the search for endophenotypes. *Nature Reviews: Neuroscience* 3: 617–628
12. Piek JP, Dyck MJ, Nieman A, Anderson A, Hay D, Smith LM, McCoy M, Hallmayer J (2004) The relationship between motor coordination, executive functioning and attention in school aged children. *Arch Clin Neuropsychol* 19: 1063–1076
13. Russell VA, Oades RD, Tannock R, Auerbach J, Killeen PR, Johansen EB, Sagvolden T (2006) Response variability in attention-deficit/hyperactivity disorder: a neuronal energetics hypothesis. *BMC Behav Brain Functn*, in press
14. Brodeur DA, Pond M (2001) The development of selective attention in children with attention deficit hyperactivity disorder. *J Abnorm Child Psychol* 29: 229–239
15. de Sonneville LMJ, Njiokiktjien C, Bos H (1994) Methylphenidate and information processing. Part 1: Differentiation between responders and nonresponders; Part 2: Efficacy in responders. *J Clin Exp Neuropsychol* 16: 877–897
16. Oades RD (2000) Differential measures of sustained attention in children with attention-deficit/hyperactivity or tic disorders: relationship to monoamine metabolism. *Psychiat Res* 93: 165–178

17. Konrad K (2006) Catecholamines and attentional function in children with ADHD. In: Oades RD (ed) *Attention-deficit/hyperactivity disorder and the hyperkinetic syndrome: current ideas and ways forward*. Hauppauge, New York: Nova Science Publishers Inc., 155–169
18. Rubia K, Smith AB, Woolley J, Nosarti C, Heyman I, Taylor E, Brammer M (2006) Progressive increases of frontostriatal brain activation from childhood to adulthood during event-related tasks of cognitive control. *Hum Brain Mapp* DOI 10.1002/hbm.20237
19. Wiersema JR, van der Meere JJ, Roeyers H (2005) ERP correlates of impaired error monitoring in children with ADHD. *J Neur Trans* 112: 1417–1430
20. Sonuga-Barke EJS, Dalen L, Remington B (2003) Do executive deficits and delay aversion make independent contributions to preschool attention-deficit/hyperactivity disorder symptoms? *J Am Acad Child Adolesc Psychiat* 42: 1335–1342
21. Moeller FG, Barratt ES, Dougherty DM, Schmitz JM, Swann AC (2001) Psychiatric aspects of impulsivity. *Am J Psychiat* 158: 1783–1793
22. Oades RD, Slusarek M, Velling S, Bondy B (2002) Serotonin platelet-transporter measures in childhood attention-deficit/hyperactivity disorder (ADHD): clinical versus experimental measures of impulsivity. *World J Biol Psychiatry* 3: 96–100
23. Nigg JT, Blaskey LG, Huang-Pollock CL, Rappley MD (2002) Neuropsychological executive functions and DSM-IV ADHD subtypes. *J Am Acad Child Adolesc Psychiat* 41: 59–66
24. Barry RJ, Clarke AR, McCarthy R, Selikowitz M, Brown CR (2006) Event related potentials in two DSM-IV subtypes of attention-deficit/hyperactivity disorder: An investigation using a combined modality auditory/visual oddball task. In: Oades RD (ed): *Attention-deficit/hyperactivity disorder and the hyperkinetic syndrome: current ideas and ways forward*. Hauppauge, New York: Nova Science Publishers Inc., 229–247
25. Brown CR, Clarke AR, Barry RJ, McCarthy R, Selikowitz M, Magee C (2005) Event related potentials in attention-deficit/hyperactivity disorder of the predominantly-innatentive type: An investigation of EEG-defined subtypes. *Int J Psychophysiol* 58: 94–107
26. Oades RD, Dittmann-Balcar A, Schepker R, Eggers C (1996) Auditory event-related potentials and mismatch negativity in healthy children and those with attention-deficit- or Tourette-like symptoms. *Biol Psychol* 43: 163–185
27. Lazzaro I, Whitmont GE, Meares R, Clarke S (2001) The modulation of late component event related potentials by pre-stimulus EEG theta activity in ADHD. *Int J Neurosci* 107: 247–264
28. Brandeis D, Banaschewski T, Baving L, Georgiewa P, Blanz B, Schmidt MH, Warnke A, Steinhausen H-C, Rothenberger A, Scheuerpflug P (2002) Multicenter P300 brain mapping of impaired attention to cues in hyperkinetic children. *J Am Acad Child Psychiat* 41: 990–998
29. Kemner C, Jonkman LM, Kenemans JL, Böcker KBE, Verbaten MN, van Engeland H (2004) Sources of auditory selective attention and the effects of methylphenidate in children with attention-deficit/hyperactivity disorder. *Biol Psychiat* 55: 776–778
30. Liotti M, Pliszka SR, Perez R, Kothmann D, Woldorff MG (2005) Abnormal brain activity related to performance monitoring and error detection in children with ADHD. *Cortex* 41: 377–388
31. Steger J, Imhof K, Steinhausen H-C, Brandeis D (2000) Brain mapping of bilateral interactions in attention deficit hyperactivity disorder and control boys. *Clin Neurophysiol* 111: 1141–1156

32. Kanemura H, Aihara M, Aoki S, Araki T, Nakazawa S (2003) Development of the (pre-frontal lobe in infants and children: a three-dimensional magnetic resonance volumetric study. *Brain Dev* 25: 195–199
33. Huttenlocher PR, de Courten C, Garey LA, van der Loos H (1982) Synaptogenesis in human visual cortex - evidence for synaptic elimination during normal development. *Neurosci Lett* 33: 247–252
34. O'Donnell S, Noseworthy MD, Levine B, Dennis M (2005) Cortical thickness of the frontopolar area in typically developing children and adolescents. *Neuroimage* 24: 948–954
35. Blanton RE, Levitt JG, Peterson JR, Sporty ML, Lee M, To D, Mormino EC, Thompson PM, McCracken JT, Toga AW (2004) Gender differences in the left inferior frontal gyrus in normal children. *Neuroimage* 22: 626–636
36. Sowell ER, Jernigan TL (1998) Further MRI evidence of late brain maturation: limbic volume increases and changing asymmetries during childhood and adolescence. *Dev Neuropsychol* 14: 599–617
37. Sowell ER, Trauner DA, Gamst A, Jernigan TL (2002) Development of cortical and subcortical brain structures in childhood and adolescence: a structural MRI study. *Dev Med Child Neurol* 44: 4–16
38. Gogtay N, Giedd JN, Lusk L, Hayashi KM, Greenstein D, Vaituzis AC, Nugent TF, Herman DH, Clasen LS, Toga AW et al. (2004) Dynamic mapping of human cortical development during childhood through early adulthood. *Proc Natl Acad Sci (USA)* 101: 8174–8179
39. Wild-Wall N, Oades RD, Juran SA (2005) Maturation processes in automatic change detection as revealed by event-related brain potentials and dipole source localization: Significance for adult AD/HD. *Int J Psychophysiol* 58: 34–46
40. Bartzokis G, Nuechterlein KH, Gitlin M, Rogers S, Mintz J (2003) Dysregulated brain development in adult men with schizophrenia: a magnetic resonance imaging study. *Biol Psychiat* 53: 412–421
41. Lambert J, Bard C (2005) Acquisition of visuomanual skills and improvement of information processing capacities in 6- to 10-year-old children performing a 2D pointing task. *Neurosci Lett* 377: 1–6
42. Rival C, Ceyte H, Olivier I (2005) Developmental changes of static standing balance in children. *Neurosci Lett* 376: 133–136
43. Rueda MR, Fan J, McCandliss BD, Halparin JD, Gruber DB, Lercari LP, Posner MI (2004) Development of attentional networks in childhood. *Neuropsychologia* 42: 1029–1040
44. Spear L (2000) The adolescent brain and age-related behavioral manifestations. *Neurosci Biobehav Rev* 24: 417–463
45. Luna B, Garver KE, Urban TA, Lazar NA, Sweeney JA (2004) Maturation of cognitive processes from late childhood to adulthood. *Child Dev* 75: 1357–1372
46. Casey BJ, Trainor RJ, Orendi JL, Schubert AB, Nystrom LE, Giedd JN, Castellanos FX, Haxby JV, Noll DC, Cohen JD et al. (1997) A developmental functional MRI study of prefrontal activation during performance of a go-no-go task. *J Cog Neurosci* 9: 835–847
47. Ladouceur CD, Dahl RE, Carter CS (2004) ERP correlates of action monitoring in adolescence. *Ann NY Acad Sci* 1021: 329–336
48. Sharma A, Kraus N, McGee TJ, Nicol TG (1997) Developmental changes in P1 and N1 central auditory responses elicited by consonant-vowel syllables. *EEG Clin Neurophysiol* 104: 540–545
49. Marshall PJ, Bar-Haim Y, Fox NA (2003) The development of P50 suppression in the auditory event-related potential. *Int J Psychophysiol* 51: 135–141

50. Freedman R, Adler LE, Waldo MC (1987) Gating of the auditory evoked potential in children and adults. *Psychophysiol* 24: 223–227
51. Oades RD, Dittmann-Balcar A, Zerbin D (1997) Development and topography of auditory event-related potentials, mismatch and processing negativity from 8 to 22 years of age. *Psychophysiol* 34: 677–693
52. Albrecht R, von Suchodoletz W, Uwer R (2000) The development of auditory evoked dipole source activity from childhood to adulthood. *Clin Neurophysiol* 111: 2268–2276
53. Polich J, Ladish C, Burns T (1990) Normal variation of P300 in children: age, memory span, and head size. *Int J Psychophysiol* 9: 237–248
54. Enoki H, Sanada S, Yoshinaga H, Oka E, Ohtahara S (1993) The effects of age on the N200 component of the auditory event-related potentials. *Cogn Brain Res* 1: 161–167
55. Oades RD, Halliday GM (1987) The ventral tegmental (A 10) system. Neurobiology I: anatomy and connectivity. *Brain Res Rev* 12: 117–165
56. Lewis DA (2003) The catecholamine innervation of primate cerebral cortex. In: Solanto MV, Arnsten AFT, Castellanos FX (eds): *Stimulant drugs and ADHD: basic and clinical neuroscience*. Oxford: Oxford University Press, 77–103
57. Loughlin SE, Foote SL, Bloom FE (1986) Efferent projections of nucleus locus coeruleus: topographic organization of cells of origin demonstrated by three-dimensional reconstruction. *Neurosci* 18: 291–306
58. Svensson TH (2003) a-Adrenoceptor modulation hypothesis of antipsychotic atypicality. *Prog Neuropsychopharmacol Biol Psychiat* 27: 1145–1158
59. Oades RD (2005) The role of norepinephrine and serotonin in ADHD. In: Gozal D, Molfese DL (eds): *Attention deficit hyperactivity disorder: from genes to animal models to patients*. Tootawa, NY: Humana Press, 97–130
60. Lewis DA, Foote SL, Goldstein M, Morrison JH (1988) The dopaminergic innervation of monkey prefrontal cortex: a tyrosine hydroxylase immunohistochemical study. *Brain Res* 449: 225–243
61. Lucas G, Spampinato U (2000) Role of striatal serotonin$_{2A}$ and serotonin$_{2C}$ receptor subtypes in the control of the in vivo dopamine outfow in the rat striatum. *J Neurochem* 74: 693–701
62. Gobert A, Rivet J-M, Audinot V, Newman-Tancredi A, Cistarelli L, Millan MJ (1998) Simultaneous quantification of serotonin, dopamine and noradrenaline levels in single frontal cortex dialysates of freely-moving rats reveals a complex pattern of reciprocal auto- and heteroceptor-mediated control of release. *Neurosci* 84: 413–429
63. De Haes JI, Bosker FJ, Van Waarde A, Pruim J, Willemsen AT, Vaalburg W, Den Boer JA (2002) 5-HT1A receptor imaging in the human brain: Effect of tryptophan depletion and infusion on [18F]MPPF binding. *Synapse* 46: 108–115
64. Wright DE, Seroogy KB, Lundgren KH, Davis BM, Jennes L (1995) Comparative localization of serotonin 1A, 1C and 2 receptor subtype mRNAs in rat brain. *J Comp Neurol* 351: 357–373
65. Kalsbeek A (1989) *The role of dopamine in the development of the rat prefrontal cortex*. Krips Repro Meppel, Amsterdam (Acad. proefschrift)
66. Meng SZ, Ozawa Y, Itoh M, Takashima S (1999) Development and age-related changes of dopamine transporter, and dopamine D1 and D2 receptors in human basal ganglia. *Brain Res* 843: 136–144
67. Seeman P, Bzowej NH, Guan HC, Bergeron C, Becker LE, Reynolds GP, Bird ED, Riederer P, Jellinger K, Watanabe S, Tourtellotte WW (1987) Human brain dopamine receptors in children and aging adults. *Synapse* 1: 399–404
68. Andersen SL, Teicher MH (2000) Sex differences in dopamine receptors and their relevance to ADHD. *Neurosci Biobehav Rev* 24: 137–141

69. Rodriguez M, Martin L, Santana C (1994) Ontogenic development of brain asymmetry in dopaminergic neurons. *Brain Res Bull* 33: 163–171
70. Lambe EK, Krimer LS, Goldman-Rakic PS (2000) Differential postnatal development of catecholamine and serotonin inputs to identified neurons in prefrontal cortex of Rhesus monkey. *J Neurosci* 20: 8780–8787
71. Oades RD, Röpcke B, Schepker R (1996) A test of conditioned blocking and its development in childhood and adolescence: relationship to personality and monoamines metabolism. *Dev Neuropsychol* 12: 207–230
72. Verney C, Milosevic A, Alvarez C, Berger B (1993) Immunocytochemical evidence of well-developed dopaminergic and noradrenergic innervations in the frontal cerebral cortex of human fetuses at midgestation. *J Comp Neurol* 336: 331–344
73. Zecevic N, Verney C (1995) Development of the catecholamine neurons in human embryos and fetuses, with special emphasis on the innervation of the cerebral cortex. *J Comp Neurol* 351: 509–535
74. Tomasini R, Kema IP, Muskiet FAJ, Meiborg G, Staal MJ, Go KG (1997) Catecholaminergic development of fetal ventral mesencephalon: characterization by high-performance liquid chromatography with electrochemical detection and immunohistochemistry. *Exp Neurol* 145: 434–441
75. Konradi C, Kornhuber J, Sofic E, Heckers S, Riederer P, Beckmann H (1992) Variations of monoamines and their metabolites in the human brain putamen. *Brain Res* 579: 285–290
76. Neddens J, Dawirs RR, Bagorda F, Busche A, Horstmann S, Teuchert-Noodt G (2004) Postnatal maturation of cortical serotonin lateral asymmetry in gerbils is vulnerable to both environmental and pharmacological epigenetic challenges. *Brain Res* 1021: 200–208
77. Arnsten AFT (2006) Noradrenergic actions in prefrontal cortex: Relevance to ADHD. In: Oades RD (ed) *Attention-deficit/hyperactivity disorder and the hyperkinetic syndrome: current ideas and ways forward.* Hauppauge, New York: Nova Science Publishers Inc., 109–129
78. Beane M, Marrocco RT (2005) Norepinephrine and acetylcholine mediation of the components of reflexive attention: implications for attention deficit disorders. *Prog Neurobiol* 74: 167–181
79. Oades RD (1985) The role of noradrenaline in tuning and dopamine in switching between signals in the CNS. *Neurosci Biobehav Rev* 9: 261–283
80. Rajkowski J, Majczynski H, Clayton E, Aston-Jones GS (2004) Activation of monkey locus coeruleus neurons varies with difficulty and performance in a target detection task. *J Neurophysiol* 92: 361–371
81. Carli M, Samanin R (2000) The 5-HT_{1A} receptor agonist 8-OH-DPAT reduces rats' accuracy of attentional performance and enhances impulsive responding in a five-choice serial reaction time task: role of presynaptic 5-HT_{1A} receptors. *Psychopharmacol* 149: 259–268
82. Dalley JW, Theobald DE, Eagle DM, Passetti F, Robbins TW (2002) Deficits in impulse control associated with tonically-elevated serotonergic function in rat prefrontal cortex. *Neuropsychopharmacol* 26: 716–728
83. Kavoussi R, Armstead P, Coccaro EF (1997) The neurobiology of impulsive aggression. *Psychiat Clin N Am* 20: 395–403
84. Winter JC, Eckler JR, Doat MM, Rabin RA (2002) The effects of acute and subchronic treatment with fluoxetine and citalopram on stimulus control by DOM. Pharmacol *Biochem Behav* 74: 95–101

85. Aglioti S, Smania N, Barbieri C, Corbetta M (1997) Influence of stimulus salience and attentional demands on visual search patterns in hemispatial neglect. *Brain Cogn* 34: 388–403
86. Malapani C, Pillon B, Dubois B, Agid Y (1994) Impaired simultaneous cognitive task performance in Parkinson's disease: a dopamine-related dysfunction. *Neurol* 44: 319–326
87. Oades RD (1997) Stimulus dimension shifts in patients with schizophrenia, with and without paranoid hallucinatory symptoms, or obsessive compulsive disorder: strategies, blocking and monoamine status. *Behav Brain Res* 88: 115–132
88. McCormick DA, Pape HC, Williamson A (1991) Actions of norepinephrine in the cerebral cortex and thalamus: implications for function of the central noradrenergic system. *Prog Brain Res* 88: 293–305
89. Devoto P, Flore G, Pira L, Longu G, Gessa GL (2004) Mirtazapine-induced corelease of dopamine and noradrenaline from noradrenergic neurons in the medial prefrontal and occipital cortex. *Eur J Pharmacol* 487: 105–111
90. Lanau F, Zenner MT, Civelli O, Hartman DS (1997) Epinephrine and norepinephrine act as potent agonists at the recombinant human dopamine D4 receptor. *J Neurochem* 68: 804–812
91. Auclair A, Cotecchia S, Glowinski J, Tassin J-P (2002) D-amphetamine fails to increase extracellular dopamine levels in mice lacking alpha 1b-adrenergic receptors: relationship between functional and nonfunctional dopamine release. *J Neurosci* 22: 9150–9154
92. Sonuga-Barke EJS (2005) Causal models of attention-deficit/hyperactivity disorder: from common simple deficits to multiple developmental pathways. *Biol Psychiat* 57: 1231–1238
93. Sergeant JA, Oosterlaan J, van der Meere JJ (1999) Information processing in attention-deficit/hyperactivity disorder. In: Quay HC, Hogan AE (eds): *Handbook of disruptive behavior disorders*. Plenum Press, New York, 75–104
94. Barkley RA (1997) Behavioral inhibition, sustained attention and executive functions: constructing a unifying theory of ADHD. *Psychol Bull* 121: 65–94
95. Quay HC (1997) Inhibition and attention deficit hyperactivity disorder. *J Abnorm Child Psychol* 25: 7–13
96. Sagvolden T, Johansen EB, Aase H, Russell VA (2005) A dynamic developmental theory of attention-deficit/hyperactivity disorder (ADHD) predominantly hyperactive/impulsive and combined subtypes. *Behav Brain Sci* 28: 397–468
97. Luman M, Oosterlaan J, Sergeant JA (2005) The impact of reinforcement contingencies on AD/HD: A review and theoretical appraisal. *Clin Psychol Rev* 25: 183–213
98. Casey BJ, Castellanos FX, Giedd JN, Marsh WL, Hamburger SD, Schubert AB, Vauss YC, Vaituzis AC, Dickstein DP, Sarfatti SE, Rapoport JL (1997) Implication of right frontostriatal circuitry in response inhibition and attention-deficit / hyperactivity disorder. *J Am Acad Child Adolesc Psychiat* 36: 374–383
99. Rubia K, Overmeyer S, Taylor EA, Brammer MJ, Williams SCR, Simmons A, Bullmore ET (1999) Hypofrontality in attention deficit hyperactivity disorder during higher-order motor control: a study with functional MRI. *Am J Psychiat* 156: 891–896
100. Smith AB, Taylor EA, Brammer M, Rubia K (2004) Neural correlates of switching set as measured in fast, event-related functional magnetic resonance imaging. *Hum Brain Mapp* 21: 247–56
101. Rubia K, Smith AB, Brammer MJ, Toone B, Taylor E (2005) Abnormal brain activation during inhibition and error detection in medication-naive adolescents with ADHD. *Am J Psychiat* 162: 1067–1075
102. Vaidya CJ, Bunge SA, Dudukovic NM, Zalecki CA, Elliiott GR, Gabrieli DE (2005) Altered neural substrates of cognitive control in childhood ADHD: Evidence from functional magnetic resonance imaging. *Am J Psychiat* 162: 1605–1613

103. Bunge SA, Dudokovic NM, Thomason MA, Vaidya CJ, Gabrieli JDE (2002) Immature frontal lobe contributions to cognitive control in children: evidence from fMRI. *Neuron* 17: 301–311
104. Aron AR, Fletcher PC, Bullmore ET, Sahakian BJ, Robbins TW (2003) Stop-signal inhibition disrupted by damage to right inferior frontal gyrus in humans. *Nature Neurosci* 6: 115–116
105. Bush G, Frazier JA, Seidman LJ, Whalen PJ, Jenike MA, Rosen BR, Biederman J (1999) Anterior cingulate cortex dysfunction in attention-deficit/hyperactivity disorder revealed by fMRI and the Counting Stroop. *Biol Psychiat* 45: 1542–1552
106. Durston S (2003) A review of the biological bases of ADHD: what have we learned from imaging studies? *Ment Retard Dev Disabil Res Rev* 9: 184–195
107. Durston S, Hulshoff Pol HE, Schnack HG, Buitelaar J, Steenhuis MP, Minderaa RB, Kahn RS, van Engeland H (2004) Magnetic resonance imaging of boys with attention-deficit/hyperactivity disorder and their unaffected siblings. *J Am Acad Child Adolesc Psychiat* 43: 332–340
108. Castellanos FX, Giedd JN, Marsh WL, Hamburger SD, Vaituzis AC, Dickstein DP, Sarfatti SE, Vauss YC, Snell JW, Lange N et al. (1996) Quantitative brain magnetic resonance imaging in attention-deficit hyperactivity disorder. *Arch Gen Psychiat* 53: 607–616
109. Aron AR, Monsell S, Sahakian BJ, Robbins TW (2004) A componential analysis of task-switching deficits associated with lesions of left and right frontal cortex. *Brain* 127: 1561–1573
110. Aron AR, Dowson JH, Sahakian BJ, Robbins TW (2003) Methylphenidate improves response inhibition in adults with attention-deficit/hyperactivity disorder. *Biol Psychiat* 54: 1465–1468
111. Cepeda NJ, Cepeda ML, Kramer AF (2000) Task switching and attention deficit hyperactivity disorder. *J Abnorm Child Psychol* 28: 213–226
112. Kramer AF, Cepeda NJ, Cepeda ML (2001) Methylphenidate effects on task switching performance in attention-deficit/hyperactivity disorder. *J Am Acad Child Adolesc Psychiat* 40: 1277–1284
113. Schweitzer JB, Faber TL, Grafton ST, Tune LE, Hoffman JM, Kilts CD (2000) Alterations in the functional anatomy of working memory in adult attention deficit hyperactivity disorder. *Am J Psychiat* 157: 278–280
114. Ernst M, Kimes AS, London ED, Matochik JA, Eldreth D, Tata S, Contoreggi C, Leff M, Bolla K (2003) Neural substrates of decision making in adults with attention deficit hyperactivity disorder. *Am J Psychiat* 160: 1061–1070
115. Schweitzer JB, Lee DO, Hanford RB, Zink CF, Ely TD, Tagamets MA, Hoffman JM, Grafton ST, Kilts CD (2004) Effect of methylphenidate on executive functioning in adults with attention-deficit/hyperactivity disorder: Normalization of behavior but not related brain activity. *Biol Psychiat* 56: 597–606
116. Volkow ND, Fowler JS, Wang GJ, Ding YS, Gatley SJ (2002) Role of dopamine in the therapeutic and reinforcing effects of methylphenidate in humans: results from imaging studies. *Eur Neuropsychopharmacol* 12: 557–566
117. Volkow ND, Wang G-J, Fowler JS, Telang F, Maynard L, Logan J, Gatley SJ, Pappas N, Wong C, Vaska P, Zhu W, Swanson JM (2004) Evidence that methylphenidate enhances the saliency of a mathematical task by increasing dopamine in the human brain. *Am J Psychiat* 161: 1173–1180
118. Volkow ND, Wang G-J, Ma Y, Fowler JS, Wong C, Ding Y-S, Hitzemann RJ, Swanson JM, Kalivas PW (2005) Activation of orbital and medial prefrontal cortex by methylphenidate

in cocaine-addicted subjects but not in controls: relevance to addiction. *J Neurosci* 25: 3932–3939

119. Volkow ND, Wang G-Y, Fowler JS, Hitzemann RJ, Gatley J, Ding Y-S, Wong C, Pappas N (1998) Differences in regional brain metabolic responses between single and repeated doses of methylphenidate. *Psychiat Res (Neuroimaging)* 83: 29–36
120. Volkow ND, Logan J, Fowler JS, Wang G-J, Gur RC, Wong C, Felder C, Gatley J, Ding Y-S, Hitzemann RJ, Pappas N (2000) Association between age-related decline in brain dopamine activity and impairment in frontal and cingulate metabolism. *Am J Psychiat* 157: 75–80
121. Ernst M, Zametkin AJ, Matochik JA, Jons PH, Cohen RM (1998) DOPA decarboxylase activity in attention deficit hyperactivity disorder adults.A [fluorine-18] fluorodopa positron emission tomography study. *J Neurosci* 18: 5901–5907
122. Ernst M, Zametkin AJ, Matochik JA, Pascualvaca D, Jons PH, Cohen RM (1999) High midbrain [18F]DOPA accumulation in children with attention deficit hyperactivity disorder. *Am J Psychiat* 156: 1209–1215
123. Bellgrove MA, Domschke K, Hawi Z, Kirley A, Mullins C, Robertson IH, Gill M (2005) The methionine allele of the COMT polymorphism impairs prefrontal cognition in children and adolescents with ADHD. *Exp Brain Res* 163: 352–360
124. Mullins C, Bellgrove MA, Gill M, Robertson IH (2005) Variability in time reproduction: difference in ADHD combined and inattentive subtypes. *J Am Acad Child Adolesc* Psychiat 44: 169–176
125. Sunohara GA, Roberts W, Malone MA, Schachar RJ, Tannock R, Basile VS, Wigal T, Wigal SB, Schuck S, Moriarty J et al. (2000) Linkage of the dopamine D4 receptor gene and attention-deficit/hyperactivity disorder. *J Am Acad Child Adolesc Psychiat* 39: 1537–1542
126. Leung PWL, Lee CC, Hung SF, Ho TP, Tang CP, Kwong SL, Leung SY, Yuen ST, Lieh-Mak F, Oosterlaan J et al. (2005) Dopamine receptor D4 (DRD4) gene in Han chinese children with attention-deficit/hyperactivity disorder (ADHD): increased prevalence of the 2-repeat allele. *Am J Med Genet* 133B: 54–56
127. Swanson JM, Oosterlaan J, Murias M, Schuck S, Spence AA, Wasdell M, Ding Y, Chi H-C, Smith M, Mann M et al. (2001) Attention deficit/hyperactivity disorder children with a 7-repeat allele of the dopamine receptor D4 gene have extreme behavior but normal performance on criticl neuropsychological tests of attention. *Proc Natl Acad Sci (USA)* 97: 4754–4759
128. Durston S, Fossella JA, Casey BJ, Hulshoff Pol HE, Galvan A, Schnack HG, Steenhuis MP, Minderaa RB, Buitelaar JK, Kahn RS, van Engeland H (2005) Differential effects of DRD4 and DAT1 genotype on fronto-striatal gray matter volumes in a sample of subjects with attention deficit hyperactivity disorder, their unaffected siblings, and controls. *Mol Psychiat* 10: 678–685
129. Durston S, Fossella JA, Casey BJ (2006) Neuroimaging as an approach to the Neurobiology of ADHD. In: Oades RD (ed) *Attention-deficit/hyperactivity disorder and the hyperkinetic syndrome: current ideas and ways forward*. Hauppauge, New York: Nova Science Publishers Inc., 173–184
130. Burnette WB, Bailey MD, Kukoyi S, Blakely RD, Trowbridge CG, Justice JB (1996) Human norepinephrine transporter kinetics using rotating disk electrode voltammetry. *Anal Chem* 68: 2932–2938
131. Devoto P, Flore, Pani L, Gessa GL (2001) Evidence for co-release of noradrenaline and dopamine from noradrenergic neurons in the cerebral cortex. *Mol Psychiat* 6: 657–664
132. Biederman J, Spencer T (1999) Attention-deficit/hyperactivity disorder (ADHD) as a noradrenergic disorder. *Biol Psychiat* 46: 1234–1242

133. Madras BK, Miller GM, Fischman AJ (2005) The dopamine transporter and attention-deficit/hyperactivity disorder. *Biol Psychiat* 57: 1397–1409
134. Bobb AJ, Addington AM, Sidransky E, Gornick MC, Lerch JP, Greenstein DK, Clasen LS, Sharp WS, Inoff-Germain G, Wavrant-De Vrie'ze F et al. (2005) Support for Association Between ADHD and Two Candidate Genes: NET1 and DRD1. *Am J Med Genet* 134B: 67–72
135. Yang L, Wang Y-F, Li JMS, Faraone SV (2004) Association of norepinephrine transporter gene with methylphenidate response. *J Am Acad Child Adolesc Psychiat* 43: 1154–1158
136. Xu X, Knight J, Brookes K, Mill J, Sham P, Craig I, Taylor E, Asherson P (2005) DNA pooling analysis of 21 norepinephrine transporter gene SNPs with attention deficit hyperactivity disorder: no evidence for association. *Am J Med Genet* 134B: 115–118
137. De Luca V, Muglia P, Jani U, Kennedy JL (2004) No evidence of linkage or association between the norepinephrine transporter (NET) gene MnlI polymorphism and adult ADHD. *Am J Med Genet* 124B: 38–40
138. Barr CL, Kroft J, Feng Y, Wigg K, Roberts W, Malone M, Ickowicz A, Schachar RJ, Tannock R, Kennedy JL (2002) The norepinephrine transporter gene and attention-deficit hyperactivity disorder. *Am J Med Genet* 114: 255–259
139. Michelson D, Adler L, Spencer T, Reimherr FW, West SA, Allen AJ, Kelsey D, Wernicke J, Dietrich A, Milton DR (2003) Atomoxetine in adults with ADHD: two randomized, placebo-controlled studies. *Biol Psychiat* 53: 211–220
140. Roman T, Schmitz M, Polanczyk GV, Eizirik M, Rohde LA, Hutz MH (2003) Is the α-2a adrenergic receptor gene (ADRRA2A) associated with attention-deficit/hyperactivity disorder? *Am J Med Genet* 120B: 116–120
141. Hawi Z, Lowe N, Kirley A, Nöthen M, Greenwood T, Kelsoe J, Fitzgerald M, Gill M (2003) Linkage disequilibrium mapping at DAT1, DRD5 and DBH narrows the search for ADHD susceptibility alleles at these loci. *Mol Psychiat* 8: 299–308
142. Park L, Nigg JT, Waldman ID, Nummy KA, Huang-Pollock C, Rappley M, Friderici KH (2005) Associations and linkage of α-2A adrenergic receptor gene polymorphisms with childhood ADHD. *Mol Psychiat* 10: 572–580
143. Wang M, Tang ZX, Li BM (2004) Enhanced visuomotor associative learning following stimulation of alpha 2A-adrenoceptors in the ventral prefrontal cortex in monkeys. *Brain Res* 1024: 176–182
144. Wang M, Ji JZ, Li BM (2004) The alpha(2A)-adrenergic agonist guanfacine improves visuomotor associative learning in monkeys. *Neuropsychopharmacol* 29: 86–92
145. Ma CL, Qi XL, Peng JY, Li BM (2003) Selective deficit in no-go performance induced by blockade of prefrontal cortical alpha 2-adrenoceptors in monkeys. *NeuroReport* 14: 1013–1016
146. Ma C-L, Arnsten AFT, Li B-M (2005) Locomotor hyperactivity induced by blockade of prefrontal cortical a_2-adrenoceptors in monkeys. *Biol Psychiat* 57: 192–195
147. Clarke AR, Barry RJ, McCarthy R, Selikowitz M (2006) EEG predictors of good response to imipramine hydrochloride in children with attention deficit/hyperactivity disorder. In: Oades RD (ed): *Attention-deficit/hyperactivity disorder and the hyperkinetic syndrome: current ideas and ways forward*. Hauppauge, New York: Nova Science Publishers, Inc., 249–267
148. Hermens DF, Williams LM, Clarke S, Kohn M, Cooper N, Gordon E (2005) Responses to methylphenidate in adolescent AD/HD. Evidence from concurrently recorded autonomic (EDA) and central (EEG and ERP) measures. *Int J Psychophysiol* 58: 21–33
149. Sangal RB, Sangal JM (2005) Attention-deficit/hyperactivity disorder: cognitive evoked potential (P300) amplitude predicts treatment response to atomoxetine. *Clin Neurophysiol* 116: 640–647

150. Klorman R, Brumaghim JT (1991) Stimulant drugs and ERPs. *EEG Clin Neurophysiol Suppl* 42: 135–141
151. Seifert J, Scheuerpflug P, Zillessen K-E, Fallgatter AJ, Warnke A (2003) Electrophysiological investigation of the effectiveness of methylphenidate in children with and without ADHD. *J Neur Trans* 110: 821–829
152. Lazzaro I, Anderson J, Gordon E, Clarke S, Leong J, Meares R (1997) Single trial varaibility within the P300 (250-500 ms) processing window in adolescents with attention defict hyperactivity disorder. *Psychiat Res* 73: 91–101
153. Jonkman LM, Kemner C, Verbaten MN, Koelega HS, Camfferman G, van der Gaag R-J, Buitelaar JK, van Engeland H (1997) Effects of methylphenidate on event-related potentials and performance of attention-deficit hyperactivity disorder children in auditory and visual selective attention tasks. *Biol Psychiat* 41: 690–702
154. Jonkman LM, Kemner C, Verbaten MN, van Engeland H, Camfferman G, Buitelaar JK, Koelega HS (2000) Attentional capacity, a probe ERP study: differences between children with attention-deficit hyperactivity disorder and normal control children and effects of methylphenidate. *Psychophysiol* 37: 334–346
155. Prichep LS, Sutton S, Hakerem G (1976) Evoked potentials in hyperkinetic and normal children under certainty and uncertainty: a placebo and methylphenidate study. *Psychophysiol* 13: 419–428
156. Broyd SJ, Johnstone SJ, Barry RJ, Clarke AR, McCarthy R, Selikowitz M, Lawrence CA (2005) The effect of methylphenidate on response inhibition and the event-related potential of children with attention deficit/ hyperactivity disorder. *Int J Psychophysiol* 58: 47–58
157. Gehlert DR, Schober DA, Hemrick-Luecke SK, Kushinski J, Howbert JJ, Robertson DW, Fuller RW, Wong DT (1995) Novel halogenated analogs of tomoxetine that are potent and selective inhibitors of norepinephrine uptake in brain. *Neurochem Int* 26: 47–52
158. Nocjar C, Roth BL, Pehek EA (2002) Localization of 5-HT(2A) receptors on dopamine cells in subnuclei of the midbrain A10 cell group. *Neurosci* 111: 163–176
159. Di Giovanni G, Di Matteo V, Di Mascio M, Esposito E (2000) Preferential modulation of mesolimbic vs. nigrostriatal dopaminergic function by serotonin$_{2c/2b}$ receptor agonists: a combined in vivo electrophysiological and microdialysis study. *Synapse* 35: 53–61
160. Di Matteo V, Cacchio M, Di Giulio C, Esposito E (2002) Role of serotonin (2C) receptors in the control of brain dopaminergic function. *Pharmacol Biochem Behav* 71: 727–734
161. Hutson PH, Barton CL, Jay M, Blurton P, Burkamp F, Clarkson R, Bristow LJ (2000) Activation of mesolimbic dopamine function by phencyclidine is enhanced by 5-$HT_{2C/2B}$ receptor antagonists: neurochemical and behavioural studies. *Neuropharmacol* 39: 2318–2328
162. Castellanos FX, Elia J, Kruesi MJP, Marsh WL, Gulotta CS, Potter WZ, Ritchie GF, Hamburger SD, Rapoport JL (1996) Cerebrospinal fluid homovanillic acid predicts behavioral response to stimulants in 45 boys with attention deficit/hyperactivity disorder. *Neuropsychopharmacol* 14: 125–137
163. Rogers RD, Blackshaw AJ, Middleton HC, Matthews K, Hawtin K, Crowley C, Hopwood A, Wallace C, Deakin JFW, Sahakian BJ, Robbins TW (1999) Tryptophan depletion impairs stimulus reward learning while methylphenidate disrupts attentional control in healthy young adults: implications for the monoaminergic basis of impulsive behaviour. *Psychopharmacol* 146: 482–491
164. Oades RD, Müller BW (1997) The development of conditioned blocking and monoamine metabolism in children with attention-deficit-hyperactivity disorder or complex tics and healthy controls: an exploratory analysis. *Behav Brain Res* 88: 95–102

165. Oades RD (2002) Dopamine may be 'hyper' with respect to noradrenaline metabolism, but "hypo" with respect to serotonin metabolism in children with ADHD. *Behav Brain Res* 130: 97–101
166. Rubia K, Lee F, Cleare AJ, Tunstall N, Fu CHY, Brammer M, McGuire PK (2004) Tryptophan depletion reduces right inferior prefrontal activation during no-go trials in fast, event-related fMRI. *Psychopharmacol* 179: 791–803
167. Castellanos FX, Elia J, Kruesi MJP, Gulotta CS, Mefford IN, Potter WZ, Ritchie GF, Rapoport JL (1994) Cerebrospinal fluid monoamine metabolites in boys with attention-deficit hyperactivity disorder. *Psychiat Res* 52: 305–316
168. Hegerl U (1998) Event-related potentials and clinical response to serotonin agonists in patients with affective disorders. *Eur Arch Psychiat clin Neurosci* 248 (Suppl. 2): S75
169. Nathan PD, O'Neill B, Croft RJ (2005) Is the loudness dependence of the auditory evoked potential a sensitive and selective *in vivo* marker of central serotonergic function? *Neuropsychopharmacol* 30: 1584–1585
170. Gallinat J, Stroehle A, Lang UE, Bajbouj M, Kalus P, Montag C, Seifert F, Wernicke C, Rommelspacher H, Rinneberg H, Schubert F (2005) Association of human hippocampal neurochemistry, serotonin transporter genetic variation, and anxiety. *Neuroimage* 26: 123–131
171. Carrilo-de-la-Pena MT (2001) One year test-retest reliability of auditory evoked potentials (AEEPs) to tones of increasing intensity. *Psychophysiol* 38: 417–424
172. Dykman RA, Holcomb PJ, Ackerman PT, McCray DS (1983) Auditory ERP augmentation-reduction and methylphenidate dosage needs in attention and reading disordered children. *Psychiat Res* 9: 255–269
173. Bruneau N, Barthelemy C, Roux S, Jouve J, Lelord G (1989) Auditory evoked potential modifications according to clinical and biochemical responsiveness to fenfluramine treatment in children with autistic behavior. *Neuropsychobiol* 21: 48–52
174. Jemel B, Achenbach C, Müller B, Röpcke B, Oades RD (2002) Mismatch negativity results from bilateral asymmetric dipole sources in the frontal and temporal lobes. *Brain Topogr* 15: 13–27
175. Fallgatter AJ, Herrmann MJ, Roemmler J, Ehlis A-C, Wagener A, Heidrich A, Ortega G, Zeng Y, Lesch KP (2005) Allelic variation of serotonin transporter function modulates the brain electrical response for error processing. *Neuropsychopharmacol* 29: 1506–1511
176. Curran S, Purcell S, Craig I, Asherson P, Sham P (2005) The serotonin transporter gene as a QTL for ADHD. *Am J Med Genet* 134B: 42–47
177. Lakatos K, Nemoda Z, Birkas E, Ronai Z, Kovacs E, Ney K, Toth I, Sasvari-Szekely M, Gervai J (2003) Association of D4 dopamine receptor gene and serotonin transporter promoter polymorphisms with infants' response to novelty. *Mol Psychiat* 8: 90–97
178. Seeger G, Schloss P, Schmidt MH (2001) Marker gene polymorphisms in hyperkinetic disorder – predictors of clinical response to treatment with methylphenidate? *Neurosci Lett* 313: 45–48
179. Uzbekov MG (2006) Hyperkinetic syndrome as a manifestation of a disturbance of metabolism and mental development. In: Oades RD (ed): *Attention-deficit/hyperactivity disorder and the hyperkinetic syndrome: current ideas and ways forward*. Hauppauge, New York: Nova Science Publishers, Inc., 133–154
180. David SP, Murthy NV, Rabiner EA, Munafo MR, Johnstone EC, Jacob R, Walton RT, Grasby PM (2005) A functional genetic variation of the serotonin (5-HT) transporter affects 5-HT_{1A} receptor binding in humans. *J Neurosci* 25: 2586–2590
181. Bantick RA, de Vries MH, Grasb PM (2005) The effect of a 5-HT_{1A} receptor agonist on striatal dopamine release. *Synapse* 57: 67–75

182. Martin-Ruiz R, Puig MV, Celada P, Shapiro DA, Roth BL, Mengod G, Artigas F (2001) Control of serotonergic function in medial prefrontal cortex by serotonin-2A receptors through a glutamate-dependent mechanism. *J Neurosci* 21: 9856–9866
183. Ferre S, Artigas F (1993) Dopamine D2 receptor-mediated regulation of serotonin extracellular concentration in the dorsal raphe nucleus of freely moving rats. *J Neurochem* 61: 772–775
184. Reneman L, De Bruin K, Lavalaye J, Guning WB, Booij J (2001) Addition of a 5-HT receptor agonist to methylphenidate potentiates the reduction of [^{123}I]FP-CIT binding to dopamine transporter in rat frontal cortex and hippocampus. *Synapse* 39: 193–200
185. Fleckenstein AE, Hanson GR (2003) Impact of psychostimulants on vesicular monoamine transporter function. *Eur J Pharmacol* 479: 283–289
186. Truong JG, Newman AH, Hanson GR, Fleckenstein AE (2004) Dopamine D2 receptor activation increases vesicular dopamine uptake and redistributes vesicular monoamine transporter-2 protein. *Eur J Pharmacol* 504: 27–32
187. Oades RD, Sadile AG, Sagvolden T, Viggiano D, Zuddas A, Devoto P, Aase H, Johansen EB, Ruocco LA, Russell VA (2005) The control of responsiveness in ADHD by catecholamines: evidence for dopaminergic, noradrenergic, and interactive roles. *Dev Sci* 8: 122–131
188. Sonuga-Barke EJS, Williams E, Hall M, Saxton T (1996) Hyperactivity and delay aversion. III: The effect on cognitive style of imposing delay after errors. *J Child Psychol Psychiat* 37: 189–194
189. Neef NA, Bicard DF, Endo S (2001) Assessment of impulsivity and the development of self-control in students with attention deficit hyperactivity disorder. *J Appl Behav Anal* 34: 397–408
190. Kuntsi J, Oosterlaan J, Stevenson J (2001) Psychological mechanisms in hyperactivity; I Response inhibition deficit, working memory impairment, delay aversion, or something else? *J Child Psychol Psychiat* 42: 199–210
191. Tripp G, Alsop B (2001) Sensitivity to reward delay in children with attention deficit hyperactivity disorder (ADHD). *J Child Psychol Psychiat* 42: 691–698
192. Winstanley CA, Theobald DEH, Cardinal RN, Robbins TW (2004) Contrasting roles of basolateral amygdala and orbitofrontal cortex in impulsive choice. *J Neurosci* 24: 4718–4722
193. Solanto MV, Abikoff H, Sonuga-Barke EJS, Schachar RJ, Logan GD, Wigal T, Hechtman L, Hinshaw S, Turkel E (2001) The ecological validity of delay aversion and response inhibition as measures of impulsivity in AD/HD: a supplement to the NIMH multimodal treatment study of AD/HD. *J Abnorm Child Psychol* 29: 215–228
194. Schultz W (2002) Getting formal with dopamine and reward. *Neuron* 36: 241–263
195. Oades RD (1999) Dopamine: Go/No-Go motivation vs. switching. Commentary on Depue & Collins "Neurobiology of the structure of personality: dopamine, facilitation of incentive motivation and extraversion." *Behav Brain Sci* 22: 532–533
196. Harris GC, Wimmer M, Aston-Jones GS (2005) A role for lateral hypothalamic orexin neurons in reward seeking. *Nature* 437: 556–559
197. Federici M, Geracitano R, Bernardi G, Mercuri NB (2005) Actions of methylphenidate on dopaminergic neurons of the ventral midbrain. *Biol Psychiat* 57: 361–365
198. Grace AA (2001) Psychostimulant actions on dopamine and limbic system function: relevance to the pathophysiology and treatment of ADHD. In: Solanto MV, Arnsten AFT, Castellanos FX (eds) *Stimulant drugs and ADHD: basic and clinical neuroscience*. Oxford University Press. Oxford, 134–157
199. Krause K-H, Dresel SH, Krause J, Kung HF, Tatsch K (2000) Increased striatal dopamine transporter in adult patients with attention deficit hyperactivity disorder: effects of

methylphenidate as measured by single photon emission computed tomography. *Neurosci Lett* 285: 107–110
200. Volkow ND, Wang G-J, Fowler JS, Logan J, Gerasimov M, Maynard L, Ding Y-S, Gatley SJ, Gifford A, Franceschi D (2001) Therapeutic doses of oral methylphenidate significantly increase extracellular dopamine in the human brain. *J Neurosci* 21: RC121 (1–5)
201. Gerasimov MR, Franceschi M, Volkow ND, Gifford A, Gatley SJ, Marsteller D, Molina PE, Dewey SL (2000) Comparison between intraperitoneal and oral methylphenidate administration: a microdialysis and locomotor study. *J Pharmacol Exp Ther* 295: 51–57
202. Volkow ND, Wang G-J, Fowler JS, Ding Y-S (2005) Imaging the effects of methylphenidate on brain dopamine: new model on its therapeutic actions for attention-deficit/hyperactivity disorder. *Biol Psychiat* 57: 1410–1415
203. Winsberg BG, Comings DE (1999) Association of the dopamine transporter gene (DAT1) with poor methylphenidate response. *J Am Acad Child Adolesc Psychiat* 38: 1474–1477
204. Roman T, Szobot C, Martins S, Biederman J, Rohde LA, Hutz MH (2002) Dopamine transporter gene and response to methylphenidate in attention-deficit/hyperactivity disorder. *Pharmacogenet* 12: 497–499
205. Rohde LA, Roman T, Szobot C, Cunha RD, Hutz MH, Biederman J (2003) Dopamine transporter gene, response to methylphenidate and cerebral blood flow in attention-deficit/hyperactivity disorder: a pilot study. *Synapse* 48: 87–89
206. Loo SK, Specter E, Smolen A, Hopfer C, Teale PD, Reite ML (2003) Functional effects of the DAT1 polymorphism on EEG measures in ADHD. *J Am Acad Child Adolesc Psychiat* 42: 986–993
207. Cheon K-A, Ryu Y-H, Kim J-W, Cho D-Y (2004) The homozygosity for 10-repeat allele at dopamine transporter gene and dopamine transporter density in Korean children with attention deficit hyperactivity disorder: relating to treatment response to methylphenidate. *Eur Neuropsychopharmacol* 15: 95–101
208. Bellgrove MA, Hawi Z, Kirley A, Futzgerald M, Gill M, Robertson IH (2005) Association between dopamine transporter (DAT1) genotype, left-sided inattention, and an enhanced response to methylphenidate in attention-deficit hyperactivity disorder. *Neuropsychopharmacol* 30: 2290–2297
209. Wienbruch C, Paul I, Bauer S, Kivelitz H (2005) The influence of methylphenidate on the power spectrum of ADHD children – an MEG study. *BMC Psychiatry* 5: 29
210. Cornish KM, Manly T, Savage R, Swanson J, Morisano D, Butler N, Grant C, Cross G, Bentley L, Hollis CP (2005) Association of the dopamine transporter (DAT1) 10/10-repeat genotype with ADHD symptoms and response inhibition in a general population sample. *Mol Psychiat* 10: 686–698
211. Simseka M, Al-Sharbatib M, Al-Adawib S, Gangulyc SS, Lawatiaa K (2005) Association of the risk allele of dopamine transporter gene (DAT1*10) in Omani male children with attention-deficit hyperactivity disorder. *Clin Biochem* 38: 739–742
212. Galvan A, Hare TA, Davidson M, Spicer J, Glover G, Casey BJ (2005) The role of ventral frontostriatal circuitry in reward-based Learning in humans. *J Neurosci* 25: 8650–8656
213. Sagvolden T, Metzger MA, Schiorbeck HK, Rugland A-L, Spinnangr I, Sagvolden G (1992) The spontaneously hypertensive rat (SHR) as an animal model of childhood hyperactivity (ADHD): changed reactivity to reinforcers and to psychomotor stimulants. *Behav Neur Biol* 58: 103–112
214. Pietras CJ, Cherek DR, Lan SD, Tcheremissine OV, Steinberg JL (2004) Effects of methylphenidate on impulsive choice in adult humans. *Psychopharmacol* 170: 390–398
215. Volkow ND, Wang GJ, Fowler JS, Logan J, Jayne M, Franceschi D, Wong C, Gatley SJ, Gifford AN, Ding YS, Pappas N (2002) "Nonhedonic" food motivation in humans

involves dopamine in the dorsal striatum and methylphenidate amplifies this effect. *Synapse* 44: 175–180
216. Bolanos CA, Barrot M, Berton O, Wallace-Black D, Nestler EJ (2003) Methylphenidate treatment during pre- and periadolescence alters behavioral responses to emotional stimuli at adulthood. *Biol Psychiat* 54: 1317–1329
217. Mateo Y, Budygin EA, John CE, Jones SR (2004) Role of serotonin in cocaine effects in mice with reduced dopamine transporter function. *Proc Natl Acad Sci (USA)* 101: 372–377
218. Budygin EA, Brodie MS, Sotnikova TD, Mateo Y, John CE, Cyr M, Gainetdinov RR, Jones SR (2004) Dissociation of rewarding and dopamine transporter-mediated properties of amphetamine. *Proc Natl Acad Sci (USA)* 101: 7781–7786
219. Ahn K-C, Pazderka-Robinson H, Clements R, Ashcroft R, Ali T, Morse C, Greenshaw AJ (2005) Differential effects of intra-midbrain raphé and systemic 8-OH-DPAT on VTA self-stimulation thresholds in rats. *Psychopharmacol* 178: 381–388
220. Dekeyne A, Gobert A, Auclair A, Girardon S, Millan MJ (2002) Differential modulation of efficiency in a food-rewarded "differential reinforcement of low-rate" 72-s schedule in rats by norepinephrine and serotonin reuptake inhibitors. *Psychopharmacol* 162: 156–167
221. Lucki I (1998) The spectrum of behaviors influenced by serotonin. *Biol Psychiat* 44: 151–162
222. Cherek DR, Lane SD, Pietras CJ, Steinberg JL (2002) Effects of chronic paroxetine administration on measures of aggressive and impulsive responses of adult males with a history of conduct disorder. *Psychopharmacol* 159: 266–274
223. Sargent PA, Williamson DJ, Pearson G, Odontiadis J, Cowan PJ (1997) Effect of paroxetine and nefazodone on 5-HT1A receptor sensitivity. *Psychopharmacol* 132: 296–302
224. Balleine B, Fletcher N, Dickinson A (1996) Effect of the $5HT_{1A}$ agonist, 8-OH-DPAT, on instrumental performance in rats. *Psychopharmacol* 125: 79–88
225. Fletcher PJ (1994) Effects of 8-OH-DPAT, 5-CT and muscimol on behaviour maintained by a DRL 20s schedule of reinforcment following microinjection into the dorsal or median raphe nuclei. *Behav Pharmacol* 5: 326–336
226. Fletcher PJ, Korth KM (1999) Activation of 5-HT1B in the nucleus accumbens reduces amphetamine induced enhancement of responding for conditioned reward. *Psychopharmacol* 142: 165–174
227. Evenden JL (1999) The pharmacology of impulsive behaviour in rats. VII: the effects of serotonergic agonists and antagonists on responding under a discrimination task using unreliable visual stimuli. *Psychopharmacol* 146: 422–431
228. Cousins MS, Vosmer G, Overstreet DH, Seiden LS (1999) Rats selectively bred for responsiveness to 5-hydroxytryptamine$_{1A}$ receptor stimulation: differences in differential reinforcement of low rate 72-second performance and response to serotonergic drugs. *J Pharmacol Exp Ther* 292: 104–113
229. Thiebot M-H, Martin P, Puech AJ (1992) Animal behavioural studies in the evaluation of antidepressant drugs. *Br J Psychiat Suppl*: 44–50
230. Bizot J-C, Le Bihan C, Puech AJ, Hamon M, Thiebot M-H (1999) Serotonin and tolerance to delay of reward in rats. *Psychopharmacol* 146: 400–412
231. Denk F, Walton ME, Jennings KA, Sharp T, Rushworth MFS, Bannerman DM (2005) Differential involvement of serotonin and dopamine systems in cost-benefit decisions about delay or effort. *Psychopharmacol* 179: 587–596
232. Pliszka SR, Maas JW, Javors MA, Rogeness GA, Baker J (1994) Urinary catecholamines in attention-deficit hyperactivity disorder with and without comorbid anxiety. *J Am Acad Child Adolesc Psychiat* 33: 1165–1173

233. Raskin LA, Shaywitz SE, Shaywitz BA, Anderson GM, Cohen DJ (1984) Neurochemical correlates of attention deficit disorder. *Pediatr Clin N Am* 31: 387–396
234. Hunt RD, Cohen DJ, Anderson G, Clark L (1984) Possible change in noradrenergic receptor sensitivity following methylphenidate treatment: growth hormone and MHPG response to clonidine challenge in children with attention deficit disorder and hyperactivity. *Life Sci* 35: 885–897
235. Shekim WO, Javaid J, Dekirmenjian H, Chapel JL, Davis JM (1982) Effects of d-amphetamine on urinary metabolites of dopamine and norepinephrine in hyperactive boys. *Am J Psychiat* 139: 485–488
236. Shekim WO, Javaid J, Davis JM, Bylund DB (1983) Urinary MHPG and HVA excretion in boys with attention deficit disorder and hyperactivity treated with d-amphetamine. *Biol Psychiat* 18: 707–713
237. Shen YC, Wang YF (1984) Urinary 3-methoxy-4-hydroxyphenylglycol sulfate excretion in seventy three schoolchildren with minimal brain dysfunction. *Biol Psychiat* 19: 861–869
238. Zametkin AJ, Karoum F, Linnoila M, Rapoport JL, Brown GL, Chuang LW, Wyatt RJ (1985) Stimulants, urinary catecholamines and indoleamines in hyperactivity: a comparison of methylphenidate and dextroamphetamine. *Arch Gen Psychiat* 42: 251–255
239. Jacobowitz D, Sroufe LA, Stewart M, Leffert N (1990) Treatment of attentional and hyperactivity problems in children with sympathomimetic drugs: a comprehensive review. *J Am Acad Child Adolesc Psychiat* 29: 677–688
240. Castellanos FX (1999) The psychobiology of attention-deficit/hyperactivity disorder. In: Quay HC, Hogan TP (eds.) *Handbook of disruptive behavior disorders*. Kluwer Academic/Plenum Publishers, New York, 179–198
241. Halperin JM, Newcorn JH, Koda VH, Pick L, McKay KE, Knott P (1997) Noradrenergic mechanisms in ADHD children with and without reading disabilities: a replication and extension. *J Am Acad Child Adolesc Psychiat* 36: 1688–1697
242. Kusaga A, Yamashita Y, Koeda T, Hiratani M, Kaneko M, Yamada S, Matsuishi T (2002) Increased urine phenylethylamine after methylphenidate treatment in children with ADHD. *Ann Neurol* 52: 371–374
243. Potter WZ, Hsiao JK, Goldman SM (1989) Effects of renal clearance on plasma concentrations of homovanillic acid. *Arch Gen Psychiat* 46: 558–562
244. Shetty T, Chase TN (1976) Central monoamines and hyperkinesis of childhood. *Neurol* 26: 1000–1002
245. Irwin M, Belendiuk K, McCloskey K, Freedman DX (1981) Tryptophan metabolism in children with attentional deficit disorder. *Am J Psychiat* 138: 1082–1085
246. Spivak B, Vered Y, Yoran-Hegesh R, Graff E, Averbuch E, Vinokurow S, Weizman A, Mester R (2001) The influence of three months of methylphenidate treatment on platelet-poor plasma biogenic amine levels in boys with attention deficit hyperactivity disorder. *Hum Psychopharmacol Clin Exp* 16: 333–337
247. Laufer MW, Denhoff E, Solomons G (1957) Hyperkinetic impulse disorder in children's behaviour problems. *Psychosom Med* 19: 38–49
248. Ashtari M, Kumra S, Bhaskar SL, Clarke T, Thaden E, Cervellione KL, Rhinewine J, Kane JM, Adesman A, Milanaik R et al. (2005) Attention-deficit/hyperactivity disorder: A preliminary diffusion tensor imaging study. *Biol Psychiat* 57: 448–455
249. Satterfield JH, Schell AM, Nicholas T (1994) Preferential processing of attended stimuli in attention-deficit hyperactivity disorder and normal boys. *Psychophysiol* 31: 1–10
250. Johnstone SJ (1999) Auditory event-related potentials in attention-deficit hyperactivity disorder: developmental and clinical aspects. University of Wollongong: PhD Thesis.

Neurotransmitter Interactions and Cognitive Function
Edited by Edward D. Levin
© 2006 Birkhäuser Verlag/Switzerland

Prepulse inhibition mechanisms and cognitive processes: a review and model

José Larrauri and Nestor Schmajuk

Duke University, Department of Psychology and Neuroscience, Durham, NC 27708, USA

Prepulse inhibition (PPI) refers to the decrease in the response to a startling stimulus when a weak pulse precedes it. The phenomenon depends mostly on the intensity of the pulse and prepulse, the time interval between them, and the intensity of background noise. In this chapter, we review and discuss studies describing the behavioral properties and the neurobiological basis of the acoustic startle response (ASR) and PPI, show how a computational model summarizes these data, and extend an existing theory of how simple and fast PPI mechanisms interact with higher and slower cognitive processes.

Acoustic startle response and prepulse inhibition

The acoustic startle response (ASR) is the reflex reaction elicited by the presentation of loud auditory stimuli (pulses). The ASR can be modified when the startling stimulus is preceded closely in time by another stimulus (prepulse). In this case, the amplitude of the ASR can be enhanced (prepulse facilitation, PPF) or decreased (prepulse inhibition, PPI), compared to the case in which the startling stimulus is not preceded by a prepulse. Furthermore, prepulses can be given in the same modality as the ASR pulse (i.e., acoustic), as well as in others (e.g., visual; [1]). When the interval between prepulse and pulse onset (referred to as lead interval) is around 100 ms, PPI exhibits a maximum. Gradually increasing the lead interval beyond this interval decreases inhibition and shorter lead intervals (below 50 ms) might result in PPF [2].

Behavioral properties of the ASR

The ASR increases with increasing pulse intensity, decreasing prepulse intensity and when the background noise level changes from a silent condition to 80 dB

Hoffman and Searle [3] (Experiment 6) analyzed the role of the prepulse intensity and its interaction with background noise level on the startle response in rats. In

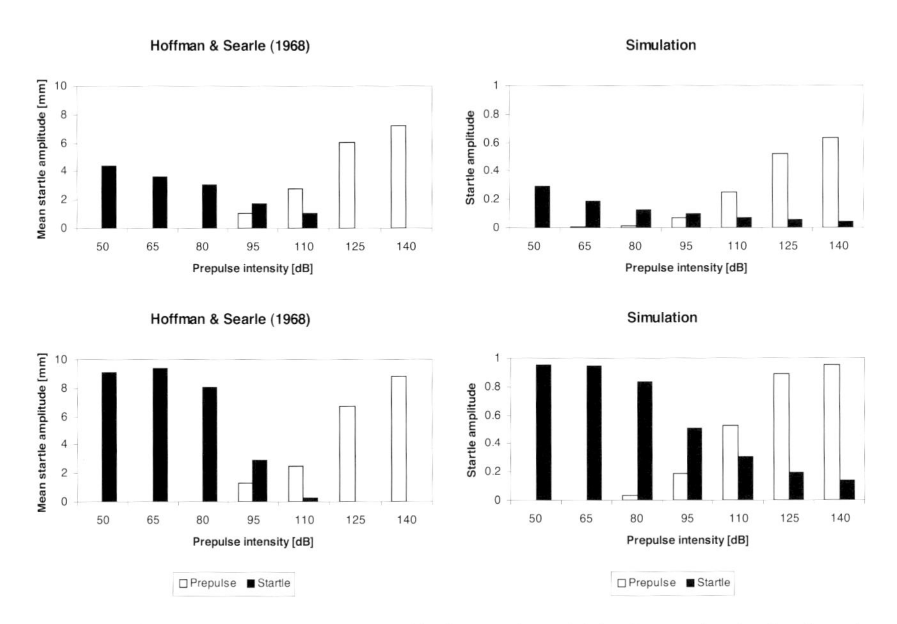

Figure 1. Effects of prepulse intensity and its interaction with background noise level on the startle response in rats. Left panels: Experimental results [3]. Right panels: Simulated results. Top panels: Silent condition. Bottom panels: 80 dB background noise.

their study, two different sound sources were used, one to generate the pulses and prepulses and another to provide the background level stimulation. The experiment consisted in presenting a 20 ms prepulse of varying intensity (50, 65, 80, 95, 110, 125 or 140 dB) preceding a 20 ms, 140 dB pulse by a 100 ms lead interval. Responses to both prepulses and pulses were measured in two different background conditions, namely, silence or 80 dB broadband noise. In order to gauge the animals' responses, rats were placed in chambers that had magnets attached to them, which in turn were located inside stationary coils. Therefore, rats' movements produced variations in the magnetic flux transversing the coil, making the instrument sensitive to sudden movements, such as startle responses. As shown in the left panels on Fig. 1, Hoffman and Searle [3] found that, in a silent background, low intensity prepulses (50 dB) could inhibit the ASR, even when they do not elicit detectable responses. When the prepulse intensity increased, the ASR was further attenuated. In the 80 dB background condition, low intensity prepulses (50, 65 and 80 dB) did not significantly inhibit the ASR, but when the prepulse intensity was increased beyond 80 dB, the ASR decreased. In both background conditions, as the prepulse intensity increased, the prepulse-elicited reactivity increased monotonically. However, the response to the 140 dB prepulse was higher in the 80 dB background condition.

The ASR is an inverted U-shape function of the background level noise

Ison and Hammond [4] (Experiment 6) extended the results reported by Hoffman and Searle [3] regarding the effects of the background noise level on the ASR. Since the latter study had only analyzed two different background conditions (silence and 80 dB), Ison and Hammond [4] studied the rats' responses to a 20 ms, 119 dB pulse using six different background noise levels (65, 70, 75, 80, 85 and 90 dB). The apparatus used in the experiment consisted in an accelerometer attached to the animal chamber, the output of which was fed into a device that converted it into millimeters of pen deflection. Experimental results shown in the left panel of Fig. 2 indicate that the ASR peaks when the background intensity is 75 dB.

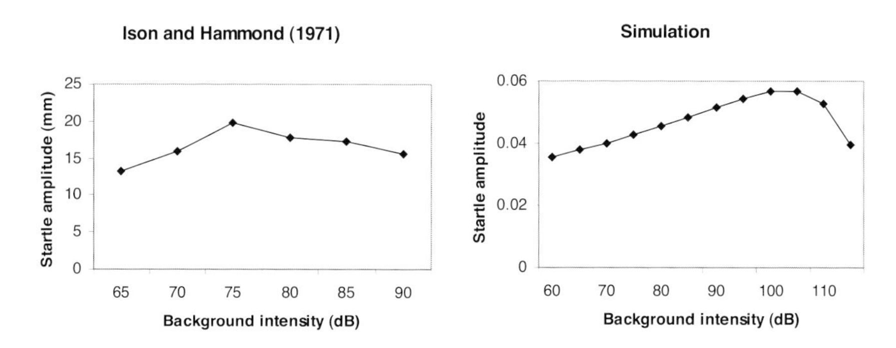

Figure 2. Effects of background noise level on the startle response in rats. Left panel: Experimental results [4]. Right panel: Simulated results.

The ASR shows different forms of plasticity

The ASR can undergo amplitude habituation [3], sensitization [5], potentiation by fear [6] and attenuation by pleasure [7].

Habituation

The decrease of the ASR after repeated stimulation in a low background can be explained by either an increase in the animal's threshold for responding or a change in the input-output slope (ratio). Pilz and Schnitzler [8] tested these hypotheses with two experiments. In the first, the rat's ASR showed a decrease in amplitude across blocks, and an increase in responding for increasing stimuli levels. Using a linear function to fit the ASR amplitude dependence on dB input, a threshold level could be determined for each block, which remained constant across blocks (around 80 dB). In the second experiment, near-threshold stimuli were used until a criterion-level response was reached (i.e., until the ASR level reached a small fixed value), intermixed with a series of intense stimuli. As expected, responding to the high amplitude stimuli decreased across the trial blocks. In addition, the amplitude of the input stimulus needed to reach the criterion level also increased, a result similar to the one obtained by Hoffman and Searle [3]. Even though Hoffman and Searle [3] interpreted these results as an indication of a change in the threshold level after repeated stimulation, Pilz and Schnitzler [8] showed that this effect was caused by a change in the slope of the response function, and not by a threshold variation. Taken together, the results of both experiments support the view that habituation of the ASR is due to a change in the input-output slope, and not a change in the threshold level. Therefore, the authors concluded that the amplitude habituation center of the ASR should be located downstream of the startle circuit.

Sensitization
Davis [5] studied the conditions under which the startle response of rats changed after repeated simulation. In one of his experiments, he found that the ASR amplitude of rats to a salient tone decreased (habituated) across blocks of trials in a 60 dB background noise. However, when the background noise level was increased to 80 dB, responding increased across trials, an effect that remained even after repeated sessions. Davis [5] showed that this sensitization effect was due to the background exposure previous to the presentation of the startling stimuli, and not to a tone-repetition effect. Furthermore, he showed that sensitization reached a maximum level (around 30 to 45 min of background noise exposure), after which habituation to the startling stimulus prevails, decreasing the amplitude of the ASR. These results show that responding to acoustic stimulation is dependent not only on the background level at the time of testing, but also on the background level previous to it.

Potentiation by fear and attenuation by pleasure
The fear-potentiated startle effect [9] refers to the amplitude increase in the ASR when the startle stimulus is delivered in the presence of another stimulus, which has been previously conditioned to an aversive US. For example, rats that are trained to fear an initially innocuous CS (such as a visual stimulus) by pairing it with a footshock, produce larger ASRs when tested in the presence of the CS.

An opposite effect to the potentiation by fear of the ASR is observed when the startling stimulus is presented paired with food or any other rewarding stimulus (attenuation by pleasure). In this case, animals produce smaller ASRs than those generated in the absence of the food. For example, Steidl, Yeomans and Li [7] gave brain-stimulation rewards to rats in the presence or absence of a light, and after this conditioning, rats tested with the reinforced CS produced smaller ASRs than those of the rats tested with the light off.

Behavioral properties of PPI

PPF is obtained for short lead intervals (below 30 ms) and PPI reaches a maximum for lead intervals ranging between 50 and 100 ms

Plappert et al. [2] (Experiment 2) studied the range of lead intervals for which PPF and PPI showed maximum values, and how the lead interval interacted with prepulse

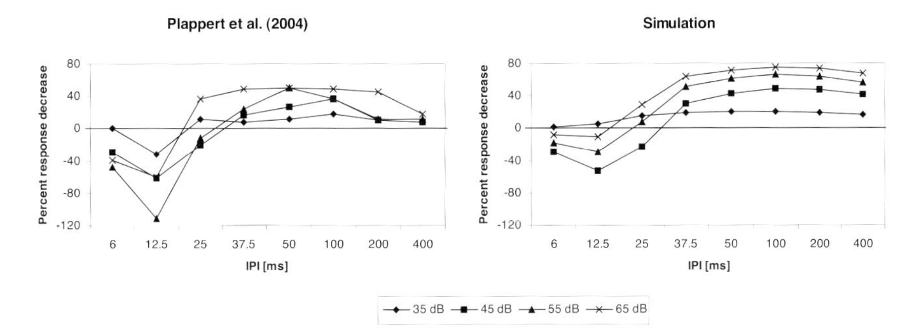

Figure 3. Effects of lead interval and its interaction with prepulse intensity on the ASR. Left panel: Experimental results [2]. Right panel: Simulated results.

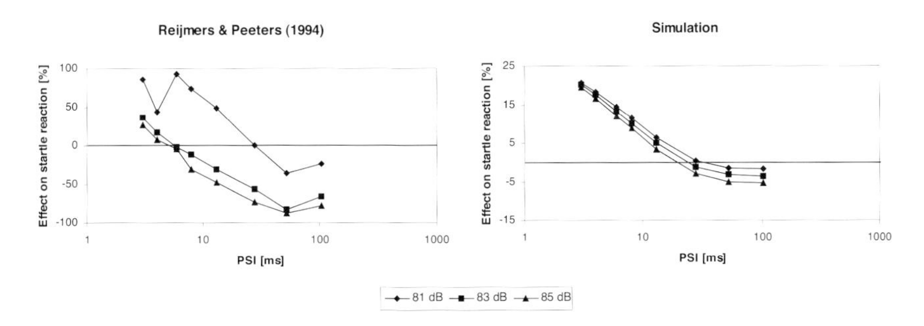

Figure 4. Effects of lead interval and its interaction with prepulse intensity on the ASR. Left panel: Experimental results [10]. Right panel: Simulated results.

intensity. In their study, mice were exposed to a 33 dB background noise level, in which a 20 ms, 110 dB startling stimulus (pulse) was presented either alone or preceded by a prepulse. The prepulse intensity was either 35, 45, 55 or 65 dB, and the lead interval took one of eight possible values: 6.25, 12.5, 25, 37.5, 50, 100, 200 and 400 ms. The experimental chamber was placed on top of a piezoelectric platform that transformed animals' movements into voltage signals. After this information was filtered and amplified, startle responses were determined as the peak-to-peak voltage difference during a 50 ms time window before and after stimulus onset. As shown in the left panel of Fig. 3, Plappert et al. [2] reported that PPI decreased for low intensity prepulses, and that PPF was obtained for shorter lead intervals (below 30 ms) and lower prepulses. In addition, PPI peaked between 50 and 100 ms.

Reijmers and Peeters [10] conducted a similar experiment with rats, in which three different 3 ms prepulses (81, 83 and 85 dB) presented on a continuous 79 dB background noise preceded a 25 ms, 119 dB startling pulse. The rats' ASRs in these conditions were compared to those obtained when the pulse was presented preceded by no prepulse to determine the occurrence of PPF or PPI. They used eight different lead intervals (3, 4, 6, 8, 13, 28, 53 and 103 ms) to analyze the interaction of prepulse intensity with lead interval. Animal responses were measured using a piezoelectric device attached to the experimental chambers, and startle was determined by averaging electrical activity elicited in a 200 ms time window after stimulus onset. Concordant to Plappert's et al. [2] results, Reijmers and Peeters [10] found greater PPF for lower prepulses (81 dB), increased PPI for more salient prepulses (85 dB), and maximum inhibition for lead intervals around 50 ms, as show on the left panel of Fig. 4.

Decrements in background level (gaps) produce PPI

Stitt et al. [11] reported the effects of decreasing the background noise level previous to the presentation of the startling stimulus. In their experiment, rats were exposed to a constant 70 dB background level, that was suddenly discontinued for a variable period of time (thus producing a gap, in which the noise level was below 30 dB) immediately followed by the presentation of a 20 ms, 125 dB pulse. Nine intervals between termination (offset) of the background signal and presentation of the pulse were used, namely, 0, 1, 4, 16, 64, 250, 1000, 4000 ms and 30 s. As in Hoffman and Searle's [3] experiment, startle responses were obtained by measuring the currents induced in a coil, caused by the movement of a magnet attached to the experimental chambers in which the rats were placed. Their results showed that increasing the gap duration up to 250 ms resulted in an increased PPI of the ASR, whereas further increases lead to smaller inhibitions. Interestingly, the ASR for the 0 ms condition (when the pulse was presented in the 70 dB background) was greater than that obtained for the 30 s condition (when the pulse was presented on a 30 dB background), in agreement with Ison and Hammond's [4] results (see section "The ASR is an inverted U-shape function of the background level noise").

An additional experiment analyzing the interaction between background noise decrement with offset lead time was conducted by Ison et al. [12]. Mice in this study were exposed to a 70 dB background noise, which was suddenly reduced in some trials, thereby producing a gap. Four different background decrements (40, 30, 20 or 10 dB) preceded the presentation of a 20 ms, 115 dB startling pulse. Seven different intervals between background decrement and pulse presentation (offset lead time) were used (1, 2, 4, 6, 8, 10 and 15 ms). Startle responses were obtained by measuring the voltage signal generated by an accelerometer connected to the mice cage. In order to determine the percentage inhibition caused by the different gaps, startle responses in these cases were compared to a control ASR, which was obtained when the startling stimulus was not preceded by a background level decrement. Ison et al. [12] reported that increasing the offset lead interval up to 15 ms resulted in increased inhibition, in agreement with Stitt et al.'s [11] results. In addition, as shown in the left panel of Fig. 5, their results showed that increasing the gap magnitude produced greater inhibition.

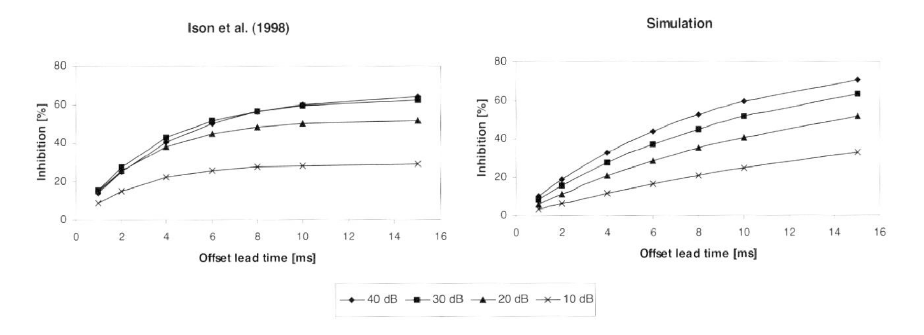

Figure 5. Effects of decreasing the background noise level and its interaction with lead interval on the ASR. Left panel: Experimental results [12]. Right panel: Simulated results.

Increasing the duration of prepulses increases PPI

Blumenthal [13] (Experiment 3) studied the effects of varying the prepulse duration, as well as its interaction with prepulse level, on the human eyeblink response. In the experiment, startle stimuli consisted of 50 ms, 85 dB pulses, whereas three different prepulse intensities (40, 50 and 60 dB) were used. Keeping the lead interval constant at 150 ms, four different prepulse durations (6, 20, 50 and 100 ms) were tested. Startle responses were determined by measuring electromyographic activity in the subjects' orbicularis oculi. Blumenthal [13] reported that increasing prepulse intensity increased PPI, as well as increasing prepulse duration from 6 to 20 ms (Fig. 6, left panel). However, when prepulse duration was increased beyond 20 ms (up to 100 ms), it did not produce a significant inhibition increase.

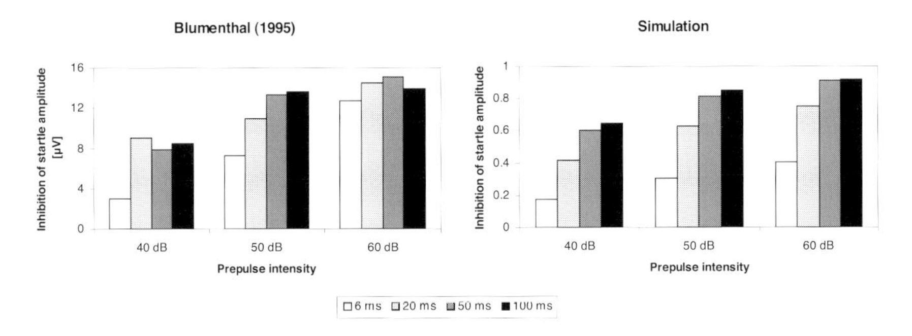

Figure 6. Effects of prepulse duration and its interaction with prepulse level on the human eyeblink response. Left panel: Experimental results [13]. Right panel: Simulated results.

Prepulses inhibit startle responses of pulses but not their ability to inhibit subsequent pulses

Swerdlow et al. [14] designed a study in order to test whether the inhibitory effect of a prepulse (s) in the ASR of a pulse (S1) also inhibited the capability of this pulse (S1) to inhibit the ASR of another pulse (S2) presented subsequently. In their experiment, 40 ms, 120 dB pulses and 85 dB, 20 ms prepulses were used, presented on a 70 dB background. Responses to startling pulses were measured under different stimuli configurations and lead intervals, namely, to pulses 1) presented alone (S1 trials), 2) preceded by another pulse (S1–S2 trials; 1 and 3 s lead intervals), 3) preceded by a pulse (S1) which had in turn been preceded by a weak prepulse (s) with a 100 ms lead interval (s–S1–S2 trials; 1 and 3 s S1–S2 intervals), and 4) preceded by a weak prepulse (s–S1 trials; 100, 1140 and 3140 ms lead interval). Both rats' and humans' responses were analyzed in this study, and similar results were obtained, although inhibition in humans was smaller and decreased faster for longer intervals between stimuli (rats' responses were determined by recording electrical activity from a piezoelectric crystal attached to the animals' chambers, whereas humans' responses were obtained through electromyographic recordings). As shown in the left panel of Fig. 7, startle responses (in rats) decreased when pulses were preceded by either pulses (S1–S2) or prepulses (s–S1). But even when prepulses decreased the motor responses of pulses (s–S1 trials, 100 ms), they did not hinder the pulses' ability to inhibit subsequent pulses, as manifested by the decreased responses to S2 pulses on the s–S1–S2 trials.

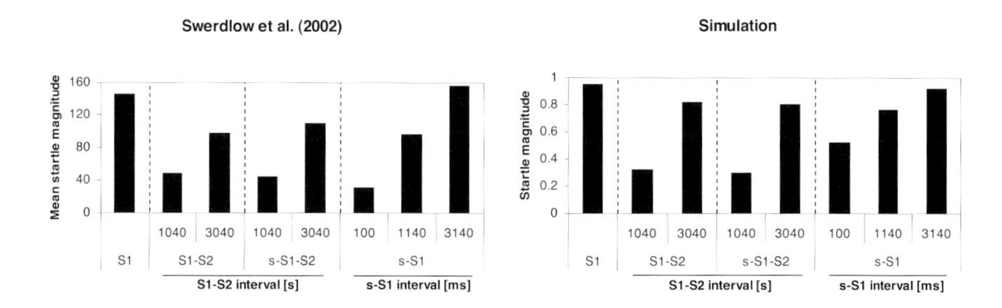

Figure 7. Inhibitory effects of prepulses and pulses, varying lead intervals and stimuli configurations. Left panel: Experimental results [14]. Right panel: Simulated results.

Increasing the startling stimulus increases the ASR but reduces PPI

Yee et al. [15] analyzed the effects of pulse intensity on the ASR and PPI. In their study, three prepulse intensities (71, 77 and 83 dB) and three pulse intensities (100, 110 and 120 dB) were used. Stimuli were presented on a 65 dB background and mice activity was obtained by measuring the voltage signal generated by a piezoelectric device attached to the experimental chambers. Startle responses were determined by integrating this electrical activity over a 65 ms time window after stimulus onset.

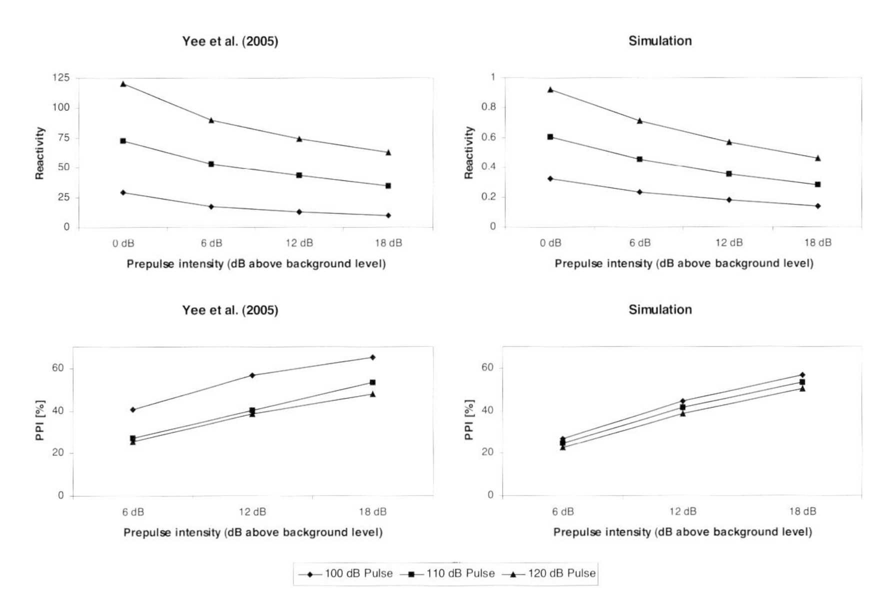

Figure 8. Effects of pulse intensity on startle and prepulse inhibition. Left panels: Experimental results [15]. Right panels: Simulated results. Upper panels: Startle responses as a function of pulse and prepulse intensity. Lower Panels: Prepulse inhibition as a function of pulse and prepulse intensity.

Yee et al.'s [15] experimental results are shown on the left panels of Fig. 8. Results in the upper panel show that increasing the pulse intensity yields increased ASRs for all the prepulse levels, and that increasing the prepulse intensity, as expected, results in decreased ASRs. However, when the ASR for the different pulse intensities are converted into percentage PPI values (as shown on the lower panel), it can be seen that the same prepulses generate greater percentage PPI when lower intensity pulses are used.

Habituation of PPI

The inhibitory effect of a prepulse on the ASR decreases after repeated presentations of the startling stimulus preceded by the prepulse [16]. This reduction in inhibition could, in theory, be explained by two different mechanisms. One possibility is that either the prepulse sensory input, or its inhibitory projection to the startle center becomes habituated after repetitive stimulation. Another alternative is that this reduction in inhibition could be explained by a decrease in the startle response, if the inhibitory potential of a prepulse does not decrease, but is instead related to the response magnitude produced by the startling stimulus. Blumenthal [17] devised an experiment to determine the mechanisms underlying the reduction of PPI observed as the test sessions progress. In his experiment, the human eye-blink response was analyzed using 95 dB noise pulses and 60 and 70 dB tone prepulses. Subjects in this study were initially exposed to one of three types of trials, namely, prepulses without pulses, pulses alone or both prepulses and pulses paired on some trials (control condition). After this initial session, all subjects were exposed to trials in which pulses were sometimes paired with prepulses. Therefore, by comparing the amount of inhibition in the first trials of the second phase, the mechanisms responsible for the attenuation of PPI could be determined. Blumenthal's [17] results showed that, compared to the control condition, PPI increased in the second session when prepulses were presented alone in the first session, and decreased when only startling stimuli were delivered in the initial session. Since repeated exposure to the prepulse did not reduce this stimulus' ability to inhibit the ASR, this result provided support for the second mechanism described above regarding decreased PPI after continuous testing. Hence, repeated presentation of stimuli that can act as inhibitors of startle do not decrease their inhibitory potential.

Attentional mechanisms involved in PPI

According to the attentional theory of PPI [18], attention to the prepulse influences the subsequent inhibition of the ASR. Dawson et al. [19] reported that using 120 ms lead intervals produced greater inhibition when the prepulse stimulus was attended compared to that observed when the prepulse was unattended. However, when the lead interval was either decreased (60 ms) or increased (240 ms), the prepulse attentional condition did not influence inhibition of the ASR. Therefore, Dawson et al. [19] suggested that even when the inhibitory mechanisms triggered by a prepulse are automatic, they can be influenced by attentional states.

Using fMRI, Hazlett et al. [20] examined the brain areas that show differential activation when attention to the prepulse is manipulated on a PPI paradigm. They reported a significant difference among conditions (attended prepulse + startle stimulus, unattended prepulse + startle stimulus, startle stimulus alone) in the right thalamus, and in the anterior and mediodorsal nuclei of the thalamus. In all cases, blood-oxygen-level-dependent (BOLD) responses were greatest in the attended condition, and weakest when the startle stimulus was presented alone. This result is in agreement with the ventral pallidum (VP)–mediodorsal thalamus (MD) circuit proposed by Kodsi and Swerdlow [21], which is believed to play a delayed mediatory role on PPI. Therefore, Hazlett et al. [20] proposed that attentional manipulations of the prepulse result in delayed cortical-thalamic activity, which converges into the ASR circuit at the pedunculopontine tegmental nucleus (PPT) level. However, a caveat in this study is that the experimental design did not include behavioral assessments of PPI, and therefore the degree to which the observed differential brain activities correlate to response inhibition cannot be determined.

In a related study, Bitsios and Giakoumaki [22] analyzed the relationship between PPI and the Rapid Visual Information Processing, Stockings of Cambridge and Stroop tests, cognitive tasks assumed to involve attentional and executive mechanisms. In the Stroop test, subjects are asked to name the color of the word they see. Individuals take longer to name words that describe a color different from the color of the text. In the Rapid Visual Information Processing test, subjects are presented with rapidly-changing number sequences and their task is to press a button when a they detect a target. In the Stockings of Cambridge test, two sets of objects are presented in different patterns, and the executive task consists in rearranging one set in the minimum number of movements to reproduce the configuration of the other. By performing all the tasks with the same pool of subjects, the authors could determine the correlation between PPI and these tasks. Among the cognitive tasks, it was hypothesized that the Stroop test (which is an "inhibition-based" paradigm), could closely resemble a PPI paradigm, since this task involves the suppression of a customary response (the word which is actually written) to selectively attend a feature (the color in which the word is written) in order to produce a correct response. No correlation between the results of the PPI experiment and the cognitive tasks reached significance in the Bitsios and Giakoumaki [22] study. However, they reported that even when the correlation coefficient between the PPI measurements and the Stroop interference test did not reach significance, it showed a trend ($p = .068$). As hypothesized, this result led the authors to suggest a possible connection between PPI and the Stroop test, since a better performance in the latter task reveals higher cognitive inhibition, which could be reflected in greater inhibition of the ASR.

Mediating circuit of the ASR and PPI

Hoffman and Ison [23] proposed a hypothetical circuit in which PPI is mediated by a fast excitatory pathway that is in turn inhibited by a slower-activated parallel pathway. In line with Hoffman and Ison's suggestion, the physiological data described below support the view that the mediating circuit is composed of those two main pathways (excitatory and inhibitory). As shown in Fig. 9, the excitatory pathway of the mediating circuit is composed of the cochlear root nucleus (CRN) that projects to the giant neurons in the caudal pontine reticular nucleus (PNC), whereas the inhibitory path includes the ventral (VCN) and dorsal cochlear nucleus (DCN), inferior (IC) and superior colliculi (SC), and the PPT.

Excitatory pathway

Experimental results provide evidence of a CRN-PNC connection, in which neural activity to acoustic stimulation exhibits latency and threshold features totally compatible to the ones observed in the ASR.

Cochlear root nucleus (CRN)
Cochlear root neurons are large cells (35 μm in diameter) in the cochlear nucleus, which receive direct input from the cochlea via the auditory nerve, and project to several areas, including the PNC, SC and lateral lemniscus (LL) [24].

Caudal pontine reticular nucleus (PNC)
Lingenhöhl and Friauf [25] studied cell populations in the reticular formation in rats, focusing especially on whether PNC neurons receive auditory input and if their

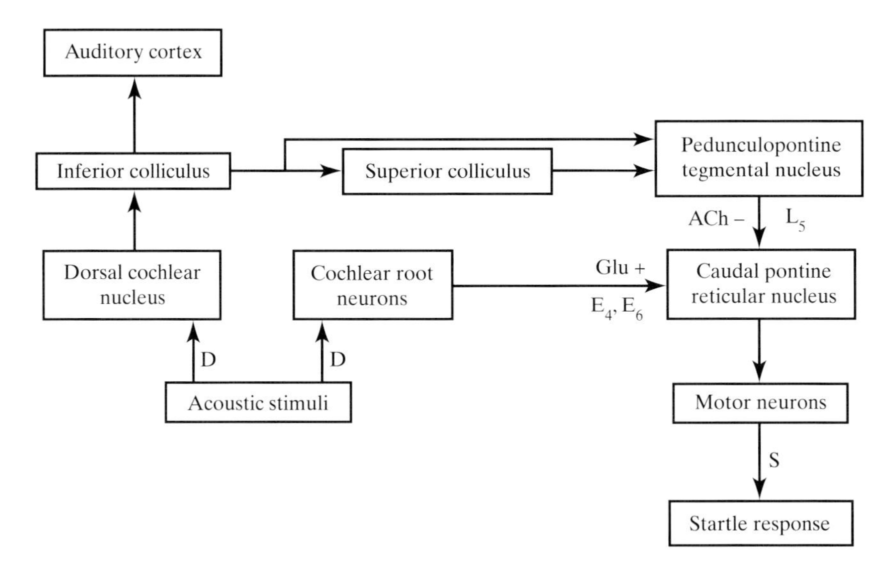

Figure 9. Schematic outline of the mediating circuit of the startle response.

potential latency is as short as the one involved in the ASR. In order to address these questions, they used a combination of intracellular electrophysiological recordings to measure acoustically-elicited activity, and injections of horseradish peroxidase to identify the anatomy of the rat's cells being analyzed. Their results showed that neurons in the PNC were excited shortly after acoustic stimulation, leading Lingenhöhl and Friauf [25] to suggest a direct connection between the CN and the PNC. They were also able to characterize the morphology of the acoustically-driven PNC neurons, reporting that these cells exhibited very large somas. These results made Lingenhöhl and Friauf [25] suggest that the "giant PNC neurons" could both mediate and modulate the ASR, since their large size allows them to rapidly transmit information to the spinal interneurons, as well as to receive information from other brain regions.

In an ensuing study, Lingenhöhl and Friauf [24] used tracing techniques to identify the sources of input to the PNC neurons. Their results showed bilateral projections from several auditory brainstem nuclei to this structure, mainly from the CN and superior olivary complex (SOC), but none from the LL. This finding simplified the acoustic startle circuit originally proposed by Davis et al. [26], which included the LL as a component of the excitatory pathway.

Further evidence for the role of the PNC neurons in the mediation of the ASR and PPI was reported by Carlson and Willott [27]. They analyzed the relationship between the ASR and neural activity in neurons of the PNC in C57BL/6J mice under three different conditions, namely, when mice responded to startle stimuli alone, on PPI trials and after the effects of high-frequency hearing loss observed in aged mice. Their results showed that action potentials evoked on PNC neurons after acoustic stimulation of the mice closely resembled the main characteristics of the measured

ASR (threshold and latency). In addition, and replicating results in cats of Wu et al. [28], Carlson and Willott [27] found neural inhibition in the PNC neurons when the startling stimulus was preceded by a weak prepulse. And lastly, old mice that exhibited high-frequency hearing loss displayed enhanced neural PNC inhibition and PPI to low-frequency tone stimulation. High-frequency hearing loss is accompanied by a shift in responding of neurons in the IC from higher to lower frequencies [29], causing the overrepresentation of low-frequency tones in the auditory pathway. Carlson and Willott [27] reported that this increased salience of low-frequency inhibitory tones caused by the hearing loss in old mice, manifested behaviorally as an increased PPI, was correlated to neural activity in the PNC, further supporting the notion that this brain region mediates the ASR.

Inhibitory pathway

Experimental results suggest a CN-IC-SC-PPT-PNC pathway that mediates inhibition of the ASR. This path runs parallel to the excitatory pathway described before, and both converge at the level of the PNC, where the resulting processing is relayed to the spinal motoneurons to produce the startle response after acoustic stimulation.

Cochlear nucleus (CN)
The CN receives input from all the axons in the auditory nerve, and constitutes a point of information divergence, since fibers from the auditory nerve project to different areas in the CN, the ventral CN (VCN) and the dorsal CN (DCN). These areas do not only differ in their location within the CN, but also in the types of cells that compose them. The VCN consists of four different types of neurons, namely, spherical bushy cells, globular bushy cells, octopus cells and multipolar/stellate cells, whereas the cells that compose the DCN are fusiform, radiate, fan, cartwheel and small stellate. The VCN projects bilaterally to the SOC, and both the VCN and DCN project contralaterally to the IC and LL.

Inferior colliculus (IC)
Carlson and Willott [30] suggested a PPI model in which the prepulse-elicited inhibition was mediated by the IC, from where information was transmitted to the PPT and converged later with the excitatory pathway of the ASR at the level of the PNC. Leitner and Cohen [31] tested the role of the IC in the inhibition of the ASR in rats. In their experiment, PPI of the ASR was initially assessed using both visual and auditory prepulses in two groups of rats (control and experimental). No differences in response amplitude or latencies were found when rats were tested using an acoustic startling stimulus alone or preceded by either an acoustic or visual prepulse. Subsequently, rats in the experimental group sustained electrolytic lesions in the IC, after which all rats were tested again with the same design used prior to the lesions. Whereas the response amplitude from rats in the control group did not significantly differ between pre- and post-lesion tests, rats in the experimental group exhibited a significant increase in the ASR. In addition, auditory prepulses were not longer able to inhibit the ASR in the lesioned rats, but visual prepulses were. Latencies

in both groups were reduced in the post-lesion tests, suggesting that different processes might govern latency and amplitude modification of the ASR. These results led Leitner and Cohen [31] to suggest a role of the IC in the ASR reduction by acoustic prestimulation. However, the fact that IC lesions resulted not only in decreased acoustic PPI, but also in increased ASR amplitude, is consistent with the view that the IC could also play a role in the inhibition of the startling stimulus itself.

Superior colliculus (SC)
Fendt [32] studied the effects of blocking GABA receptors of the SC on PPI of the ASR, by using microinjections of picrotoxin. As a result of this manipulation, the SC was moderately stimulated (though not to the point to elicit motor reactions), and PPI was significantly enhanced, without changing the ASR baseline responding. Therefore, Fendt [32] argued that the SC might be a component of the circuit mediating PPI of the ASR. Furthermore, since the SC receives and integrates information from different sensory modalities (acoustic, visual and tactile; [33]), it could also represent the entry point of other sensory prepulse information into the inhibitory pathway of the ASR. However, the fact that the ASR baseline responding was not affected after picrotoxin injections (as was the case after IC electrolytic lesions, see above) could be explained in terms of the low doses used (to avoid behavioral responses), or to parallel IC-PPT projections [30].

Pedunculo pontine tegmental nucleus (PPT)
Since the PPT receives projections from the SC [34] and also presents a short latency activation after acoustic stimulation (13 ms; [35]) this structure is a likely component of the mediating circuit of the ASR. Therefore, Koch et al. [36] studied the way in which the PPT influences PPI of the ASR in rats, specifically through cholinergic neurons that innervate the PNC. Using retrograde tracing techniques, they found that the only sources of cholinergic input to the PNC were provided by the PPT and laterodorsal tegmental nucleus (LDT). Acetylcholine (ACh) agonists (AMCH and carbachol) both increased and decreased the evoked activity of neurons in the PNC. However, a large number of acoustic neurons (those that respond to acoustic stimulation) in the PNC were inhibited (58.5% when the agonist used was AMCH, and 78% when it was carbachol), whereas a similar amount of non-acoustic neurons were equally excited and inhibited. Koch et al. [36] attributed the lack of inhibition of all acoustic neurons in the PNC after ACh agonist administration to the fact that the PNC receives cholinergic input also from the LDT, which might exert a different function. In addition, lesions of the PPT resulted in reduced PPI, further suggesting a role of this region in the mediation of PPI. Koch et al. [36] also reported that these lesions have no effect on the habituation of the ASR. Finally, Swerdlow and Geyer [37] also analyzed the effects of PPT lesions on the ASR, and in agreement with Koch's et al. [36] results, found a decreased PPI in lesioned rats. However, Swerdlow and Geyer's [37] results also showed an increased ASR responding in the rats that underwent PPT lesions, suggesting a role of this region in the inhibition of the startling stimulus itself.

Neuronal activity in the mediating circuit

In this section, we present neural recordings of the brain regions involved in the generation of the ASR, described in Fig. 9.

Excitatory pathway

Cochlear root neurons

Through extracellular recordings, Sinex et al. [38] were able to characterize CRNs' responses. They found that single units exhibited a marked first-spike to noise onset, followed by a short (2 to 3 ms) refractory period, after which the cells continued to fire, but with a lower rate than the one observed at the tone onset. These features are shown on Fig. 10, for both individual (left panel) and averaged (right panel) trials.

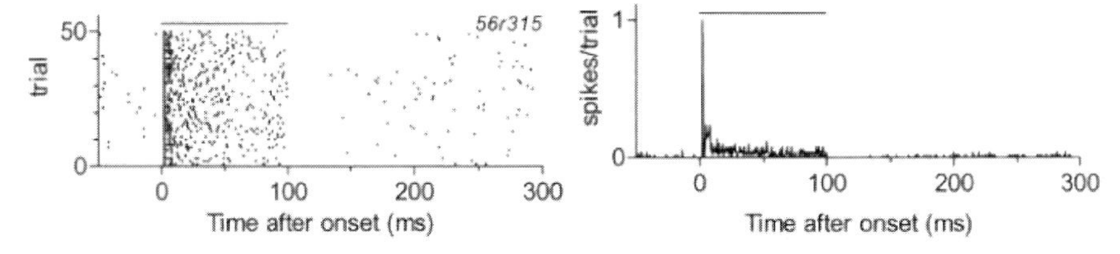

Figure 10. Discharge rates for a CRN, shown as dot rasters (Left panel) and peristimulus time histogram (Right panel) (reproduced from [38], with permission). The solid line indicates the duration of the acoustic stimulus.

Sinex et al. [38] also analyzed the CRN frequency dependence and found that these neurons exhibit high characteristic frequencies (CF), around 30 KHz (i.e., the threshold for responding is the lowest for this frequency).

Caudal pontine reticular nucleus

Lingenhöhl and Friauf [24, 25] characterized PNC neurons responses in rats using intracellular recordings. In agreement with behavioral data showing a threshold around 80 dB in order to generate acoustic startle responses [3], Lingenhöhl and Friauf [24] reported that tones below 80 dB, although capable of producing excitatory post-synaptic potentials (EPSPs) on PNC neurons, were less likely to generate action potentials. Fig. 11 shows the number of firing neurons in this brain region as a function of input noise level.

It is worthwhile noticing that the PNC is an area where signals from other regions relevant to acoustic stimulation converge, such as the PPT [30, 36]. Hence, in order to better determine the CRN-PNC pathway connectivity (i.e., without interference from other brain areas) a study similar to those performed by Lingenhöhl and Friauf could be carried out, disrupting the influence of the PPT on the PNC either by electrolytic lesions or drug manipulations (via cholinergic antagonists).

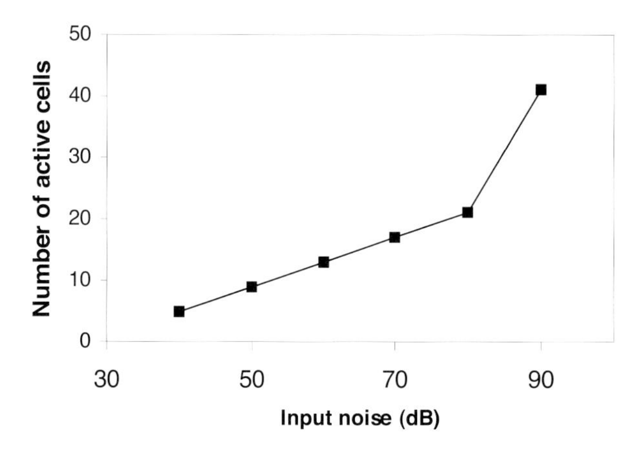

Figure 11. Number of active PNC cells as a function of sound intensity [24].

Inhibitory pathway

Inferior colliculus
Moller and Rees [39] recorded single cells' responses extracellularly in the inferior colliculus of the rat in order to investigate if these could be described using a linear model. They found that all cell measurements showed an almost linear relationship between responding and sound intensity, with different slopes.

After comparing the experimental data with the results provided by a linear model, Moller and Rees [39] inferred that neural discharge rates for increments and decrements in stimulus intensity should be asymmetrical. Other experimental results also seem to support the view that increments in sound intensity are more efficient in inhibiting than equal decrements [40].

Pedunculopontine tegmental nucleus
Reese, Garcia-Rill and Skinner [35, 41] studied the evoked potentials and unit responses of PPT neurons after acoustical stimulation. One of their main findings was that, within the PPT, there seems to be two different groups of neurons, with different latencies and threshold responses. The short-latency group (around 6 ms) exhibits also a lower threshold (from 50 to 60 dB), whereas the long-latency group (around 14 ms) evidences a higher threshold (from 70 to 80 dB). The existence of these two cell groups in the PPT has led some researchers to propose that the facilitatory effect of the prepulse for short lead intervals might be controlled by the short-latency cells, whereas inhibition could be regulated by the long-latency group [42].

Neural basis of ASR habituation and sensitization

In order to study the neural basis of habituation, Leaton, Casella and Borszcz [43] compared the ASR from decerebrate rats and normal animals. Their results showed that rats with incisions from near the intercollicular junction to the pontine-mesencephalic junction exhibited short-term habituation of the ASR, consistent with the view that this process takes place in the basic stimulus-response pathway, through some form of synaptic depression. Since in their surgical procedure all animals in the experimental group sustained damage to the IC (which, as described above, is a main component of the inhibitory pathway in the startle circuit), this damage might be responsible for the increased responding. However, decerebrate rats showed no long-term habituation of the ASR, expressed as a non-changing responding over days, in contrast to the decreased responding observed in control animals. These results suggest two different mechanisms of habituation, a short-term one occurring within the stimulus-response pathway, and a long-term mechanism involving brain areas more rostral to the locus of the lesion.

Based on the assumption that habituation and sensitization involve different neural processes [44], Davis et al. [45] investigated if eliciting electrical startle responses from the CN and the PNC could separate these mechanisms. According to their results, since increased startle occurred after stimulation of the CN or the PNC, sensitization should take place in the later parts of the circuit (PNC or motor neurons). By the same token, since responding decreases after stimulation of the CN but not the PNC, habituation should occur in the early stages of the circuit (before the PNC).

Since synapses from auditory afferents arising from the CN or CRN are the most likely candidates to provide the cellular basis for short-term habituation, Weber, Schnitzler and Schmid [46] extracellularly stimulated neurons in the lateral superior olive in order to excite traversing fibers originating in the CN, and projecting to the PNC. By applying presynaptic current bursts in these cells (mimicking the high frequency firing during auditory stimulation), they were able to measure excitatory post-synaptic currents (EPSCs) in the PNC neurons. Their results showed an exponential decay of EPSCs following repeated presynaptic stimulation over trials. The parameters of this homosynaptic depression (HSD) matched several others from short-term plasticity, e.g., facilitation of the second EPSC in a paired pulse paradigm for short interstimulus intervals. This result is of particular interest, for it provides support to the hypothesis that prepulse facilitation occurs in the CRN-PNC pathway, and not through decreased activity in the inhibitory path. In addition, Weber et al. [46] provided evidence that PNC neurons might receive inputs from cochlear nucleus cells via group III metabotropic glutamate receptors, since specific antagonists of these receptors blocked the HSD.

Modulatory circuit

The above-described mediating circuit of the ASR receives influence from other brain regions, which modulate its behavior [47]. These modulatory areas, which are schematized in Fig. 12, include the VP, nucleus accumbens (NAC), entorhinal cortex, medial prefrontal cortex (MPFC), hippocampus and the amygdala.

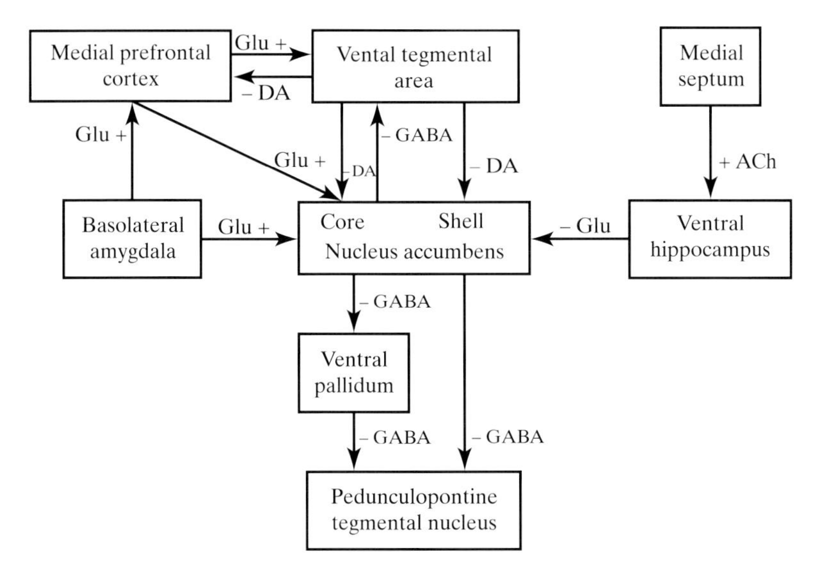

Figure 12. Idealized circuit showing the influence of higher brain structures in the modulation of the ASR and PPI.

Ventral pallidum (VP)

Dopamine infusion in the NAC reduces PPI in rats [48]. This disruption is believed to be mediated by GABA-ergic projections from the NAC to the VP, since this effect can be eliminated by infusions of GABA agonists (such as muscimol) into the VP [49]. In order to determine which pallidal efferents influence the ASR, Kodsi and Swerdlow [21] studied the three major areas to which the VP projects to, using both lesions and infusion techniques. These areas include the PPT, MD and Subthalamic Nucleus (STN). Their results showed that quinolinic lesions of the PPT significantly reduced PPI, but did not alter the amplitude of the ASR. Intra-PPT infusion of the GABA agonist muscimol reduced PPI when the drug dose was 10 ng or higher (when compared to a saline control), but did not alter the startle amplitude for any dose. In contrast, quinolinic lesions of the STN did not produce any statistical change on PPI regulation and also failed to modify the forebrain dopaminergic regulation of PPI. The effects of the MD lesions on PPI were more complex, since when rats were initially tested, lesioned animals showed the same inhibition than controls. However, subsequent testing using saline solution in both lesioned and control rats, showed a decreased PPI in the former group. Kodsi and Swerdlow [21] conducted further studies to determine if these results were caused by a delayed quinolinic acid action, a delayed circuit modification or an experimental design effect. They concluded that the most likely cause was an excitatory projection from the MD neurons to the MPFC that forms a cortico (MPFC) – striato (NAC) – pallidum (VP) – thalamic (MD) circuit. Therefore, Kodsi and Swerdlow [21] suggested that the VP influences PPI through a main VP-PPT projection (via GABAergic transmission) and a VP-MD connection, which indirectly modulates PPI through the cortical loop.

Nucleus accumbens

Even though dopamine agonist infusions in the NAC modulate PPI through a VP-PPT projection (see above), Kretschmer and Koch [50] proposed that the PPI disruption caused by intra-NAC glycine NMDA antagonists is not regulated by the VP. To test their hypothesis, they analyzed the effects of systemic injections of both a dopamine agonist (apomorphine) and a non-competitive NMDA receptor antagonist (dizocilpine) on PPI and the ASR amplitude. Their results showed that whereas in sham lesioned animals both drugs reduced PPI, this effect was disrupted in VP lesioned animals when apomorphine was infused, but not when dizocilpine was used. Regarding the drugs' effect on the ASR amplitude, only the dizocilpine infused rats significantly increased their responding, and this effect was preserved in animals with VP lesions. Similar results were obtained with intra-NAC drug infusions, when dopamine or 7-CLKYN (glycine-site NMDA antagonist) was used. In this case, PPI in sham lesioned rats is disrupted in both dopamine and 7-CLKYN infused animals, but this effect is only present in 7-CLKYN infused animals with VP lesions. These data support the hypothesis presented by Kretschmer and Koch [50] that NMDA mediated PPI disruption is not regulated by the VP, as is the case in the dopamine disruption of PPI, but instead could be controlled by a direct NAC-PPT GABA-ergic projection.

Entorhinal cortex (EC)

Using microdyalisis techniques, Goto, Ueki, Iso and Morita [51] studied the effects of bilateral EC lesions on dopamine release in the NAC in rats during acoustic stimulation. They reported that lesions in the EC decreased PPI, but did not alter significantly the startle amplitude, or its habituation across trial blocks. In addition, they found that concentration of extracellular dopamine in the NAC was higher in the lesioned group, even without acoustic stimulation, suggesting that this projection could be acting as a tonic regulator of dopamine in the NAC. In sum, these results are in agreement with physiological data showing that the NAC is a region interconnected with other brain dopaminergic areas (such as the ventral tegmental area, VTA), and therefore, when one of the NAC's afferents is lesioned (such as the EC), dopamine level changes can arise.

Medial prefrontal cortex and ventral hippocampus

Based on neurophysiological data indicating that schizophrenic patients exhibit GABA-ergic deficits in the PFC and hippocampus, and on behavioral data showing a decreased PPI of the ASR on these subjects, Japha and Koch [52] studied the influence of these brain regions in the modulation of PPI in rats. Animals injected with a GABA antagonist (picrotoxin) in the MPFC exhibited a dose-dependent PPI reduction, an effect that could be reversed by the intraperitoneal injection of haloperidol (a dopamine antagonist). Similar results, but of weaker magnitude, were found when the picrotoxin was injected into the ventral hippocampus. The combined administration of picrotoxin and haloperidol in both areas, however, did not restore the ASR level to the control value, indicating a wearing effect of the drug combination in the motor system. When picrotoxin was infused in the lateral PFC, no PPI reduction was observed. These experimental results suggest that the MPFC and ventral hippocampus influence PPI via dopaminergic modulations.

Dopamine depletion of MPFC neurons reduces PPI in rats
Bubser and Koch [53] studied the effects of reducing the concentration of prefrontal cortex dopamine on PPI, by administering two different doses of 6-hydroxydopamine hydrobromide (6-OHDA), 3.0 and 6.0 μg/μl. Even when neither concentration affected the amplitude of the ASR, the higher dose of 6-OHDA significantly reduced PPI. Bubser and Koch [53] explained these results in terms of the inhibitory effects of dopamine in MPFC neurons. When dopamine concentration in this region is reduced, glutamatergic projections from this area to the NAC and VTA are strengthened, increasing dopamine activity in the NAC. However, the mechanisms underlying this outcome are not completely identified, and two alternatives have been proposed. The first one suggests that increases in dopamine activity in the NAC could be the consequence of a presynaptic glutamatergic connection from the MPFC to the VTA-NAC dopamine projection. The second alternative proposes that an increased glutamatergic stimulation of the VTA neurons by the MPFC results in an increased dopamine release from the VTA into the NAC. However, these mechanisms are not exclusive, and therefore they could occur simultaneously.

Effects of noncompetitive NMDA antagonists (dizocilpine) in limbic regions
Bakshi and Geyer [54] studied the effects of microinfusions of dizocilpine in different brain areas involved in the regulation of PPI. Their results showed that only high doses (6.25 micrograms) of dizocilpine statistically reduced PPI in the amygdala and dorsal hippocampus, and found a similar trend towards significance for higher doses in the MPFC. In the other regions analyzed (ventral hippocampus, NAC and MD), PPI did not decrease after administration of dizocilpine. Startle amplitude, however, increased when dizocilpine was infused into the amygdala, dorsal hippocampus, NAC and MD. These results seem to indicate a role from the amygdala, dorsal hippocampus and to a lesser degree, from the MPFC in the regulation of PPI (which reflects sensory-motor gating), and that different brain regions are responsible for the startle magnitude changes and PPI decreasing effects observed after the systemic administration of NMDA antagonists.

Hippocampus
Pouzet et al. [55] analyzed the effects of different types of lesions in the hippocampus. Electrolytic and aspiration lesions of the dorsal hippocampus did not affect the startle response or its habituation, but rats that sustained dorsal hippocampus aspiration exhibited decreased PPI. Based on this last result, Pouzet et al. [55] decided to study the effects of selective excitotoxic (NMDA) lesions in the dorsal, medial and complete hippocampus, but their results showed no differences in startle amplitude, habituation or PPI. Since excitotoxic lesions do not damage axons going through the brain region under analysis, the authors decided to lesion the fimbria-fornix (FF), which is the main path connecting the hippocampus to the NAC (a main modulatory component of PPI, as explained above). Lesions to the FF did not affect the startle amplitude,

habituation or inhibition of the response. However, when systemic apomorphine (a dopamine agonist) was subsequently administered, the FF lesioned rats showed a larger PPI decrease than controls (reversing facilitation for low prepulses), indicating that these hippocampus lesions might be necessary for the disruption of PPI, but not for its manifestation.

Carbachol infusion into the dentate gyrus of the rat's hippocampus disrupts PPI
Caine, Geyer and Swerdlow [56] reported that infusion of the cholinergic agonist carbachol into the dentate in rats resulted in the disruption of PPI. Intra-hippocampal doses of carbachol administered bilaterally produced a dose-dependent effect in both the startle magnitude and PPI. Both low (up to .4 μg) and high (up to 1.6 μg) drug doses reduced the ASR, but this effect was not statistically significant in the first case. In contrast, PPI was disrupted even with low doses. In order to assess the specificity of the region involved in this effect, Caine et al. [56] infused carbachol into the cortex area surrounding the hippocampus of different rats, and tested their responding to the same stimuli. Their results showed that in this case neither the ASR nor PPI were affected by the drug administration, inducing the authors to conclude that carbachol acts on the hippocampus. In addition, Caine et al. [56] pretreated rats with spiperone (D2 dopamine receptor antagonist, administered subcutaneously), and found that this manipulation could not reverse the carbachol-induced PPI disruption. In contrast, if apomorphine (administered subcutaneously) was used to reduce PPI (instead of carbachol), Caine et al. [56] reported that the spiperone pretreatment could reinstate inhibition to control levels. Finally, these authors found that carbachol infusion into the hippocampus had the same PPI disrupting effects if the acoustic startling stimulus was replaced with an airpuff, suggesting that this modulation is not modality-specific. Swerdlow, Geyer and Braff [47] suggested that this hippocampus regulation of PPI could reflect septum-hippocampus projections, since AMPA activation of the septal nucleus reduces PPI, an effect that can be reversed by infusion of scopolamine (a muscarinic antagonist) into the hippocampus [57].

Amygdala
Campeau and Davis [6] studied the role of different regions of the amygdala in fear-potentiated startle. In order to assess the changes in the ASR due to fear, rats were initially trained in a conditioned suppression paradigm, where either a tone or a visual CS was paired with an electric shock (US). When exposed to a startling auditory stimulus, responding increased in the presence of the previously conditioned CS. Campeau and Davis [6] showed that post-training electrolytic or ibotenic acid lesions of the central nucleus of the amygdala completely eliminated the potentiation by fear of the ASR, when both visual and auditory CSs were conditioned to an aversive US (electric shock) in the same rat. In addition, similar results were obtained for post-training electrolytic or NMDA lesions of the basolateral complex of the amygdala, as well as for pre-training NMDA lesions of this region. Since pre-training NMDA lesions of the central nucleus (but not the basolateral complex) did not hinder the potentiation by fear of startle, and post-training lesions of both regions disrupted it, Campeau and Davis [6] argued that the central nucleus of the amygdala should be a final relay for

the expression of fear conditioning, whereas the basolateral complex should act as a relay of information from cortical areas to the central nucleus of the amygdala.

Lesions to the amygdala in rats lead to diminished conditioned emotional responses and avoidance, suggesting a role of this area in fear. Furthermore, electrical stimulation of the central and basolateral nucleus in cats produces threat or defensive rage. Therefore, Rosen and Davis [58] analyzed the role of the amygdala in the modulation of the ASR by electrically stimulating different areas within this region in rats' brains. They found that pairing a startle stimulus with low pulses of current ranging from 200 to 400 μA (which by themselves did not elicit any behavioral responses) could increase the ASR twofold, the central nucleus of the amygdala being the most effective site of stimulation (i.e., that it required lower currents to increase the response). Other regions of the amygdala that potentiated the ASR under electrical stimulation were the medial area, and the medial and basolateral nuclei. Interestingly, the ventral amygdalofugal pathway (VAF), which sends projections to the brain stem (including the PNC) and originates in the medial area of the amygdala, exhibited the lower threshold for ASR increase. Hence, Rosen and Davis [58] proposed that the central nucleus of the amygdala modulates the ASR, through the descending VAF.

Even when low electrical stimulation of the amygdala nuclei does not elicit behavioral activity, high currents can evoke startle-like responses. Yeomans and Pollard [59] analyzed the thresholds of response generation in the VAF, midbrain areas and medulla sites using a one pulse electrical stimulation, and studied their refractory periods. In addition, they examined the neural connectivity of these different regions by delivering pairs of conditioning-test C-T pulses at different intervals. The collision test of the midbrain and medulla yielded symmetric C-T intervals, providing evidence of bidirectional action potential conduction (axonal) between these two areas. In contrast, the VAF-midbrain collision test produced an asymmetric C-T interval curve, as well as the VAF-medulla test, indicating an indirect (synaptic) transmission between the VAF and the midbrain or medulla. To further assess the role of the midbrain areas involved in the electrically-produced startle responses on the potentiation of the ASR, Yeomans and Pollard [59] used a classical conditioning paradigm, where a light was conditioned to an aversive US (footshock). After conditioning, when the light was presented along with an acoustic stimulus, the startle response was twice as large as the one produced by the acoustic stimulus alone, or by electrically stimulating the VAF or medulla sites. Electrolytical lesions of the midbrain area did not affect the response to the acoustic stimulus, but disrupted the fear-potentiation effect of the startle response when the acoustic stimulus was presented along with the light. In addition, the midbrain-lesioned rats also showed a decreased startle-like response when either the VAF or medulla sites were electrically stimulated. However, electrical stimulation of the VAF after the midbrain lesion caused a complete elimination of the startle response, whereas the reduction of the ASR after stimulation of the medulla was only partial. Since an amygdala-PNC circuit would predict no disruption of a medulla electrically-evoked startle response after a midbrain lesion, Yeomans and Pollard [59] proposed that the midbrain might be the target of the VAF pathways which project beyond the PNC to the medulla in parallel to the primary acoustic startle circuit.

Different dopamine receptors in the basolateral amygdala (BLA) in the rat regulate PPI in opposite ways, but do not affect latent inhibition (LI) in a conditioned taste-aversion (CTA) paradigm
LI refers to the delayed conditioning of a target stimulus (CS) to an unconditioned stimulus (US), when non-reinforced presentations of the CS precede the conditioning trials. Based on experimental results showing that (a) dopamine transmission in the MPFC and NAC is regulated by a BLA dopamine-controlled mechanism [60], (b) PPI can be disrupted by dopamine increases in both the MPFC and NAC [50, 53] (see above), (c) dopamine release in the NAC is reduced in LI [61], and (d) BLA lesions, as described above, decrease PPI [54], Stevenson and Gratton [62] designed a study to examine the role of BLA dopamine in the regulation of both PPI and LI. In their experiments, two drugs were used to block specific dopamine receptors in the BLA, namely SCH 23390 (D1 receptor blocker) and raclopride (D2/D3 receptor blocker). Infusion of dopamine blockers into the BLA did not affect the startle response, but had distinct effects on PPI. Whereas SCH 23390 enhanced PPI for the lowest and highest prepulse intensities tested, raclopride caused a dose-dependent reduction of PPI. The LI experiment was conducted in a CTA paradigm, in which rats in the preexposed group had access to sucrose (CS) for three 30-min sessions, whereas rats in the control condition had access to water. In order to determine the effects of BLA dopamine in this task, 5 min before the preexposure and conditioning sessions began, Stevenson and Gratton [62] injected vehicle, SCH 23390 or raclopride to the rats' BLA. Their results showed that none of the dopamine blockers affected LI. Taken together, these findings led Stevenson and Gratton [62] to conclude that PPI can be modulated by BLA dopamine and that this modulation (either increasing or decreasing) depends on the type of dopamine receptor activation. Moreover, since infusion of SCH 23390 into the BLA increases NAC dopamine [63], it would be expected for this drug manipulation to reduce PPI [48] (see above), but the opposite result was obtained by Stevenson and Gratton [62], who suggested that the BLA dopamine regulation of PPI should be independent of the NAC and MPFC. Finally, since none of the drugs analyzed had an effect on LI, Stevenson and Gratton [62] concluded that dopamine in the BLA (which is not a part of the LI circuit [64]) does not influence LI.

Pharmacology of PPI

Systemic injections of different drugs, sometimes combined with brain lesions, have been used to characterize the circuit controlling PPI.

Differential effects of NMDA-receptor antagonists and apomorphine-induced PPI disruption on startle response
Yee et al. [65] reported that intraperitoneal infusion of dopaminergic agonists (apomorphine) not only disrupted PPI, but also increased the prepulse-elicited reactivity. According to Yee et al. [65], this result contradicted the hypothesis set forth by Davis et al. [66], by which the diminished prepulse inhibition was due to a decreased detectability of the prepulse. In a similar way, Yee et al. [15] analyzed the effects of

non-competitive NMDA-receptor anatgonists (dizocilpine) on the reactivity elicited by prepulses, and found that this reactivity increased for low doses (0.1 mg/kg), and decreased for high doses (3 mg/kg). PPI disruption also exhibited a dose-dependent behavior, increasing for higher doses. These findings showing a negative correlation between prepulse reactivity (decrease for high doses) and PPI disruption after the administration of dizocilpine in mice, led Yee et al. [15] to propose a model in which the detection of a prepulse activates two different mechanisms, one triggering the information gating process (i.e., PPI), and the other responsible for the prepulse-generated reactivity. They assumed different modulatory factors on these mechanisms and proposed differential drug effects on them. Since apomorphine increased the prepulse-elicited reactivity and disrupted PPI, Yee et al. [15] suggested an excitatory effect of this drug on the second mechanism (prepulse-reactivity), and an inhibitory one on the first one (PPI). Analogously, because administration of dizocilpine decreased both prepulse-elicited reactivity and PPI, they proposed an inhibitory modulation of this drug in both mechanisms.

Effects of typical and atypical antipsychotics on early ventral hippocampus lesions
Since neonatal ventral [67] and adult hippocampal lesions [68] in rats produce post-puberty abnormal behaviors analogous to those observed in schizophrenic patients (such as high responsiveness to stress or novelty situations), Le Pen and Moreau [69] studied the ability of typical antipsychotics (haloperidol) and atypical antipsychotics (clozapine, olanzapine and risperidone) to counteract the PPI deficits observed in rats that sustained neonatal hippocampal lesions. Haloperidol, known to reinstate LI following hippocampal lesions [68], was unable to reinstate PPI in lesioned animals, regardless of the dose or prepulse level. Clozapine, on the other hand, could reverse the PPI deficits caused by the hippocampal lesions, in a dose-dependent way, for the three higher prepulse intensities. Intermediate doses of olanzapine (3 mg/kg) could reinstate PPI for intermediate prepulse intensities, and risperidone could reverse the deficits at high prepulse levels in a dose-dependent way. Regarding the effects of the antipsychotics on the responding amplitude, all the drugs tested in the study reduced the ASR in a dose-dependent way, and no effect of neonatal ventral hippocampal lesion was obtained. These results show that atypical antipsychotics seem to reinstate PPI (and therefore, sensory-motor gating abilities) in rats better than typical antipsychotics, which is in agreement with the results obtained in schizophrenic patients [70]. Le Pen and Moreau [69] argued that atypical and non-typical antipsychotics could reinstate PPI due to their agonist effect on the glutamatergic system.

Interactions between NMDA glutamatergic receptor blockade and nicotinic cholinergic agonists in PPI
Levin et al. [71] analyzed the ASR of rats after the administration of different doses of nicotine (cholinergic agonist) and dizocilpine (NMDA antagonist). Their results showed an increase of PPI with increasing nicotine doses, probably acting on the septal-hippocampal excitatory cholinergic projection in the modulatory circuit (see Fig. 12). Dizocilpine subcutaneous infusions resulted in a dose-dependent disruption of PPI. This PPI deficit caused by dizocilpine was enhanced when an intermediate (0.2

to 0.4 mg/kg) nicotine dose was simultaneously used. In a later experiment, Levin et al. [71] studied the interactions of these drugs with the atypical antipsychotic clozapine, and found that this drug by itself could not reverse the PPI disruption effects caused by dizocilpine. However, when clozapine was combined with nicotine, both drugs together could reinstate PPI in the rats, leading them to suggest an interaction effect of these drugs, which has a practical importance in the treatment of patients with sensory-motor gating deficits, such as schizophrenics. Since clozapine can reinstate PPI after hippocampal lesions through the glutamatergic system [69] and nicotine might act on the septal-hippocampal excitatory cholinergic connection (see above), in terms of the model shown in Fig. 12, the combined action of both drugs results in an increased excitation of the NAC. This activation, in turn, generates more inhibition of the PNC by the PPT, and reinstates PPI, as reported by Levin et al. [71].

Neuropsychiatric disorders and PPI

Prepulse inhibition is affected in subjects with specific neuropsychiatric disorders, such as Huntington's disease, Parkinson's disease, Tourette syndrome, and schizophrenia. Huntington's disease deficits in PPI are related to deterioration of GABAergic cells in the striatum, which project to the VP and also regulate PPI [72]. Diminished PPI in Parkinson's disease is thought to be influenced by dopamine receptors in the striatum [73], whereas the reduction in PPI observed in Tourette syndrome patients is believed to involve the striatum or cortical-striatum projections [74]. Braff et al. [75] compared the responses of acute and chronic schizophrenic patients to 50 ms, 104 dB tones preceded by continuous, 71 dB prepulses to those of control subjects. In their study, Braff et al. [75] used short lead intervals (30 to 120 ms), and found that schizophrenic patients showed weaker PPI. This result led these authors to hypothesize that schizophrenia disrupts the preattentive sensory filtering (sensory gating), which could cause information overload [76].

Kumari et al. [77] conducted an fMRI experiment designed to examine the brain areas involved in PPI. Even when this study used tactile stimuli as both pulses and prepulses, there is evidence that the brain areas that regulate PPI are the same as those involved when acoustic stimuli are used [72]. In their study, both healthy subjects and schizophrenic patients received a 40 ms, 30 psi airpuff (pulse) preceded in some trials by a 20 ms, 10 psi airpuff (prepulse), with a 120 ms lead interval. Startle responses were obtained measuring the electromyographic activity elicited by eye blinks. Kumari et al. [77] reported that in trials where pulses and prepulses were presented, healthy subjects showed increased BOLD responses bilaterally in the striatum, which extended to the hippocampus and thalamus. Schizophrenic patients also showed activation in these areas, but of a lower magnitude, as well as reduced (albeit not statistically significant) PPI compared to healthy subjects. In addition, all subjects showed a linear relationship between BOLD responses and PPI, consistent with the notion that the striatum, hippocampus and thalamus are relevant structures in PPI, as described above.

A computational model of ASR and PPI

Recently, Schmajuk and Larrauri [78] presented a real-time model of acoustic PPI and PPF in animals and humans. The assumptions introduced in the model were derived from behavioral experiments in which similar independent variables produced non-conflicting dependent values [2, 3, 11].

The assumptions were:

1. The ASR is controlled by the positive value of changes in an exponential function of the intensity of the input noise expressed in dB. Under this assumption, the model correctly describes experimental data showing that the startle response (a) grows as a nonlinear function of the input noise [3], and (b) is elicited by increments, but not decrements, in the background noise (Blumenthal, personal communication). Instead, had a linear function of the intensity of the input been assumed, the model would wrongly predict that increments and decrements of identical absolute value produce responses of the same strength.
2. The ASR decreases (and PPI increases) with the absolute value of changes in a linear function of the input noise expressed in decibels. Under this assumption, the model is able to describe experimental results showing that both increments [3] and decrements [11] in noise intensity produce PPI. In addition, the model properly addresses Hoffman and Searle's [3] data showing that when a 110 dB prepulse almost completely inhibits the startle response to a 140 dB pulse, a 50 dB prepulse also produces some degree of inhibition, a combined result unattainable under the assumption that inhibition is controlled by an exponential function of the input expressed in dB. Notice that, because we assumed that PPI increases with the absolute value of changes in the input noise, the model captures the idea [79–81] that PPI increases as the ratio between prepulse intensity and background intensity ratio (signal to noise ratio) increases, either by increasing the prepulse without changing the background intensity or by decreasing the background without changing the prepulse intensity.

Because, according to (1.) the ASR increases as an exponential function of the difference between the intensity of the pulse and the intensity of the background noise, and according to (2.) the ASR decreases as a linear function of the same difference, the ASR to a given pulse should be an inverted-U function of the background level noise. Therefore, the two above-mentioned assumptions correctly account for the results reported by Ison and Hammond [4] (Experiment 6) described in the section "The ASR is an inverted U-shape function of the background level noise."

Notice that because both pulses and prepulses have access to both excitatory and inhibitory pathways, the model expects that (a) a pulse will inhibit itself, and (b) a prepulse will generate a weak startle response.

3. Facilitation of the startle is controlled by the positive change in an exponential function of the input noise expressed in dB. Under this assumption, the model correctly describes experimental data showing facilitation of the startle for short

lead intervals and weak prepulses [2] and the fact that facilitation remains constant in the case of decrements in the level of background noise that last until the presentation of the startle stimulus [11].

The descriptions of the Schmajuk-Larrauri [78] model were confirmed by applying it to the original data, as shown in the right panels of Figs. 1–8: 1) PPI increases with prepulse intensity [3], 2) PPI and PPF depend on the duration of the lead interval [2], and 3) PPI can be produced by a decrease in the background noise [11]. In addition, the model is able to describe the following results: 4) PPI produced by a decrease in the background noise depends on the intensity decrement and the duration of the lead interval [12], 5) PPI increases with prepulse duration [13], and 6) A prepulse does not inhibit the inhibitory power of a pulse [14]. The model also describes experimental results showing that lesions of the PPT enhance the strength of the startle response and impair PPI [37]. In addition, the model correctly predicted that PPI decreases with increasing pulses intensity [15].

Figure 13 shows a diagram of the model, which includes 1) an excitatory pathway with output E_4 activated by the positive values of changes in an exponential function of the input noise, 2) a facilitatory pathway with output E_6 activated by the same positive values, and 3) an inhibitory pathway with output L_5 activated by the absolute values of changes in a linear function of the input noise.

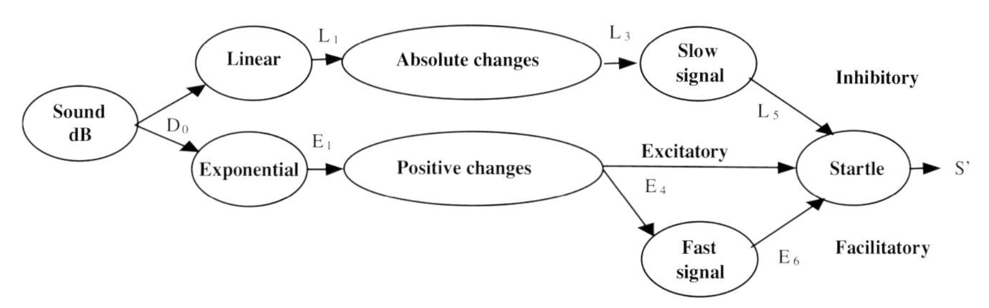

Figure 13. Diagram of the model, showing the excitatory, facilitatory and inhibitory pathways involved in the generation of simulated startle responses.

Schmajuk and Larrauri [78] mapped different parts of the model onto the brain circuits controlling ASR and PPI (see Fig. 9). Interestingly, even though the mapping assumptions were based on the behavioral data previously mentioned, neural activity seems to provide some support for our conjectures.

1. As mentioned, we assumed that the ASR is controlled by (a) the positive value of changes in an (b) exponential (sigmoid) function of the intensity of the input noise expressed in dB. In the case of an increment in the ambient noise level, this positive value is found at the onset of the increment. Likewise, neurons in the CRN, part of the excitatory pathway, show (a) responding to the onset of a pulse, and (b) firing rate that is a sigmoid function of the sound intensity [38].

2. As mentioned, we assumed that the threshold for generating the ASR is at around 80 dB. Similarly, the number of active neurons in the PNC, part of the excitatory pathway, rapidly increases for inputs above 80 dB [24].
3. As mentioned, we assumed that PPI increases (a) with the absolute value of changes (b) in a linear function of the input noise expressed in dB. In the case of an increment in the ambient noise level, these absolute values are found at both the onset and offset of the increment. Correspondingly, neurons in the inferior [82] and superior colliculi [83], both parts of the inhibitory path, show (a) responding to both the onset and offset of a pulse, and (b) response amplitude that is a linear function of the sound intensity.

Cognitive significance of prepulse inhibition

What is the survival value of inhibiting the startle response? According to Graham's [16] protection-of-processing hypothesis, the prepulse triggers a gating mechanism attenuating the startle response to allow the perceptual processing of the prepulse. In her view, the startle would disrupt normal perceptual processing. This idea is supported by data showing that perception of the prepulse is linked to its ability to inhibit startle [84–87, 40] (see also [88]).

Fendt et al. [33] extended Graham's theory and specified how brain areas activated during PPI improve perceptual processing and assessment of the prepulses. They suggested that the startle response (which includes eye closing and contraction of the whole body) would seriously hinder visual exploration of the environment. Even though startle responses might protect from attacks, PPI would allow the generation of exploratory responses that benefit sensory processing.

According to Fendt et al. [33], 1) activation of the SC contributes to perceptual processing by inducing orienting toward and foveation of, the prepulse stimulus, via the tectoreticulospinal pathway, 2) activation of the PPT (and LDT) enhances perceptual processing by cholinergic activation of thalamo-cortical systems, via direct PPT projections to thalamus, and 3) activation of the PPT results in the exploration of novel and rewarding stimuli through the activation of mesolimbic dopamine neurons. Furthermore, 4) activation of the PPT might be involved in attentional and learning processes through that activation of the thalamus, basal forebrain and basal ganglia [89]. Fendt et al. [33] proposed that 1) the startle reflex is organized in the hindbrain to maximize speed (CN, PNC, motoneurons), 2) more complex responses of orienting, approach and avoidance are organized in the midbrain (IC, SC, PPT), and 3) the fuller processing of stimuli occurs at forebrain levels (VTA, substantia nigra, thalamus), as shown in Fig. 14. In sum, according to the protection-of-processing view, the prepulse quickly inhibits the startle response in the hindbrain while allowing further processing of the prepulse in the forebrain.

What are the cognitive benefits and, therefore, evolutionary advantages of inhibiting the inhibition of the startle response? We suggest that if the prepulse is determined to be novel (cannot be recognized or predicted by the stimuli that precede it) in the VTA [90], the NAC is activated by the DA input from VTA [91], the PPT is inhibited through the GABA projection from the NAC, which results in a decrement in the

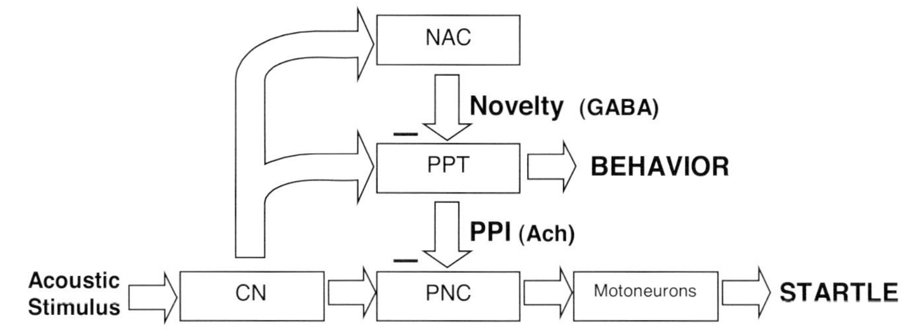

Figure 14. Circuit showing the inhibition of behavior and PPI by the NAC when novelty is detected in the environment.

cholinergic inhibition on the PNC. Therefore, the inhibition produced by the prepulse is reduced and the startle is released (see Schmajuk [64] for a complete description of the circuit). That is, if the prepulse is recognized, the appropriate response is produced; otherwise the animal startles and protects itself [92].

Discussion

In this chapter we present a comprehensive description of the behavioral properties of the ASR and PPI, and their anatomical bases.

This circuit mediating ASR, PPI and PPF includes an excitatory pathway from specialized neurons in the CRN, innervated directly by the auditory nerve, which projects to the PNC. In turn, PNC neurons stimulate motoneurons in the spinal chord, producing the startle response [24–26]. Other projections from the auditory nerve reach neurons in the CN, which exert an inhibitory effect on the PNC through IC-SC-PPT-PNC projections [27, 28, 30]. Since the inhibitory pathway contains more synapses to reach the PNC, its dynamics are slower than those of the excitatory pathway, as evidenced by the greater latency of the PPT to fire after acoustic stimulation compared to the latency of muscular responses observed in rats [26]. PPI can be also influenced by a modulatory circuit composed by several structures (VP, NAC, MPFC, hippocampus, amygdala) that exert their regulatory influence on the PPT [42]. Several lesion and drug manipulations, as well as neuropsychiatric disorders, can influence both the ASR and PPI.

We showed that a real-time model introduced by Schmajuk and Larrauri [78] is able to describe most of the data described in this chapter, including the properties of PPI and PPF, neural activity in different regions of the mediating circuit, and the effect of specific brain lesions on ASR and PPI. The model summarizes in a set of differential equations the large amount of data presented in our review.

Finally, we proposed to extend the protection-of-processing view of PPI [16, 33] that suggests that the prepulse quickly inhibits the startle response in the hindbrain while allowing the further processing of the prepulse in the forebrain. We suggest that if the prepulse is determined to be novel in the VTA and the NAC, the PPT is inhibited, which results in a decrement in PPI and the release of the startle response. In other words, if the prepulse was recognized as signaling another event, then the startle would stay inhibited and an appropriate response would be produced. If the prepulse was not recognized – i.e., determined to be novel – the inhibition would be inhibited and the animal would startle to protect itself.

References

1. Buckland G, Buckland J, Jamieson C, Ison J (1969) Inhibition of startle response to acoustic stimulation produced by visual prestimulation. *J Comp Physiol Psychol* 67: 493–496
2. Plappert C, Pilz P, Schnitzler H (2004) Factors governing prepulse inhibition and prepulse facilitation of the acoustic startle response in mice. *Behav Brain Res* 152: 403–412
3. Hoffman H, Searle J (1968) Acoustic and temporal factors in the evocation of startle. *J Acoust Soc Am* 43: 269–282
4. Ison J, Hammond G (1971) Modification of the startle reflex in the rat by changes in the auditory and visual environments. *J Comp Physiol Psychol* 75: 435–452
5. Davis M (1974) Sensitization of the rat startle response by noise. *J Comp Physiol Psychol* 87: 571–581
6. Campeau S, Davis M (1995) Involvement of the central nucleus and basolateral complex of the amygdala in fear conditioning measured with fear-potentiated startle in rats trained concurrently with auditory and visual conditioned stimuli. *J Neurosci* 15: 2301–2311
7. Steidl S, Yeomans J, Li L (2001) Conditioned brain-stimulation reward attenuates the acoustic startle reflex in rats. *Behav Neurosci* 115: 710–717
8. Pilz P, Schnitzler H (1996) Habituation and sensitization of the acoustic startle response in rats: Amplitude, threshold, and latency measures. *Neurobiol Learn Mem* 66: 67–79
9. Brown J, Kalish H, Farber I (1951) Conditioned fear as revealed by magnitude of startle response to an auditory stimulus. *J Exp Psychol* 41: 317–328
10. Reijmers L, Peeters B (1994) Effects of acoustic prepulses on the startle reflex in rats: A parametric analysis. *Brain Res* 661: 174–180
11. Stitt C, Hoffman H, Marsh R (1973) Modification of the rat's startle reaction by termination of antecedent acoustic signals. *J Comp Physiol Psychol* 84: 207–215
12. Ison J, Agrawal P, Pak J, Vaughn W (1998) Changes in temporal acuity with age and with hearing impairment in the mouse: A study of the acoustic startle reflex and its inhibition by brief decrements in noise level. *J Acoust Soc Am* 104: 1696–1704
13. Blumenthal T (1995) Prepulse inhibition of the startle eyeblink as an indicator of temporal summation. *Percept Psychophys* 57: 487–494
14. Swerdlow N, Shoemaker J, Stephany N, Wasserman L, Ro H, Geyer M (2002) Prestimulus effects on startle magnitude: Sensory or motor? *Behav Neurosci* 116: 672–681
15. Yee BK, Chang T, Pietropaolo S, Feldom J (2005) The expression of prepulse inhibition of the acoustic startle reflex as a function of three pulse stimulus intensities, three prepulse stimulus intensities, and three levels of startle responsiveness in C57BJ/6L mice. *Behav Brain Res* 163: 265–276
16. Graham F (1975) The more or less startling effects of weak prestimulation. *Psychophysiology* 12: 238–248
17. Blumenthal T (1997) Prepulse inhibition decreases as startle reactivity increases. *Psychophysiology* 34: 446–450

18. Dawson M, Schell A, Swerdlow N, Filion D (1997) Cognitive, clinical, and neurophysiological implications of startle modification. In: Lang P, Simons R, Balaban M (eds): *Attention and orienting: Sensory and motivational processes*. Erlbaum, Hillsdale, New Jersey, 257–279
19. Dawson M, Hazlett E, Filion D, Nuechterlein K, Schell A (1993) Attention and schizophrenia: Impaired modulation of the startle reflex. *J Abnorm Psychol* 102: 633–641
20. Hazlett E, Buchsbaum M, Tang C, Fleischman M, Wei T, Byne W, Haznedar M (2001) Thalamic activation during an attention-to-prepulse startle modification paradigm: A functional MRI study. *Biol Psychiatry* 50: 281–291
21. Kodsi M, Swerdlow N (1997) Regulation of prepulse inhibition by ventral pallidal projections. *Brain Res Bull* 43: 219–228
22. Bitsios P, Giakoumaki S (2005) Relationship of prepulse inhibition of the startle reflex to attentional and executive mechanisms in man. *Int J Psychophysiol* 55: 229–241
23. Hoffman H, Ison J (1980) Reflex modification in the domain of startle: I. Some empirical findings and their implications for how the nervous system processes sensory input. *Psychol Rev* 87: 175–189
24. Lingenhöhl K, Friauf E (1994) Giant neurons in the rat reticular formation: A sensorimotor interface in the elementary acoustic startle circuit? *J Neurosci* 14: 1176–1194
25. Lingenhöhl K, Friauf E (1992) Giant neurons in the caudal pontine reticular formation receive short latency acoustic input: An intracellular recording and HRP-study in the rat. *J Comp Neurol* 325: 473–492
26. Davis M, Gendelman D, Tischler M, Gendelman P (1982) A primary acoustic startle circuit: lesion and stimulation studies. *J Neurosci* 6: 791–805
27. Carlson S, Willott J (1998) Caudal pontine reticular formation of C57BL/6J mice: Responses to startle stimuli, inhibition by tones, and plasticity. *J Neurophysiol* 79: 2603–2614
28. Wu M, Suzuki S, Siegel J (1988) Anatomical distribution and response patterns of reticular neurons in relation to acoustic startle. *Brain Res* 457: 399–406
29. Willott J, Parham K, Hunter K (1991) Comparison of the auditory sensitivity of neurons in the cochlear nucleus and inferior colliculus of young and aging C57BL/6J and CBA/J mice. *Hear Res* 53: 78–94
30. Carlson S, Willott J (1996) The behavioral salience of tones as indicated by prepulse inhibition of the startle response: Relationship to hearing loss and central neural plasticity in C57BL/6J mice. *Hear Res* 99: 168–175
31. Leitner D, Cohen M (1985) Role of the inferior colliculus in the inhibition of the acoustic startle in the rat. *Physiol Behav* 34: 65–70
32. Fendt M (1999) Enhancement of prepulse inhibition after blockade of GABA activity within the superior colliculus. *Brain Res* 833: 81–85
33. Fendt M, Li L, Yeomans JS (2001) Brain stem circuits mediating prepulse inhibition of the startle reflex. *Psychopharmacology* 156: 216–224
34. Redgrave P, Mitchell I, Dean P (1987) Descending projections from the superior colliculus in rat: A study using orthograde transport of wheatgerm-agglutinin conjugated horseradish peroxidase. *Exp Brain Res* 68: 147–167
35. Reese N, Garcia-Rill E, Skinner R (1995) Auditory input to the pedunculopontine nucleus: I. Evoked potentials. *Brain Res Bull* 37: 257–264
36. Koch M, Kungel M, Herbert H (1993) Cholinergic neurons in the pedunculopontine tegmental nucleus are involved in the mediation of prepulse inhibition of the acoustic startle response in the rat. *Exp Brain Res* 97: 71–82
37. Swerdlow N, Geyer M (1993) Prepulse inhibition of acoustic startle in rats after lesions of the pedunculopontine tegmental nucleus. *Behav Neurosci* 107: 104–117

38. Sinex D, López D, Warr B (2001) Electrophysiological responses of cochlear root neurons. *Hear Res* 158: 28–38
39. Moller A, Rees A (1986) Dynamic properties of the responses of single neurons in the inferior colliculus of the rat. *Hear Res* 24: 203–215
40. Blumenthal T (1999) Short lead interval startle modification. In: Dawson M, Schell A, Böhmelt A (eds): *Startle modification: Implications for neuroscience, cognitive science, and clinical science*. Cambridge University Press, Cambridge, 51–71
41. Reese N, Garcia-Rill E, Skinner R (1995). Auditory input to the pedunculopontine nucleus: II. Unit responses. *Brain Res Bull* 37: 265–273
42. Swerdlow N, Geyer M (1999) Physiology of short lead interval inhibition. In: Dawson M, Schell A, Böhmelt A (eds): *Startle modification: Implications for neuroscience, cognitive science, and clinical science*. Cambridge University Press, Cambridge, 114–133
43. Leaton R, Casella J, Borszcz G (1985) Short-term and long-term habituation of the acoustic startle response in chronic decerebrate rats. *Behav Neurosci* 99: 901–912
44. Groves P, Thompson R (1970) Habituation: A dual-process theory. *Psychol Rev* 77: 419–450
45. Davis M, Parisi T, Gendelman D, Tischler M, Kehne J (1982) Habituation and sensitization of startle reflexes elicited electrically from the brainstem. *Science* 218: 688–690
46. Weber M, Schnitzler H, Schmid S (2002) Synaptic plasticity in the acoustic startle pathway: the neuronal basis for short-term habituation? *Eur J Neurosci* 16: 1325–1332
47. Swerdlow N, Geyer M, Braff D (2001) Neural circuit regulation of prepulse inhibition of startle in the rat: current knowledge and future challenges. *Psychopharmacology* 156: 194–215
48. Swerdlow N, Braff D, Masten V, Geyer M (1990) Schizophrenic-like sensorimotor gating abnormalities in rats following dopamine infusion into the nucleus accumbens. *Psychopharmacology* 101: 414–420
49. Swerdlow N, Braff D, Geyer M (1990) GABAergic projection from nucleus accumbens to ventral pallidum mediates dopamine-induced sensorimotor gating deficits of acoustic startle in rats. *Brain Res* 532: 146–150
50. Kretschmer B, Koch M (1998) The ventral pallidum mediates disruption of prepulse inhibition of the acoustic startle response induced by dopamine agonists, but not by NMDA antagonists. *Brain Res* 798: 204–210
51. Goto K, Ueki A, Iso H, Morita Y (2004) Involvement of nucleus accumbens dopaminergic transmission in acoustic startle: Observations concerning prepulse inhibition in rats with entorhinal cortex lesions. *Psychiatry Clin Neurosci* 58: 441–445
52. Japha K, Koch M (1999) Picrotoxin in the medial prefrontal cortex impairs sensorimotor gating in rats: Reversal by haloperidol. *Psychopharmacology* 144: 347–354
53. Bubser M, Koch M (1994) Prepulse inhibition of the acoustic startle response of rats is reduced by 6-hydroxydopamine lesions of the medial prefrontal cortex. *Psychopharmacology* 113: 487–492
54. Bakshi V, Geyer M (1998) Multiple limbic regions mediate the disruption of prepulse inhibition produced in rats by the noncompetitive NMDA antagonist dizocilpine. *J Neurosci* 18: 8394–8401
55. Pouzet B, Feldon J, Veenman C, Yee BK, Richmond M, Nicholas J, Rawlins P, Weiner I (1999) The effects of hippocampal and fimbria-fornix lesions on prepulse inhibition. *Behav Neurosci* 113: 968–981
56. Caine S, Geyer M, Swerdlow N (1991) Carbachol infusion into the dentate gyrus disrupts sensorimotor gating of startle in the rat. *Psychopharmacology* 105: 347–354
57. Koch M (1996) The septohippocampal system is involved in prepulse inhibition of the acoustic startle response in rats. *Behav Neurosci* 110: 468–477

58. Rosen J, Davis M (1988) Enhancement of acoustic startle by electrical stimulation of the amygdala. *Behav Neurosci* 102: 195–202
59. Yeomans J, Pollard B (1993) Amygdala efferents mediating electrically evoked startle-like responses and fear potentiation of acoustic startle. *Behav Neurosci* 107: 596–610
60. Simon H, Taghzouti K, Gozlan H, Studler J, Louilot A, Herve D, Glowinski J, Tassin J, Le Moal M (1988) Lesion of dopaminergic terminals in the amygdala produces enhanced locomotor response to D-amphetamine and opposite changes in dopaminergic activity in prefrontal cortex and nucleus accumbens. *Brain Res* 447: 335–340
61. Young A, Joseph M, Gray J (1993) Latent inhibition of conditioned dopamine release in rat nucleus accumbens. *Neuroscience* 54: 5–9
62. Stevenson C, Gratton A (2004) Role of the basolateral amygdala in modulating prepulse inhibition and latent inhibition in the rat. *Psychopharmacology* 176: 139–145
63. Hurd Y, McGregor A, Ponten M (1997) In vivo amygdala dopamine levels modulate cocaine self-administration behaviour in the rat: D1 dopamine receptor involvement. *Eur J Neurosci* 9: 2541–2548
64. Schmajuk N (2005) Brain-behaviour relationships in latent inhibition: a computational model. *Neurosci Biobehav Rev* 29: 1001–1020
65. Yee BK, Chang D, Feldon J (2004) The Effects of dizocilpine and phencyclidine on prepulse inhibition of the acoustic startle reflex and on prepulse-elicited reactivity in C57BL6 mice. *Neuropsychopharmacology* 29: 1865–1877
66. Davis M, Mansbach R, Swerdlow N, Campeau S, Braff D, Geyer M (1990) Apomorphine disrupts the inhibition of acoustic startle induced by weak prepulses in rats. *Psychopharmacology* 102: 1–4
67. Lipska B, Jaskiw G, Weinberger D (1993) Postpubertal emergence of hyperresponsiveness to stress and to amphetamine after neonatal excitotoxic hippocampal damage: a potential animal model of schizophrenia. *Neuropsychopharmacology* 9: 67–75
68. Schmajuk N, Christiansen B, Cox L (2000) Haloperidol reinstates latent inhibition impaired by hippocampal lesions: data and theory. *Behav Neurosci* 114: 659–670
69. Le Pen G, Moreau J (2002) Disruption of prepulse inhibition of startle reflex in a neurodevelopmental model of schizophrenia: Reversal by clozapine, olanzapine and risperidone but not by haloperidol. *Neuropsychopharmacology* 27: 1–11
70. Bakshi V, Geyer M (1995) Antagonism of phencyclidine-induced deficits in prepulse inhibition by the putative atypical antipsychotic olanzapine. *Psychopharmacology* 122: 198–201
71. Levin E, Petro A, Caldwell D (2005) Nicotine and clozapine actions on pre-pulse inhibition deficits caused by N-metyl-D-aspartate (NMDA) glutamatergic receptor blockade. *Prog Neuropsychopharmacol Biol Psychiatry* 29: 581–586
72. Swerdlow N, Paulsen J, Braff D, Butters N, Geyer M, Swenson M (1995) Impaired prepulse inhibition of acoustic and tactile startle response in patients with Huntington's disease. *J Neurol Neurosurg Psychiatry* 58: 192–200
73. Morton N, Chaudauri R, Ellis C, Gray NS, Toone BK (1995) The effects of apomorphine and L-dopa challenge on prepulse inhibition in patients with Parkinson's disease. *Schizophr Res* 15: 181–182
74. Swerdlow N, Benbow C, Zisook S, Geyer M, Braff D (1993) A preliminary assessment of sensorimotor gating in patients with obsessive compulsive disorder (OCD). *Biol Psychiatry* 33: 298–301
75. Braff D, Stone C, Callaway E, Geyer M, Glick I, Bali L (1978) Prestimulus effects on human startle reflex in normals and schizophrenics. *Psychophysiology* 15: 339–343

76. Cadenhead K, Braff D (1999) Schizophrenia spectrum disorders. In: Dawson M, Schell A, Böhmelt A (eds): *Startle modification: Implications for neuroscience, cognitive science, and clinical science*. Cambridge University Press, Cambridge, 231–244
77. Kumari V, Gray J, Geyer M, ffytche D, Soni W, Mitterschiffthaler M, Vythelingum G, Simmons A, Williams S, Sharma T (2003) Neural correlates of tactile prepulse inhibition: A functional MRI study in normal and schizophrenic subjects. *Psychiatry Res Neuroimaging* 122: 99–113
78. Schmajuk N, Larrauri JA (2005) Neural nework model of prepulse inhibition. *Behav Neurosci* 119: 1546–1562
79. Gewirtz J, Davis M (1995) Habituation of prepulse inhibition of the startle reflex using an auditory prepulse close to background noise. *Behav Neurosci* 109: 388–395
80. Miyazato H, Skinner R, Garcia-Rill E (1999) Sensory gating of the P13 midlatency auditory evoked potential and the startle response in the rat. *Brain Res* 822: 60–71
81. Flaten MA, Nordmark E, Elden A (2005) Effects of background noise on the human startle reflex and prepulse inhibition. *Psychophysiology* 42: 298–305
82. Walton J, Frisina R, O'Neill W (1998) Age-related alteration in processing of temporal sound features in the auditory midbrain of the CBA mouse. *J Neurosci* 18: 2764–2776
83. Fukuda Y, Iwama K (1978) Visual receptive-field properties of single cells in the rat superior colliculus. *Jpn J Physiol* 28: 385–400
84. Perlstein WM, Fiorito E, Simons RF, Graham FK (1989) Pre-stimulation effects on reflex blink and evoked potentials in normal and schizotypal subjects. *Psychophysiology* 26: S48
85. Perlstein WM, Fiorito E, Simons RF, Graham FK (1993) Lead stimulation effects on reflex blink, exogenous brain potentials, and loudness judgments. *Psychophysiology* 30: 347–358
86. Filion DL, Ciranni M (1994) The functional significance of prepulse inhibition: A test of the protection of processing theory. *Psychophysiology* 31: S46
87. Norris C, Blumenthal T (1996) A relationship between inhibition of the acoustic startle response and the protection of prepulse processing. *Psychobiol* 24: 160–168
88. Braff D, Grillon C, Geyer M (1992) Gating and habituation of the startle reflex in schizophrenic patients. *Arch Gen Psychiatry* 34: 25–30
89. Steckler T, Inglis W, Winn P, Sahgal A (1994) The pedunculopontine tegmental nucleus: a role in cognitive processes? *Brain Res Brain Res Rev* 19: 298–318
90. Schott BH, Sellner DB, Lauer CJ, Habib R, Frey JU, Guderian S, Heinze HJ, Düzel E (2004) Activation of midbrain structures by associative novelty and the formation of explicit memory in humans. *Learn Mem* 11: 383–387
91. Legault M, Wise RA (2001) Novelty-evoked elevations of nucleus accumbens dopamine: dependence on impulse flow from the ventral subiculum and glutamatergic neurotransmission in the ventral tegmental area. *Eur J Neurosci* 13: 819–828
92. Yeomans JS, Li L, Scott BW, Frankland PW (2002) Tactile, acoustic and vestibular systems sum to elicit the startle reflex. *Neurosci Biobehav Rev* 26: 1–11

Index

The EXS-Series
Experientia Supplementa

Experientia Supplementa (EXS) is a multidisciplinary book series originally created as supplement to the journal *Experientia* which appears now under the cover of *Cellular and Molecular Life Sciences*. The multi-authored volumes focus on selected topics of biological or biomedical research, discussing current methodologies, technological innovations, novel tools and applications, new developments and recent findings.

The series is a valuable source of information not only for scientists and graduate students in medical, pharmacological and biological research, but also for physicians as well as practitioners in industry.

Forthcoming titles:

Plant Systems Biology, EXS 97, S. Baginsky, A.R. Fernie (Editors), 2006

Published volumes:

Proteomics in Functional Genomics, EXS 88, P. Jollès, H. Jörnvall (Editors), 2000
New Approaches to Drug Development, EXS 89, P. Jollès (Editor), 2000
The Carbonic Anhydrases, EXS 90, Y.H. Edwards, W.R. Chegwidden, N.D. Carter (Editors), 2000
Genes and Mechanisms in Vertebrate Sex Determination, EXS 91, G. Scherer, M. Schmid (Editors), 2001
Molecular Systematics and Evolution: Theory and Practice, EXS 92, R. DeSalle, G. Giribet, W. Wheeler (Editors), 2002
Modern Methods of Drug Discovery, EXS 93, A. Hillisch, R. Hilgenfeld (Editors), 2003
Mechanisms of Angiogenesis, EXS 94, M. Clauss, G. Breier (Editors), 2004
NPY Family of Peptides in Neurobiology, Cardiovascular and Metabolic Disorders: from Genes to Therapeutics, EXS 95, Z. Zukowska, G.Z. Feuerstein (Editors), 2006
Cancer: Cell Structures, Carcinogens and Genomic Instability, EXS 96, L.P. Bignold (Editor), 2006

Where quality meets scientific research...

Birkhäuser

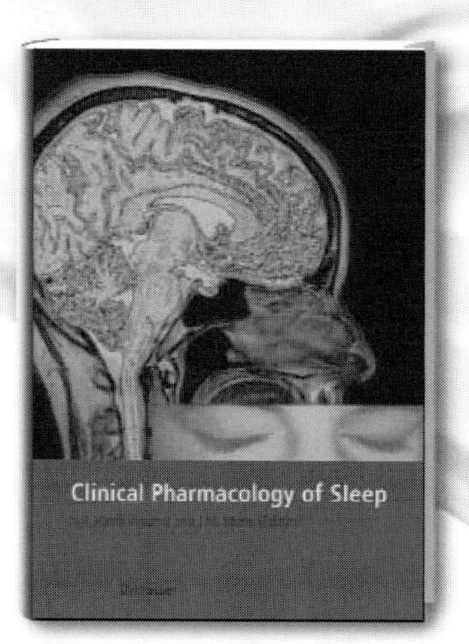

Pandi-Perumal, S.R., Comprehensive Center for Sleep Medicine, NY, USA / Monti, J.M., Montevideo, Uruguay (Eds)

Clinical Pharmacology of Sleep

2006. XII, 239 p. 6 illus. Hardcover
ISBN 3-7643-7262-1

The field of sleep medicine is rapidly growing, and the clinical pharmacology of sleep is gaining much attention from sleep physicians. Many new drugs are in the process of being developed and tested for their possible usefulness in the treatment of various sleep disorders. The opportunity for further advancement is this field is very promising. This volume covers the clinical and pharmacological treatment of several important sleep disorders such as insomnia, sleep apnea, narcolepsy, restless legs syndrome, and periodic limb movement syndrome. It further addresses the use of sleep medications in children, adolescents, and in the elderly. It offers a comprehensive overview of the currently available hypnotic medications and covers aspects of chronopharmacology and its implications for the pharmacology of sleep.

For orders originating from all over the world except USA and Canada:
Birkhäuser Customer Service
c/o SDC
Haberstrasse 7, D-69126 Heidelberg
Tel.: +49 / 6221 / 345 0
Fax: +49 / 6221 / 345 42 29
e-mail: orders@birkhauser.ch

For orders originating in the USA and Canada:
Birkhäuser
333 Meadowland Parkway
USA-Secaucus
NJ 07094-2491
Fax: +1 201 348 4505
e-mail: orders@birkhauser.com